T0191984

Flora Neotropica

Volume 123

Since 1968, these monographic volumes have provided taxonomic treatments of families or other groups of plants or fungi growing in the Americas between the Tropic of Cancer and the Tropic of Capricorn. The monographs are intended to be comprehensive, so most of them also include information on economic botany, conservation, phylogenetic relationships, taxonomic history, ecology, cytology, anatomy, and phytochemistry, among other topics. Each volume is illustrated with line drawings, black and white photographs, and distribution maps. This series is the official publication of the Organization for Flora Neotropica (OFN), an organization founded in 1964 with the mission of producing a published botanical inventory of the American tropics. The current OFN Executive Director is Wm. Wayt Thomas.
Flora Neotropica volumes 1–14 (1968–1974) were published by Hafner Publishing Company. Volumes 15–121 (1975 – 2019) were published by The New York Botanical Garden. All in-print volumes 1—121 are available exclusively through NYBG Shop: https://www.nybgshop.org/nybg-press/books-in-series/flora-neotropica. Volumes 122 and going forward are available exclusively from Springer.

More information about this series at http://www.springer.com/series/16365

Ghillean T. Prance

Humiriaceae

Ghillean T. Prance
Royal Botanic Gardens, Kew
Richmond, Surrey, UK

ISSN 0071-5794 ISSN 2330-202X (electronic)
Flora Neotropica
ISBN 978-3-030-82361-0 ISBN 978-3-030-82359-7 (eBook)
https://doi.org/10.1007/978-3-030-82359-7

Cover illustration: Project Flora da Reserva Ducke INPA/DFID

This Springer imprint is published by the registered company Springer Nature Switzerland AG
The registered company address is: Gewerbestrasse 11, 6330 Cham, Switzerland

This monograph is dedicated to Jose Cuatrecasas (1903–1996) pioneer in the study of the Humiriaceae

Acknowledgements

I thank Andrew Brown for the drawings of each genus (Fig. 10.1, 10.3, 10.5, 10.9, 10.11, 10.16, 10.18, 10.21, 10.24, and 10.26) and Michella del Rei and Matrisabel Adrianzén for Fig.10.12. I thank Rachel Crosby for much help with compiling the database for mapping. I am grateful to the herbarium of the Royal Botanic Gardens Kew for facilities to carry out this work and to Elizabeth Gjieli for the preparation of the distribution maps. I thank Thomas Zanoni and Charles Zartman for most helpful reviews of the original manuscript, and Ana Maria Ruiz for editorial assistance. Thanks to the IUCN Red List Unit in Cambridge, UK and to Barbara Goettsch who worked with me to determine the conservation status for each species. I thank The International Cosmos Prize for the funding that enabled this study.

Abstract

Ghillean T. Prance (Royal Botanic Gardens, Kew, Richmond, Surrey, TW9 3AB, UK). Humiriaceae. Flora Neotropica Monogr. 122. A taxonomic monograph of the Humiriaceae is presented. The family is almost exclusively neotropical with a single species in West Africa (which is included here). Humiriaceae is a member of the order Malpighiales close to the Erythroxylaceae, Ctenolophonaceae and Linaceae. Sixty-five species in 8 genera are treated as well as 14 subspecific taxa. Two new species are added, *Humiriastrum purusensis* and *Vantanea aracaensis*. Introductory material on the wood anatomy and pollen and on the systematic position of Humiriaceae is provided. Full descriptions, nomenclature, identification keys, at least one illustration of each genus, and distribution maps of all species are provided.

Resumo

Ghillean T. Prance (Royal Botanic Gardens, Kew, Richmond, Surrey, TW9 3AB, UK). Humiriaceae. Flora Neotropica Monogr. 122. Esta monografia presenta os oito gêneros da família Humiriaceae. Todas espécies, menos uma da África, são neotropicais, distribuidos de Nicaragua até o sul do Brasil. Humiriaceae pertence à Malpighiales perto à Erythroxylaceae, Ctenophonaceae e Linaceae. Há 65 espécies em oito gêneros. São descritas duas espécies novas, *Humiriastrum purusensis*, e *Vantanea aracaensis*. São incluidas descrições, nomenclatura, sinonímia, chaves para identificação, e pelo menos uma illustração de cada gênero, além mapas de distribuição de todas as espécies.

Contents

About the Author

Ghillean T. Prance was born in Suffolk in 1937 and was educated at Malvern College and Keble College Oxford where he obtained a BA in botany and a DPhil in 1963. His career began at the New York Botanical Garden in 1963 as a research assistant and subsequently B. A. Krukoff Curator of Amazonian Botany, director and vice president of research and finally senior vice president for science. His exploration of Amazonia included 39 expeditions in which he collected over 350 new species of plants. He was director of the Royal Botanic Gardens, Kew, from 1988 to 1999. He was McBryde Professor at the National Tropical Botanical Garden in Hawaii 2001–02 and is currently McBryde Senior Fellow there. Until recently he was a trustee of the Eden Project in Cornwall and is visiting professor at Reading University. He is author of 24 books including 9 volumes of *Flora Neotropica* Chrysobalanaceae (1972), Dichapetalaceae (1972), Rhabdodendraceae (1972) and Caryocaraceae (1973) and co-authored Lecythidaceae (1979 and 1990 with Scott A. Mori), Chrysobalanaceae Supplement (1989), Proteaceae (2008 with K. S. Edwards) and Rhizophoraceae (2018). He has published over five hundred eighty scientific and general papers in taxonomy, ethnobotany, economic botany, conservation and ecology. Sir Ghillean holds 15 honorary doctorates and in 1993 received the International COSMOS Prize and was elected a Fellow of the Royal Society. He was knighted in July 1995 and received the Victoria Medal of Honour from the Royal Horticultural Society in 1999. He received the David Fairchild Medal for plant exploration jointly with his wife Anne in 2000, and the Allerton Award in 2005. In 2000 he was made a Commander of the Order of the Southern Cross by the President of Brazil and in 2012 received the Order of the Rising Sun from Japan. He continues to be active with research in plant systematics and in conservation of the tropical rainforest. He chairs the Mass Extinction Memorial Observatory (MEMO), and was chairman of the Brazilian Atlantic Rainforest Trust (1999–2012) and A Rocha International (2008–2013). He is president of The Wildflower Society, Nature in Art and the International Tree Foundation and the patron of several other organisations.

Chapter 1
Introduction

Over the years, I have received many specimens of Humiriaceae mistaken for Chrysobalanaceae because there is a superficial resemblance of some sterile and fruiting material. There are many differences in both the flowers and the fruits of these two families. They both belong to the order Malpighiales, but they are not very closely related. Receiving these specimens of Humiriaceae by mistake often led to my identifying the specimens and as a result gradually accumulating information and interest in the Humiriaceae. The aim here is to put together that information and a lot more from recent collections to supplement the last monograph of the family by Cuatrecasas (1961). Cuatrecasas cited the 889 herbarium specimens that he studied. This revision is now based on 5397 specimens, an increase of 4508, indicating how much collecting has taken place in the Neotropics over the past 60 years. It is encouraging that 70% of the additional collections were made by resident Latin American collectors. This indicates the vast increase in local collecting efforts, especially in Brazil and Colombia.

G. T. Prance, *Humiriaceae*, Flora Neotropica 123,
https://doi.org/10.1007/978-3-030-82359-7_1

Chapter 2
Taxonomic History of Humiriaceae

A more detailed species-by-species taxonomic history of the family is given in Cuatrecasas (1961) and only the major contributions before that date are highlighted here. The first genera of *Humiriaceae* were described and illustrated by Aublet (1775), *Houmiri* Aubl., with the single species *H. balsamifera* (later Latinized to *Humiria*) and *Vantanea guianensis* Aubl. Aublet placed these genera in the class Polyandria and order Monogynia of the Linnaean system. Nees and Martius (1824) described the genus *Helleria*, which was later placed in synonymy with *Vantanea* Aubl. (Bentham 1853). Martius (1827) described the genus *Sacoglottis* with a single species *S. amazonica* and also *Humirium floribundum* (now a variety of *Humiria balsamifera*). He considered these genera to be related to Meliaceae, Symplocaceae, and Styracaceae, an opinion supported by Endlicher (1840). Jussieu (1829) was the first to treat the group at the family level including two genera: *Humirium* with two species and *Helleria* of Nees and Martius (1824) also with two species that are now in *Vantanea*. He also recognized *Sacoglottis* as belonging in this family, so in effect he brought together all three genera of the family that had been described by 1829. Planchon (1848) compared the *Humiriaceae* with *Erythroxylaceae* and the *Linaceae* due to morphological affinities to *Erythroxylum* P. Browne and *Roucheria* Planch, respectively. Bentham (1853) published a concise summary of the family treating the three genera *Humiria*, *Sacoglottis*, and *Vantanea*. The separation of these genera was mainly based on the number of stamens, *Vantanea* with numerous stamens, *Humiria* with 20 stamens, and *Sacoglottis* with 10 stamens. He also noted the number of ovules, two in *Vantanea* and one in *Sacoglottis*, but *Humiria* with one or two ovules. Baillon (1873) united all known Humiriaceae into the single genus *Houmiri*, which included 15 species placed in five sections: *Humirium* (20 stamens), *Aubrya* (10 free stamens), *Saccoglottis*[1] (10 stamens + 10 staminodia), *Vantanea* (20–60 stamens), and *Vantaneoides* (never formally published), and the genus was placed in his *Linaceae*.

[1] The spelling of *Sacoglottis* in this part is as it was in each manuscript.

G. T. Prance, *Humiriaceae*, Flora Neotropica 123,
https://doi.org/10.1007/978-3-030-82359-7_2

Baillon (1874) repeated the same division of *Humiriaceae* and placed his series *Humiriées* in the *Linaceae*. The next major contribution to the family was by Urban (1877) in Martius' *Flora brasiliensis.* Urban recognized three genera: *Humiria* with three species, *Saccoglottis* with nine species, and *Vantanea* with four species. *Saccoglottis* was grouped into three subgenera: *Eusaccoglottis* with two Brazilian species but recognizing that the African *Aubrya gabonensis* of Baillon (1862) belonged here; subgenus *Humiriastrum* with three species; and subgenus *Schistostemon* with four species. Urban divided the family into two groups distiguished by the number of stamens and number of cells in the thecae of the anthers: *Vantanea* with thecae with two locules and many stamens and *Humiria* and *Saccoglottis* with unilocular thecae and fewer stamens. The subgenera were divided by staminal characters: subg. *Eusaccoglottis* with 10 stamens, subg. *Humiriastrum* with 20 undivided stamens, and subg. *Schistostemon* with 20 stamens, the latter with 5 tridentate ones bearing 3 anthers. Urban (1893) later described *Sacoglottis glaziovii* Urb. based on material sent to him from Brazil, and this was transferred to *Humiriastrum* by Cuatrecasas (1961). Hallier (1921) followed Baillon in keeping the *Humiriaceae* and *Linaceae* united into a single family. Ducke (1922–1933; 1935–1937) in a series of important papers about Amazon plants published four new species of *Vantanea* and five of *Sacoglottis,* as well as several new varieties. Two of Ducke's new species of *Sacoglottis* were later given generic status by Cuatrecasas (1961) one named appropriately *Duckesia* (after Adolpho Ducke whose work was a major contribution to the taxonomy of the Amazonian species of *Humiriaceae*) and the other *Hylocarpa*. Ducke (1938) summarized much of his work on *Humiriaceae* and provided a key to the genera *Sacoglottis* and *Vantanea*, placing them in Linaceae. Winkler (1931) in the second edition of *Pflanzenfamilien* followed Baillon and Hallier in including the family in Linaceae, but as the subfamily *Humirioideae*. Otherwise Winkler (1931) followed Urban's system without any changes.

The seminal work on the family is the monograph of Cuatrecasas (1961) who recognized eight genera. He decribed *Duckesia*, *Hylocarpa*, and *Endopleura* all based on former species of *Sacoglottis* and he elevated Urban's *Sacoglottis* subgenera *Schistostemon* and *Humiriastrum* of *Sacoglottis* to generic rank. He described new species of *Vantanea, Humiria, Humiriastrum, Schistostemon*, and *Sacoglottis* and made numerous new nomenclatural combinations to constitute the new genera. Below the species rank there was excessive splitting into varieties and forms, for example *Humiria balsamifera* was divided into 14 subspecific taxa. Cuatrecasas studied many more fruit specimens than previous authors, and they are well illustrated throughout the monograph. He considered the *Humiriaceae* as a natural and well-defined group, separate from the *Linaceae*, and closely related to *Ixonanthes* Jack, *Ochthocosmos* Benth., and *Ctenolophon* Oliv. Cuatrecasas also divided the family into two tribes: (1) the Vantaneoideae[2]

[2] Cuatrecasas termed his division of the family into two as the "tribes" Humirioideae and Vantaneoideae using the Latin ending *oideae*, which would be correct for a subfamily. I have maintained his designation of tribe for these taxa and changed to the correct endings Humireae and Vantaneae.

consisting of the single genus *Vantanea* and characterized by the stamens arranged in three or four circles with tetrasporangiate bilocular thecae and (2) the Humirioideae comprising all other genera defined by both tetrasporangiate (*Endopleura*, *Duckesia*) and bisporangiate (*Humiria*, *Sacoglottis*, *Schistostemon*, and *Hylocarpa)* stamens in fewer numbers with unilocular thecae. He used the term tribes but used the correct ending "*oideae*" for subfamilies. More recently, Herrera et al. (2010) questioned the status of *Schistostemon* because there are no differences in fruit between it and *Sacoglottis*. They proposed returning *Schistostemon* to subgeneric status in *Sacoglottis*. However, the 10 undivided stamens of *Sacoglottis* compared to the 20 stamens, 5 being longer and episepalous and trifurcate at the apex bearing three anthers, is a sufficiently adequate character to maintain *Schistostemon* as a genus.

Cuatrecasas and Huber (1999) in the *Flora of Venezuelan Guayana* treated the six genera that occur in that region and continued to recognize 11 infraspecific taxa of *Humiria balsamifera* and did not make any significant changes to the taxonomy of Cuatrecasas (1961). Following his monograph, Cuatrecasas continued to be active with the taxonomy of the Humiriaceae and described several additional species of Humiriaceae: *Humiriastrum spiritu-sancti* from Bahia, Brazil, and *Sacoglottis holdridgei* in Cuatrecasas (1964), the latter an interesting endemic from Isla del Coco of Costa Rica. Cuatrecasas (1964) also added yet another forma to *Humiria balsamifera*. Cuatrecasas (1972) described *Sacoglottis trichogyna* from Heredia, Costa Rica. Cuatrecasas (1990) described *Humiriastrum ottohuberi* from Amazonas, Venezuela, *Schistostemon fernandezii* from Bolívar, Venezuela, and two species of *Vantanea* from Bahia, Brazil, *V. bahiensis* and *V. morii*. Cuatrecasas (1991) described *Humiriastrum liesneri* from Amazonas, Venezuela in preparation for the account in the Flora of Venezuelan Guayana, and in 1993 described *Humiriastrum mussunungense* from Espírito Santo, Brazil. *Humiriastrum liesneri* was later transferred to *Duckesia* by Wurdack and Zartman (2019). Giordano and Bove (2008) modified and corrected the description of *Humiriastrum spiritu-sancti*, and later Wurdack and Zartman (2019) transferred it to *Vantanea*. Rodrigues (1982) described *Vantanea deniseae* from the vicinity of Manuas, Brazil based on the tomentose floral disc, at that time only known in *V. parvifolia*. Sabatier (2002) described *Vantanea ovicarpa* from French Guiana, also based on the tomentose disc. Sabatier (1987) described *Schistostemon sylvaticum,* also from French Guiana; Gentry (1990) described *Vantanea spichigeri* from Loreto, Peru; and Engel and Sabatier (2018) described *Vantanea maculicarpa* from French Guiana.

Recently, Wurdack and Zartman (2019) published an important paper focussed on *Sacoglottis* (Wurdack and Zartman 2019). The authors described the new species, *Sacoglottis perryi*, and transferred two species of Cuatrecasas' post-1961 species of *Humiriastrum* into different genera, *H. liesneri* to *Duckesia* and *H. spiritu-sancti* to *Vantanea*. They also produced some elegant SEM pictures of stamen structure and extensive details about the extrafloral nectaries and the types of leaf margin in a survey of almost all species of the family.

Also, over the last few decades, the several local treatments of Humiriaceae have been published, for example: Humiriaceae of the Arboretum Jenaro Herrera in Amazonian Peru (Spichiger et al. 1983,); Panama (Gentry 1975, 5 species in 3 genera); Nicaragua (Gentry 2001, 3 species), the illustrated field guide to the Reserva Ducke near Manaus (Sothers et al. 1999, 10 species), and Costa Rica (Zamora 2007, 7 species).

Chapter 3
Relationships and Phylogeny

There has been surprisingly little doubt about the relationships of the Humiriaceae being close to the Erythroxylaceae and Linaceae, but its exact position in the Malpighiales is still unclear. The taxonomic history above shows that from early times, the Humiriaceae was associated with the Erythroxylaceae or the Linaceae, in which it was often included. Hutchinson (1973) placed Humiriaceae in his order Malpighiales next to the Malpighiaceae and the Linaceae. This order also included the Ixonanthaceae, Erythroxylaceae, and Ctenolophonaceae, all of which had previously been associated with Humiriaceae. Cronquist (1968, 1981) placed the Humiriaceae in his order Linales with the Erythroxylaceae and the Linaceae. Takhtajan (1969) placed Humiriaceae in his order Geraniales which also included Linaceae, Ixonanthaceae, Malpighiaceae, and 15 other families. Takhtajan (1997) was closer to Cronquist's system and placed Humiriaceae in his order Linales together with Hugoniaceae, Linaceae, Ctenolophonaceae, and Ixonanthaceae. Molecular data has not changed this placement of the Humiriaceae. In the various versions of the Angiosperm Phylogeny Group (APGIII), the family is placed in the Rosid order Malpighiales, which includes all the families mentioned above, such as Ixonanthaceae, Erythroxylaceae, and Linaceae. Nevertheless, the exact position of the Humiriaceae within the Malpighiales is not yet clear. In the molecular analysis of Wurdack and Davis (2009), the Humiriaceae is in a polytomy with Ctenolophonaceae, Caryocaraceae, Irvingiaceae, Ixonanthaceae, Linaceae, and the Pandaceae, confirming some of the earlier placements of the family. The phylogeny of Xi et al. (2012), however, differed considerably from the traditional placement of Humiriaceae. The family was placed within a major subclade as sister to the parietal clade including the families Achariaceae, Violaceae, Passifloraceae, and Salicaceae. In Xi's system, the Linaceae and the Erythroxylaceae were both distant from Humiriaceae and not close to each other.

There have been two attempts to propose a phylogeny within the Humiriaceae based on morphology (Bove 1997; Herrera et al. 2010; Figs. 3.1 and 3.2). Both these phylogenies are based on morphological and other characters, but not

G. T. Prance, *Humiriaceae*, Flora Neotropica 123,
https://doi.org/10.1007/978-3-030-82359-7_3

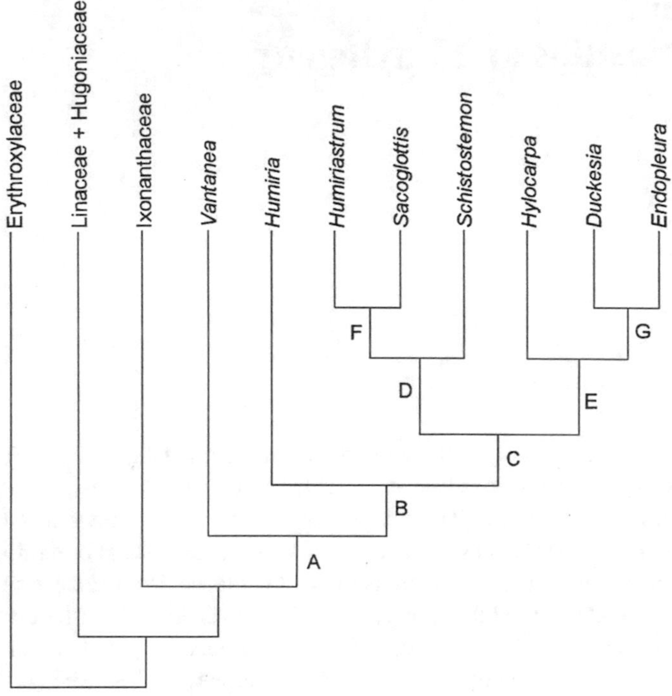

Fig. 3.1 Cladogram of Humiriaceae genera after Bove (1997)

molecular, and both used Ixonanthaceae, Linaceae, and Erythroxylaceae as the outgroups. Herrera et al. (2010) also included the Caryocaraceae as the outgroup and placed considerable emphasis on fruit characters. They both strongly confirm the monophyly of Humiriaceae and recognize all eight genera. The phylogeny of Bove (1997; Fig. 3.1) was based on characters from anatomy, palynology, ontogeny, and chemotaxonomy. Both Bove (1997) and Herrera et al. (2010) support the hypothesis of Cuatrecasas (1961) that *Vantanea* is the sister group of the other genera, and they also support the division of the family into the two tribes, Vantanoideae and Humirioideae. Both phylogenies divide the Humirioideae into two clades with *Humiria* sister to the other six genera. The division of the remaining six genera is quite different in the two systems, perhaps indicating the closeness of these genera compared with *Humiria* and *Vantanea*. In Bove (1997), *Humiriastrum, Sacoglottis*, and *Schistostemon* form one clade, and *Hylocarpa*, *Duckesia*, and *Endopleura* form the other. *Duckesia* and *Endopleura* are sister taxa based on the presence of anthers with four thecae and a foraminate endocarp (the latter character also occurring in *Humiriastrum*). A sister–group relationship between *Humiriastrum* and *Sacoglottis* is supported by the presence of dicolporate pollen. The cladogram from Herrera et al. (2010) is based on morphological characters with considerable emphasis on those of the fruit. It has well supported *Vantanea* and *Humiria* clades and the remaining six genera in a sister clade to *Humiria* divided into three pairs of sister

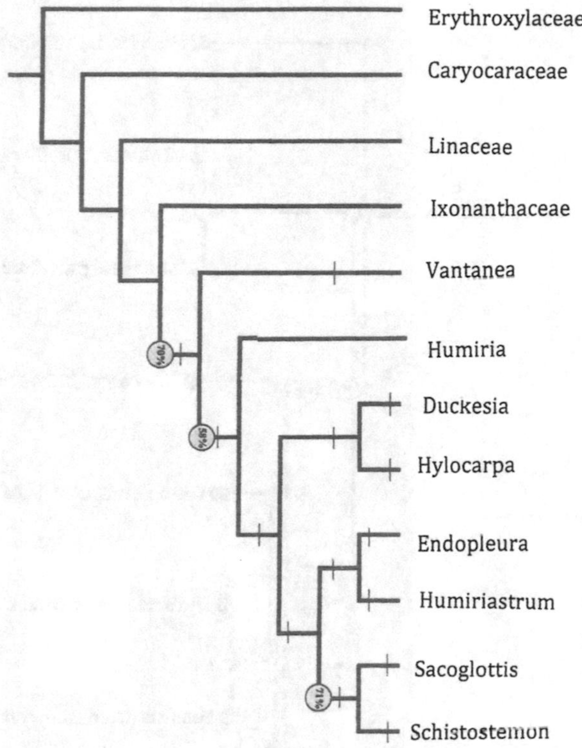

Fig. 3.2 Cladogram after Herrera et al. (2010)

taxa: *Duckesia* and *Hylocarpa*; *Endopleura* and *Humiriastrum*; and *Sacoglottis* and *Schistostemon*. The sister group relationship of *Sacoglottis* and *Schistostemon* is well supported by morphological characters.

A more recent molecular phylogeny is that of Chave (pers. comm., Fig. 3.3). The main difference from the morphological phylogenies discussed above is that *Humiria* is sister to the rest of the family rather than *Vantanea*. An important synapomorphy of *Humiria* and *Vantanea* is that these genera have 2-ovulate carpels in contrast to the other six genera, which have uniovulate carpels. A shared result, however, among all three phylogeneis consistently has *Vantanea* and *Humiria* as separated from the other six genera rather than considering *Vantanea* alone in the tribe Vantaneoideae. For this reason, I have not recognized the tribes of Cuatrecasas (1961). Instead, the family divides phylogenetically into the *Humiria* clade together with *Vantanea*, and the other six genera belong to the *Sacoglottis* clade. In the Chave (pers. comm.) cladogram, *Schistostemon* falls within *Sacoglottis* suggesting the possibility that these two taxa may be recombined into one. However, the staminal differences are significant, so I have maintained them herein as separate genera. A study with more species sampled is needed to make any conclusion about the status of *Schistostemon*.

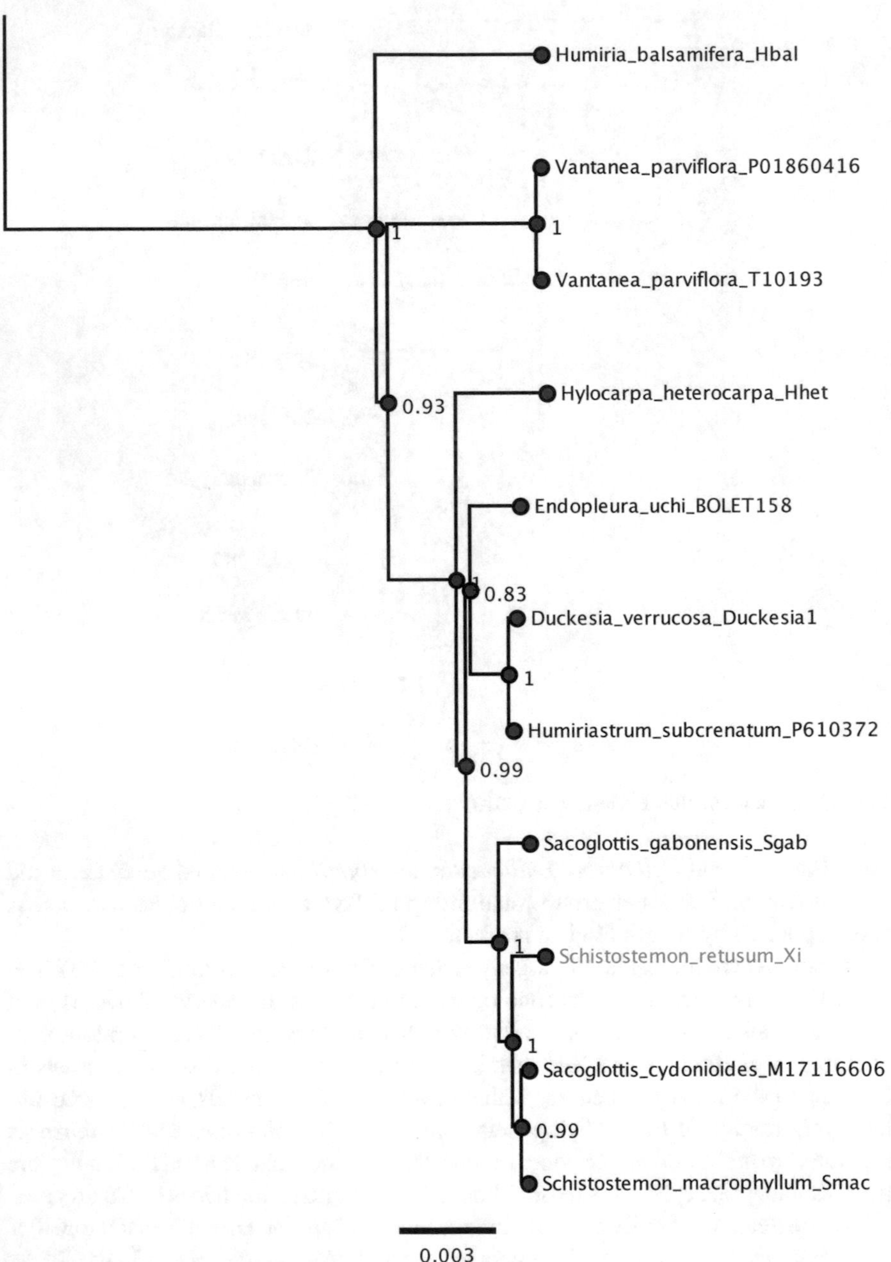

Fig. 3.3 Cladogram of Humiriaceae after Chave (pers.comm.)

Chapter 4
Fossils

There are a good number of fossil records of Humiriaceae, mainly of the woody endocarps of the fruit, which preserve well. The first record of a humiriaceous fossil is that of Berry (1922), who described *Sacoglottis tertiaria* from a Pliocene deposit in Bolivia. Berry (1924a, b) described and illustrated another fossil, *Saccoglottis cipaconensis*, based on well-preserved endocarps of the late Eocene or Oligocene of Cipacón at 2640 m altitude in Colombia. Cuatrecasas (1961) pointed out that the six seeds described by Berry (1924b) are in fact the valves of the endocarp of a *Vantanea*. However, Berry's (1924a, b) *Vantanea colombiana* is in fact a *Sacoglottis*. Berry (1929a, b, c) described three additional Humiriaceae fossils: *Saccoglottis cipaconensis* var. *peruvianus* and *Vantanea compressiformis* from the early Tertiary of Belén, Peru, and *Vantanea sheppardi* from the Eocene of Ida Seca in Ecuador. Reid (1933) described *Sacoglottis costata* from the Tertiary in Colombia. Selling (1945) described *Humiria bahiensis* from the Miocene or Pliocene of Bahia, Brazil, which Cuatrecasas (1961) suggested is probably a *Vantanea*. According to Cuatrecasas (1961) and Bove and Melhem (2000), neither of the European records, *Sacoglottis germanica* of Weyland (1938) nor *S. kayseri* of Kirchheimer (1951), belong to the Humiriaceae. This was supported by Herrera et al. (2010), who also rejected *Vantanea sheppardi* Berry, which they assigned to the category of *incertae sedis*. Wijninga and Kuhry (1990) reported a fossil endocarp of *Humiriastrum* from Colombia, but Herrera et al. (2010) reported that the specimens of this have been lost according to Wijninga. In a summary of palaeobotany in Colombia, Hooghiemstra et al. (2006) reported records of *Sacoglottis*, *Humiriastrum*, and *Vantanea*. Herrera et al. (2010) re-evaluated the fossil record of Humiriaceae and added a new record of *Sacoglottis tertiaria* endocarps from the Cucaracha Formation of Panama, which was well illustrated by a color plate. They transferred *S. cipaconensis* to *Vantanea cipaconensis* (Berry) Herrera and agreed with Cuatrecasas that *Humiria bahiensis* is probably a species of *Vantanea*. The fact that all of the reliable known fossils of Humiriaceae are from South America supports the opinion of Cuatrecasas (1961) that the family is an old tropical American one. The center of

G. T. Prance, *Humiriaceae*, Flora Neotropica 123,
https://doi.org/10.1007/978-3-030-82359-7_4

diversity was the somewhat elevated land surrounding the Amazon basin before the uplift of the Andes. This stock evolved and diversified throughout tropical America into many different habitats from the flooded forest to high in the tepuis of the Guayana Highland. In a later paper, Herrera et al. (2014) expanded their study of the fossil record of Humiriaceae and described *Duckesia berryi* Ferrera from the Oligocene of the Pacific coastal region of Peru. This is particularly interesting record as the only extant *Duckesia* is a Central Amazonian species. The fruit of *Duckesia* is quite distinct, so this is likely to be a reliable record. Herrera et al. (2014) also documented an earlier record of *Sacoglottis tertiaria* from the early Oligocene (ca. 33.9–28.4 Ma) and a new record and a detailed description of *Vantanea cipaconensis* fruits from the late Miocene of Panama.

The first report of fossil wood of Humiriaceae was that of Stern and Eyde (1963) from Ocú Panama, which they attributed to *Vantanea*. The Ocú region was later studied by Herrera et al. (2014), who described the fossil wood, as *Humiriaeeoxylon ocuensis* Herrera, Manchester, Vélez-Juarbe & Jaramillo, which they dated as Eocene (ca. 37.2–33.9 Ma). There are several records of fossil wood of Humiriaceae from Amazonian Brazil and Peru (Jupiassú, 1970; Pons and Franceschi, 2007; Kloster et al. 2012), and Herrera et al. (2014) confirmed that the identification of these woods was correct.

Bove and Melhem (2000) stated that there are no fossil pollen records for the Humiriaceae, but this was later shown to be untrue as Lorente (1986) had assigned *Psilabrevitricolporites devriesii* (Lorente) Silva-Caminha et al. to the Humiriaceae. This designation was backed up by Silva-Caminha et al. (2010) and Herrera (2010). The fossil record of *P. devriesii* is widely distributed in the Neotropics and has been observed since the early Miocene in Panama, Colombia, Venezuela, and western Amazonia according to Hoorn (1994; appendix A). The fossil record confirms that the Humiriaceae were present in central Amazonia at least since the early to mid-Miocene.

Chapter 5
Wood Anatomy

There are general descriptions of the wood of Humiriaceae in Metcalf and Chalk (1950), Elias de Paula and Alves (1977), Araujo and Mattos-Filho (1985), Bhikhi et al. (2016), and in many other places. The wood was also described in detail by Herrera et al. (2014) in order to aid the study of the fossil woods of Humiriaceae. The wood of Humiriaceae as a family is easily distinguished from other angiosperms, but it is very similar among the different genera, which means that it is of little taxonomic use below the family level.

Growth rings are absent or indistinct. Vessels are diffuse porous, exclusively solitary (rarely arranged in a radial diagonal pattern), 80–150 (–250)µm diameter, 6–14 per sq mm; perforation plates are scalariform, usually with 8–15 bars in mature samples and with up to 40 in immature wood, often with gums and other deposits in heartwood vessels; intervessel pits are alternate or rarely opposite (*Humiria* and *Humiriastrum*), usually 4–10 µm in size; vessel-ray pitting has reduced borders and distinct vessel-ray pits; fibers are very thick-walled, at least 1500 µm long, nonseptate, with distinctly bordered pits on both radial and tangential walls; apotracheal parenchyma absent/diffuse to diffuse-in-aggregates, paratracheal parenchyma ranges from scanty to short-winged: rays are 1 or 2 (rarely 3) seriate, heterogeneous (Kribs Type II) with procumbent cells in the body of the ray with two or usually more than four rows of upright/square marginal cells and abundant prismatic crystals in chambered axial parenchyma in both upright and procumbent ray cells. The sieve tube plastids contain protein and starch, and the nodes are 5-lacunar.

Wood anatomy clearly supports the relationship of Humiriaceae to the Linaceae. Heimsch (1942) and Metcalfe and Chalk (1950) both placed Humiriaceae next to Linaceae and Erythroxylaceae based on wood anatomy. They stated that "Humiriaceae forms a homogenous group, distinct from the Linaceae, but more nearly related to it than to any other family; the greatest resemblance is to *Ctenolophon* which has closer affinities with the Humiriaceae than with the Linaceae."

G. T. Prance, *Humiriaceae*, Flora Neotropica 123,
https://doi.org/10.1007/978-3-030-82359-7_5

Chapter 6
Pollen

The pollen of Humiriaceae was described in great detail in Bove and Melhem (2000), who studied 81 taxa from all eight genera. The description below is taken from Bove and Melhem. Before Bove and Melhem (2000), there were detailed studies by Oltmann (1971) and Suryakanta (1974).

Pollen grains in isopolar monads, radially symmetrical, angulaperturate (2)3(4)-zonocolporate, suboblate to subprolate; P: 20–52 (19–56) μm, E: 23–42 (18–48) μm, P/E 0.78–1.31. Amb triangular to circular for tricolporate pollen grains, elliptical for dicolporate grains. The mesocolpia straight to gently convex or rarely concave. In polar view, margins not protruding, polar outline angulaperturate. Colpi short or long, extending from 1/3 to 4/5 of the length of the polar axis, narrow to wide, with rounded to pointed apices, delimited by a more or less distinct flattened sexinous margin and by foot layer thicker near the apertures; colpus membrane finely to coarsely granular, apocolpium index 0.23–0.70. Endoapertures 2.5–7.0 × 2.5–8.0 μm in diameter, lalongate, circular to rectangular, or quadrangular, sometimes surpassing the colpi limits or a complex H-shaped thinning with arms of thinning up to ¾ length of polar axis. Exine 1.5–3.5 μm thick, with conspicuous thickenings parallel to the colpi. Columellar infratectum very thin in some taxa. Sexine tectate or semitectate, thinner than nexine. Tectum psilate, perforate, microreticulate, rarely reticulate, striato-reticulate, or rugulate.

The family is stenopalynous with only two main types of pollen based on colpus length and shape and exine stratification. The pollen of *Humiria* differs from all the other genera. The *Humiria*-type has ectocolpus short, with a thin columellar infratectum, and in the *Vantanea*-type, the ectocolpus is long, with a thick columellar infratectum. Bove recognized three pollen subtypes in *Humiria* based on the exine sculpture:

- *Humiria balsamifera* var. *balsamifera* subtype: pollen grains reticulate
- *Humiria balsamifera* var. *floribunda* subtype: pollen grains psilate with perforations in the apocolpium region

G. T. Prance, *Humiriaceae*, Flora Neotropica 123,
https://doi.org/10.1007/978-3-030-82359-7_6

- *Humiria fruticosa* subtype pollen grains reticulate

The fact that pollen morphology does not support the tribal division of Cuatrecasas (1961) led Suryakanta (1974) to reject these tribes. However, all other fields of evidence support the separation of *Vantanea* into a different basal tribe or subfamily than the other genera. This is now supported by molecular data as stated above.

Chapter 7
Chemistry

The family is ellagic acid-rich as in the closely related Ixonanthaceae. Bergenin has been isolated from *Endopleura uchi* (Da Silva et al. 2009). *Humiria balsamifera* produces a balsam-type phenolic resin in the trunk that is similar to the resin of the leguminous genus *Myroxylon* (Langenheim 2003). Silva et al. (2004) studied the chemistry of stem and leaves of *Humiria balsamifera* (Aubl.) St. Hil. and found a wide array of terpenes and other chemicals: the triterpenes arjunolic acid, 2β,3α-dihydroxy-D:A-friedooleanane, friedelin, α and β-amyrins and lupeol; the flavonol quercetin; bergenin; the diterpene phytol; two phytosteroids (sitosterol and stigmasterol); three sesquiterpenoids (caryophyllene oxide, humulene epoxide II and trans-isolongifolenone; and four aliphatic esters (methyl tetra-, hexa-, octadecanoate and ethyltetradecanoate). The fruit of *Endopleura uchi* (Huber) Cuatr. has been shown to have a high oleic acid fat content (Pinto 1956; Marx et al. 2002). Marx et al. (2002) showed that the fruit also contains phytosterols, a notable amount of vitamin C, vitamin E, and minerals. They identified 42 volatile compounds including 3,3-dimethyl-s-butanol, eugenol and methyl, and ethyl esters of fatty acids. Costa Abreu (2013) analyzed the bark of *Endopleura uchi,* and the dichloromethane fractions yielded the pentacyclic terpenes pseudotaraxasterol, lupeol, a-amyrin, botulin, and methyl 2β,3β-dihydroxy-urs-12-en-28-oate, as well as a mixture of the steroids sitosterol and stigmasterol.

G. T. Prance, *Humiriaceae*, Flora Neotropica 123,
https://doi.org/10.1007/978-3-030-82359-7_7

Chapter 8
Chromosomes

There is a single record of chromosome number for the family and that is of the African species *Sacoglottis gabonensis* (Baill.) Urb., 2n = 12 (Kubitzki 2013).

G. T. Prance, *Humiriaceae*, Flora Neotropica 123,
https://doi.org/10.1007/978-3-030-82359-7_8

Chapter 9
Morphology of Humiriaceae

9.1 Habit

The Humiriaceae consists exclusively of woody tropical plants that are mainly large trees and a few smaller shrubs or sprawling shrubs in rocky places. In some of the larger trees, buttresses may occur.

9.2 Leaves

The leaves are simple, alternate, and often distichous. The margins are serrate, dentate, or at least slightly crenulate in most species of all genera except *Vantanea*, which has entire margins. The venation is exclusively brochidodromous. Stipules are absent or minute. The stomata are anomocytic except in *Vantanea* where they paracytic.

9.3 Vestiture

The leaves of many species are glabrous. Young branches are often pubescent. Where hairs occur; they are simple and unicellular.

G. T. Prance, *Humiriaceae*, Flora Neotropica 123,
https://doi.org/10.1007/978-3-030-82359-7_9

9.4 Inflorescence

The inflorescences are commonly axillary but may be pseudoterminal or rarely terminal. They are paniculate and often corymbose, dichasial, and trichotomous, but often reduced to dichotomous or alternate branching. The flowers have short articulate pedicels. Bracts and bracteoles may be persistent or early deciduous.

9.5 Calyx

The calyx is 5-lobed, the lobes thick and fleshy at the base, thinner toward margins, suborbicular or triangular, connate into tube or cupule to varying degrees, glabrous or pubescent on the exterior, the margins always ciliate, sometimes with marginal or dorsal glands; aestivation is quincuncial or imbricate.

9.6 Corolla

The petals are free and generally rather thick, and usually white except in *Vantanea guianensis* with red petals. The aestivation is contorted, cochlear, or quincuncial.

9.7 Stamens

The stamens are in two or more whorls and vary from 10 to 230. Most species of *Vantanea* have more than 35 stamens, and the other genera have from 10 to 30. The stamens are united into a tube at the base. The anthers are fleshy and usually attached by an elongated connective; they have two bilocular thecae in *Vantanea* and are unilocular in the other genera. In *Duckesia* and *Endopleura,* there are four thecae, and in the other genera, there are two thecae. Androecial morphology is described in detail in Wurdack and Zartman (2019) including information on the vasculature of the anthers.

9.8 Disc

In all species, there is a disc surrounding the ovary. It varies from a fused cupular or tubular circle that is dentate or lobate to one divided into up to 20 free scales. It may be firm and subcoriaceous or membranaceous.

9.9 Gynoecium

The ovary is superior in all genera. The ovary is surrounded by a free disc that may be entire or consist of free or united scales. The locules are uni- or bi-ovulate. The ovules are epitropous with the micropyle facing upwards. Each carpel has a single style usually with a capitate stigma.

9.10 Fruit

The fruit (Fig. 9.1) is a drupe with a single stone derived from a syncarpous ovary with biovular cells as in *Vantanea* and *Humiria*, or uniovular in all other genera. The exocarp is a fleshy layer outside the endocarp and can be soft or hard, thick, or thin. It is usually easily removable at maturity or may be closely adnate to the stone, and hard, granulose, or coriaceous when dry. Usually the surface is smooth but sometimes it is rugate or tuberculate. The endocarp is very hard, woody, with thick walls. Rarely, it has five (also four, six, or seven) locules, but generally only one or two of the ovary cavities are fertile and develop in the fruit. The cavities are normally monospermous in genera with uniovular ovary cells. The endocarp often has numerous resin-filled cavities (Fig. 9.4).

The most interesting feature of the fruit is the germinal dehiscence of the endocarps which have as many longitudinal opercula or valves as carpels, which may open or be pushed away by the emerging embryo at germination of seed inside the fruit. In *Duckesia*, *Humiria*, and *Vantanea,* the endocarp has conspicuous longitudinal lines stretching from the apex almost to the base or shallow furrows that define the potential valves for germination. On germination, the valves of seed-containing locules are pushed away by the developing embryo leaving the ribs or costae intact. In *Humiriastrum,* the opercula are smaller and confined to the upper part of the endocarp. This is very similar to the germination opercula of the genus *Parinari* (Chrysobalanaceae). Hill (1933) described this type of opercular dehiscence for two species of *Sacoglottis*. The valves and furrows are much less conspicuous in *Sacoglottis* and *Schistostemon* until ready for germination. The endocarps of *Sacoglottis, Schistostemon*, *Duckesia*, and a few species of *Humiriastrum* have a spongy appearance in cross-section (Fig. 9.4a) due to cavities that are filled with resinous material. This provides buoyancy that enables the seeds to float and drift for long periods and explains the presence of *Sacoglottis* in Africa. Van der Ham et al. (2015) report on a possible fruit of an unidentified species of *Sacoglottis* found in drift material of the northern Atlantic

Fig. 9.1 Fruits of Humiriaceae. (**a**) *Schistostemon retusum, Ducke RB 30131*. (**b**) *Hylocarpa heterocarpa, Ducke RB 30137*. (**c**) *Vantanea tuberculata, Ducke RB 30134*. (**d**) *Vantanea obovata, Guedes 684*. (**e**) *Schistostemon macrophyllum, Ferreira 130*

9.11 Seeds

The seeds are oblong with double testa, with the exterior often adherent to the endocarp, and the inner seed coat membranaceous and thin. There are one or two seeds per locule. The embryo is straight or slightly curved, with oblong or ovate cotyledons that are often subcordate at base, the radicle half as long, and fleshy or oily endosperm.

Chapter 10
Systematic Treatment of Humiriaceae

Humiriaceae A. Juss., Fl. Bras. Merid. (Saint-Hilaire) 2: 87. 1829; Benth. & Hook., Gen. Pl. 1: 246–247; Urban, Fl. Bras. (Martius) 12, 2: 425–454. 1877; Reiche, Nat. Pflanzenfam. (Engler) 3,4: 35–37.1890; Cuatrecasas, Contr. U.S. Natl. Herb. 35, 2: 25–209. 1961; Zamora, Manual de Plantas de Costa Rica 6: 10–15. 2007; Kubitzki, Gen. Sp. Fl. Pl. 11: 223–228. 2013.

Houmircae Baill., Adansonia 10: 368, 371. 1873; Hist. Pl. (Baillon) 5: 51, 56. 1874.

Linaceae subfamily Humirioideae Winkler, Nat. Pflanzenfam. (Engler & Harms) 19a: 126–130. 1931.

Small shrubs, treelets or large trees; wood hard, aromatic, often with balsamic sap, heartwood reddish, alburnum yellow or yellowish; bark smooth, striate or fissured. Leaves alternate, simple, distichous, coriaceous or subcoriaceous, penninerved, brochidodromous, from small to large, margins entire, crenulate, dentate or slightly serrate, petiolate or rarely sessile, sometimes decurrent along branches, often punctate-glandulose (nectariferous glands) near margins (or basal) on underside; stipules very small, geminate, often deciduous. Inflorescences (synflorescences) axillary, pseudo terminal or rarely terminal, cymose-paniculate (or thysoid) or often corymbiform, of dichasial type and trichotomous, but through reduction with dichotomous or alternate (cincinnate) branching; branchlets often with incrassate ends, articulate; pedicels short, articulate; bracts and bracteoles persistent or deciduous, small, amplectant. Flowers hermaphroditic, actinomorphic, slightly aromatic; sepals 5, persistent, thick and carnose at base, thinner toward margins, suborbicular or triangular, more or less connate in tube or cupule in varying degrees, glabrous, pubescent or tomentose outside, margins always ciliate, sometimes with marginal or dorsal glands; aestivation quincuncial or imbricate; petals 5, deciduous or sometimes persistent, free, thick or membranaceous, usually 3–5-veined, oblong, linear or oblong-lanceolate, acute to obtuse, 1.5–16 mm long, exceptionally 30–40 mm long, rarely with gland at top, margins smooth, sometimes with tooth at

G. T. Prance, *Humiriaceae*, Flora Neotropica 123,
https://doi.org/10.1007/978-3-030-82359-7_10

one side near apex, above glabrous, below glabrous or pilose, white, greenish white, or yellowish white, rarely red or purple; aestivation contorted, cochlear or quincuncial; stamens monadelphous, numerous and pluriseriate or of variable number, 1–2-seriate; filaments filiform (when numerous), slender and flexuose, or thick, complanate, linear, acute at apex, straight and glabrous or papillose; connate at base in a more or less long tube, alternating in different lengths, sometimes five alternating with petals are trifurcate at apex and triantheriferous; sometimes with additional staminodial filaments; anthers dorsifixed or subbasifixed; thecae 2, bilocular, laterally attached, ellipsoid-oblong and each cell dehiscing by longitudinal slit, or 4 unilocular, rounded or ellipsoid disjunct thecae (2 lateral and 2 basal), dehiscing by detachment, or two unilocular, disjunct, basal, dehiscing by detachment; connective thick, fleshy, ovoid or lanceolate, obtuse at apex or most commonly produced in apiculum or linguiform appendage; pollen shed as isopolar monads; intrastaminal free disc surrounding ovary, membranaceous or subcoriaceous, nectariferous, annular, tubular or cupular, dentate, lobate, laciniate or composed of 10–20 free scales; style single, entire, columnar, erect, as long as stamens (1.2–12 mm long, exceptionally 30 mm long or shorter (0.3–0.9 mm long), rarely very short and rather thick or longer; stigma narrowly or broadly capitate, 5-lobate or 5-radiate; ovary superior, ovoid or ellipsoid, sessile, syncarpous, (4-)-5(−7) septate with axile placentation, locules uniovulate or biovulate; ovules anatropous, epitropous with 2 integuments, pendant at inner angle of ovary cells, micropyle pointing upward, raphe ventral; when 2 ovules present in each cell, superposite and lower one hanging from longer funiculus. Fruit a drupe (drupoid), from small (not exceeding 16 mm) to large, black, blackish, reddish, yellow, or orange when mature, usually aromatic; exocarp with smooth surface, glabrous, or pilose; mesocarp hard-fleshy varying from pulpy to fibrous, subcoriaceous texture, often aromatic and edible; endocarp woody, usually very hard, compact or with many resin-filled, round cavities, rarely spongy-woody, 5 septate, commonly with only 1–2(−5) seeds developed; surface smooth, bullate, rugose, or tuberculate, slightly striate or strongly costate; with dehiscence germinal, provided with as many longitudinal opercula or valves as carpels, which may open or be pushed away by emerging embryo at germination of seed inside fruit; often subapical foramina present in *Duckesia, Endopleura, Humiria*, and *Humiriastrum*. Seeds oblong, with double testa, exterior often adherent to endocarp, inner membranaceous, thin; one or two per locule; embryo straight or slightly curved, cotyledons oblong or ovate, often subcordate at base, radicle half as long, endosperm fleshy and oily.

A predominantly Neotropical family consisting of eight genera and 65 species distributed from Nicaragua to southern Brazil, with a single widespread species in West Africa.

Key to Genera of Humiriaceae

1. Anthers with 2 bilocular thecae; stamens usually 50–180 (rarely 15–18 in *V. depleta*); style 2–-2 mm long; carpels 2-ovulate........................8. **Vantanea**

1. Anthers with 2 or 4 unilocular free thecae; stamens 10–30; style 0.3–4 mm long; carpels uniovulate except in *Humiria*.

   2. Anthers with 4 unilocular thecae; connective acute; carpels opposite sepals, 1-ovulate.

      3. Petioles 1–3 mm long; leaves 2.2–7.5 × 1–3.3 cm; endocarp, evenly costate with long, lingulate valves, texture spongy-lignose (unknown in *D. liesneri*) resinous-lacunose.; stamens 20–25; style columnar, 1.5 mm long..1. **Duckesia**
      3. Petioles 10–30 mm long; leaves 8–33 × 2.3–8.5 cm; endocarp prominently, sharply costate and furrowed with shorter, inconspicuous valves at bottom of furrows; texture compact-woody, not resinous-lacunose...2. **Endopleura**

   2. Anthers with 2 unilocular thecae; connective obtuse or acute; carpels 1-ovulate except in *Humiria*.

      4. Anthers with thick very obtuse connective; thecae basal, glabrous; endocarp strongly costate, valvate at furrows, compact woody, not resinous lacunose..3. **Hylocarpa**
      4. Anthers with attenuate acute connective (rarely obtuse); thecae as below, glabrous or pilose; costae or shallowly furrowed, resinous lacunose or not.

         5. Thecae of anthers basal, pilose; carpels opposite petals, 2-ovulate; endocarp woody, striate, evenly costate-valvate, 5-foraminate at apex, not resinous-lacunose, valves linear, oblong or ligulate; stamens 20...4. **Humiria**
         5. Thecae of anthers basal or inferolateral, glabrous; carpels opposite sepals, 1-ovulate; endocarp as below; stamens 10 or 20.

            6. Stamens 10; thecae of anthers inferolateral; endocarp shallowly or inconspicuously furrowed, not foraminate at apex, resinous-lacunose, valve broad, adjacent, alternating ribs thin, inconspicuous...6. **Sacoglottis**
            6. Stamens 20; thecae and endocarp as below.

               7. Episepalous stamens 5, longer, trifurcate at apex, triantheriferous; epipetalous stamens 5, medium-sized, entire, monantheriferous; ten shorter alternate stamens, monantheriferous; thecae inferolateral; endocarp shallowly or inconspicuously furrowed, not foraminate at apex, resinous-lacunose, valves broad, adjacent, alternating ribs inconspicuous; leaf margins entire or only slightly crenulate..7. **Schistostemon**
               7. All 20 stamens monantheriferous, 10 episepalous and epipetalous longer than alternating ones; thecae basal; endocarp 5-foraminate at apex, with 5 alternating, descending, oblong

and short opercular valves, not resinous lacunose; leaf margins usually distinctly serrate-crenulate...........5. **Humiriastrum**

Key to Genera of Humiriaceae Based on Fruits

1. Fruit small, 0.9–1.6 cm long... 4. **Humiria**
1. Fruit much larger, at least 2.5 cm long.

    2. Endocarp resinous lacunose.

        3. Endocarp strongly verrucose, 5-foraminate....................1. **Duckesia**
        3. Endocarp nor strongly verrucose, not foraminate...6. **Sacoglottis**, 7. **Schistostemon**

    2. Endocarp not resinous lacunose.

        4. Endocarp distinctly and sharply costate.

            5. Endocarp 4–6 × 2–3.8 cm, with short inconspicuous val ves..2. **Endopleura**
            5. Endocarp 9.5 × 4.5 cm, valvate along furrows..........3. **Hylocarpa**

        4. Endocarp not strongly costate.

            6. Endocarp with lingulate valves..............................8. **Vantanea**
            6. Endocarp with short opercular valves...............5. **Humiriastrum**

10.1 Duckesia

Duckesia Cuatr., Contr. U.S. Natl. Herb. 35(2): 76–80. 1961. Type. *Duckesia verrucosa* (Ducke) Cuatr.

Trees or shrubs. Leaves alternate, thin-coriaceous, short-petiolate, lamina margins serrate, the teeth terminated by a deciduous gland. Inflorescences axillary, cymose-paniculate, furcate below, upward with alternate branching; bracts persistent or deciduous. Sepals 5, suborbicular, imbricate, united into a cup. Petals 5, free, aestivation contorted or cochlear. Stamens 20(−25), biseriate, glabrous, filaments united at base, complanate, subulate, papillose or smooth, alternating in three dimensions, 20 fertile or with 5 fertile stamens antepetalous, alternating with a group of 4–5 sterile ones with thinner filaments one of which is longer and antesepalous, occasionally substituted by some shorter staminodia; anthers linear-lanceolate, glabrous, dorsifixed, usually only 5 fertile, the connective lanceolate, fleshy and thick in fertile anthers, thecae 4, unilocular, dissociated or close together, subglobose or ellipsoid, two attached at the base, the other two laterally at the middle, dehiscing by flap-like valves. Disc of 10 subulate, free scales surrounding the ovary.

Ovary glabrous, 5-locular, the locules uniovulate; carpels opposite sepals; style erect, thick, equalling or shorter than the stamens in height, stigmas 5-lobed. Drupe large, ovoid, almost smooth, exocarp thick, coriaceous and fragile when dry; endocarp strongly verrucose, spongy-lignose, resinous-lacunose, 5-foraminate at apex, laterally with five longitudinal plates or opercula for germination. Seeds oblong.

Distribution. Until recently, a unispecific genus of central Amazonian Brazil, but based on a second collection, Wurdack and Zartman (2019) transferred *Humiriastrum liesneri* to this genus.

Differs from *Sacoglottis* in the anthers with a long connective and four unilocular thecae, two of them at the base and two higher up and lateral and in the valvate endocarp.

Key to Species of Duckesia

1. Leaves lanceolate; sepals glabrous except for margin; petals glabrous; filaments papillose; pistil longer than ovary........................ **2. D. verrucosa**

1. Leaves elliptic to obovate; sepals hispid; petals hirsute; filaments smooth; pistil shorter than ovary..**1. D. liesneri**

1. **Duckesia liesneri** (Cuatrecasas) K. Wurdack & C.E. Zartman, PhytoKeys 124: 101. 2019; *Humiriastrum liesneri* Cuatrecasas, Phytologia 71: 165. 1991; Cuatrecasas & Huber, Fl. Venez. Guayana (Steyermark et al.) 5: 632. 1999. Type. Venezuela, Amazonas: Río Negro, Cerro Aracamuni summit, 1400 m, 28 Oct 1987 (fl), *R. L. Liesner & G. Carnevali 22,589* (holotype, US 00409970, isotypes, MO 249111, NY 003299209, US 00512570, VEN 22589).

Shrubs to small trees 2–4 m tall, the young branches densely hirtellous, glabrescent. Leaves with petioles 1–3 mm long, thickened at base into a glandular pulvinus; laminas elliptic to obovate-rounded, 3.5–4.8 × 2.2–3.3 cm and some smaller, 2.2–3 × 1.5–2.8 cm, rigidly coriaceous, glabrous above, minutely hirtellous when young beneath, surface densely covered with minute granular glands, base broadly cuneate, apex retuse to bluntly attenuate, margins slightly crenate and revolute, setae present; midrib impressed above, primary veins ca, 6 pairs prominulous or inconspicuous on both surfaces, reticulate venation conspicuous beneath. Inflorescences axillary and subterminal not exceeding leaves in length, corymbose-paniculate, 2–4 × 3–4 cm, peduncle ca. 1 cm long, branches and pedicels densely minutely hirtellous; pedicels 0.2–1 mm long; bracts deciduous. Sepals 5, thick, 1.5–1.8 × 1.2–1.3 mm in buds, elliptic apex rounded, margins ciliate, densely hispid-hirsute on abaxial surface. Petals 5, 2.1–2.3 mm long in bud, thick, elliptic, medium line of dorsal surface hirsute the rest glabrous, margin smooth. Stamens 20; filaments connate to $\frac{1}{3}$ of length, glabrous, smooth; anthers adpressed 3-seriate in bud; connectives thick and more so in fertile stamens; two larger thecae inserted dorsally at level of filament insertion, two smaller ones basal. Ovary ovate, sulcate, glabrous, 5-locular, locule uniovular; style thick and shorter than ovary, stigma 5-lobed. Drupe not seen.

Distribution and habitat. This distinctive species is only known from two collections on the summits of Cerros Aracamuni and Unturián at 1150–1400 m in upland cloud forest (Fig. 10.2).

Phenology. Flowering in February and October.

Conservation status. Data deficient (DD) only known from two collections, but LC because of lack of threats.

Additional specimen examined. **VENEZUELA. AMAZONAS**: Sierra de Unturián, 1150 m, 1.55, −65.2, 3 Feb 1989 (fl), *Henderson 933* (NY, US).

This species was originally described from a single poor specimen. The second collection, cited above, enabled Wurdack and Zartman (2019) to observe the four separate sporangia, and thereby transfer it from *Humiriastrum* to *Duckesia*.

2. **Duckesia verrucosa** (Ducke) Cuatr., Contr. U.S. Natl. Herb. 35(2): 78. 1961. *Sacoglottis verrucosa* Ducke, Arch. Jard. Bot. Rio de Janeiro 3: 177. 1922. Type: Brazil, Pará, Óbidos, 11 Aug 1916 (fl), *A. Ducke MG 16325* (lectotype of Cuatrecasas (1961: 79), MG 016325; isolectotypes, BM 000796028, BM 000796029, INPA, MG 016764, P 01903244, P 01903245, NY 00388411, R 000002406, RB 00539052, S-R9884, US 101213, US 1101191). Fig. 10.1.

Large trees, the young branches minutely hirtellous pubescent. Leaves subsessile with petioles 0–1.5 mm, pubescent; laminas thin-coriaceous, lanceolate, 3–9 × 1–3 cm, cuneate at base, acuminate or cuspidate at apex, margins conspicuously crenulate and with setae, glabrous above, glabrous beneath except for midrib; midrib prominulous above, prominent beneath, minutely pubescent; primary veins 12–14 pairs, prominulous, arcuate-anastomosing near margin, venation obscure. Inflorescence axillary, shorter than the leaves, to 1.5 m long, cymose-paniculate, divaricate, branchlets thin, peduncles and branches minutely hirtellous pubescent; bracts persistent, ovate-lanceolate, clasping, 0.5–1 mm long, puberulous; pedicels thick, 0.4–0.5 mm long, articulate with short peduncles or sessile. Sepals 0.6 mm long, orbicular, glabrous and minutely ciliate on margins. Petals linear-oblong, 3.5 × 1 mm, glabrous. Stamens 20–25, glabrous, filaments 1.7–2.3 mm long, subulate, minutely papillose, united at base, the 5 opposite the petals larger; anthers linear-lanceolate, glabrous, 1 mm long, dorsifixed, 4–5 fertile, 15–20 sterile, thecae 4, two short-ellipsoid, basal, the other two oblong-ellipsoid, lateral on the middle, connective lanceolate, subtriquetrous, carnose. Disc of 10 free subulate scales, 0.5 mm long. Ovary ovoid or subglobose, 1 mm high, furrowed, 5-locular, the locules uniovulate, ovules ellipsoid-oblong, 0.6 mm long; style columnar, glabrous, 1.5 mm long, stigma subcapitate-pyramidal, slightly 5-lobed. Drupe ovoid or subglobose, 7 × 6 cm, exocarp smooth on exterior, 10–13 mm thick; endocarp ovoid abruptly acute at apex, spongy-lignose, densely resinous-lacunose with tiny cavities, verrucose, 5-valvate lengthwise, 4–5-locular, with 5 apical foramina.

Distribution and habitat. Forest on *terra firme* in Amazonian Brazil (Fig. 10.2).

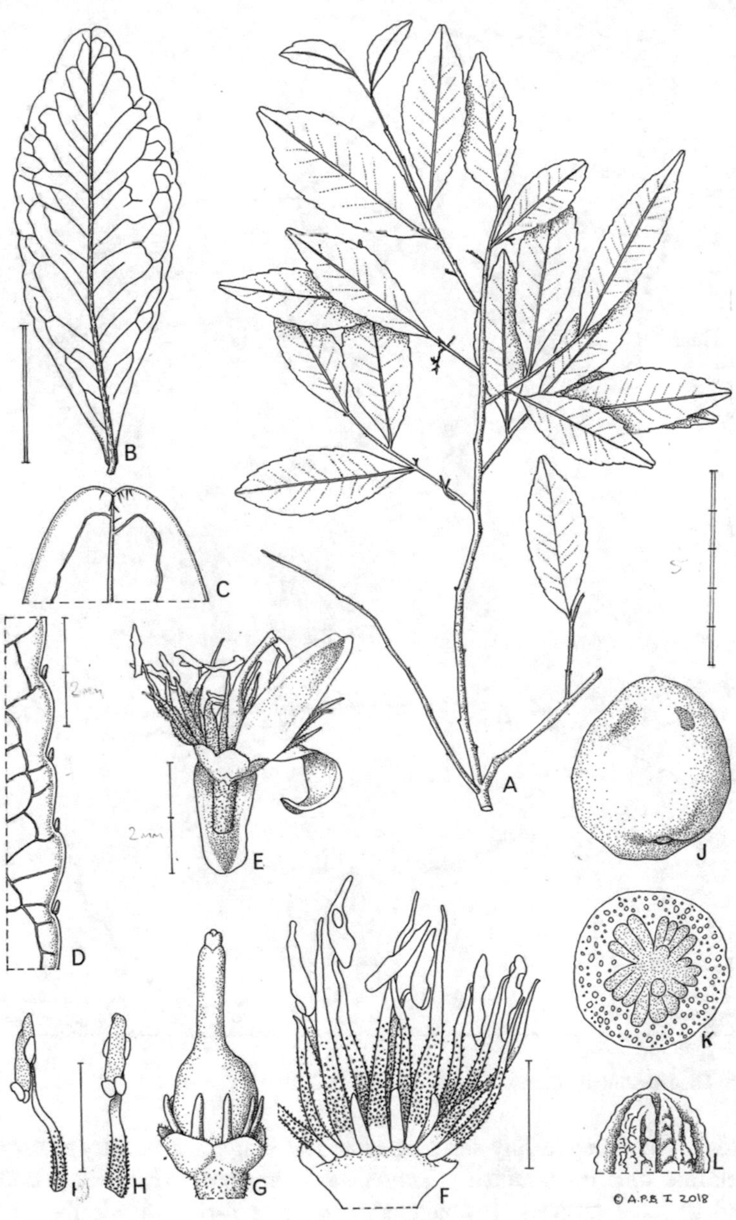

Fig. 10.1 *Duckesia verrucosa*. (**a**) Habit. (**b**) Leaf, abaxial surface. (**c**) Leaf apex, abaxial surface. (**d**) Leaf margin, abaxial surface. (**e**) Flower with two petals removed. (**f**) Part of androecium. (**g**) Ovary, style, stigma and disc appendages. (**h**) Short stamen. (**i**) Short stamen, side view. (**j**) Fruit. (**k**) Fruit section. (**l**) Upper half of seed, side view. Single bar = 1 mm, graduated single bar = 2 mm, double bar = 1 cm, graduated double bar = 5 cm. (A–D, from *Ducke 2108*; E–G, I, from *Menandro 47*; J, from *Vincentini* et al. *447*; H, L from *Sothers 847*)

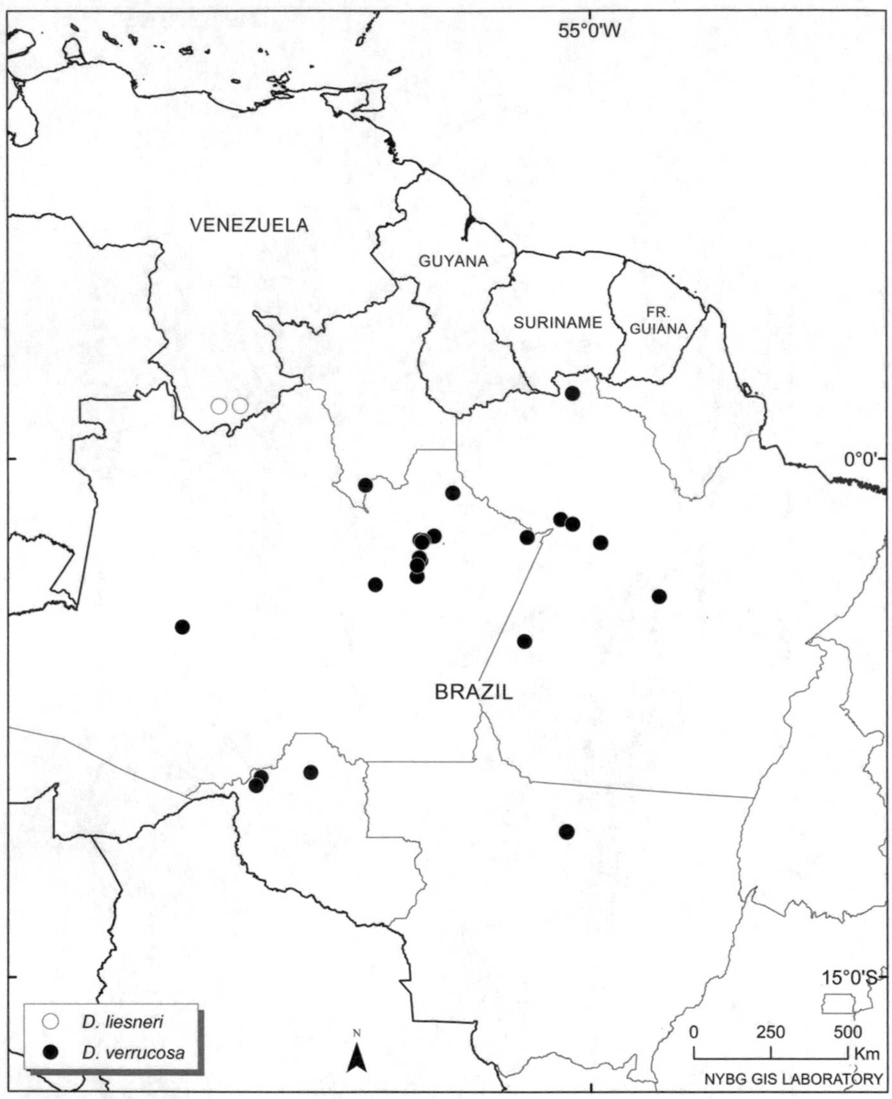

Fig. 10.2 Distribution of *Duckesia verrucosa* and *D. liesneri*

Phenology. Flowering in July and August and fruiting from January to April.

Local names and uses. Brazil: *Uchi-curuá, uchi-corôa*. The fruit is edible and found in local markets. It contains a lauric acid-rich oil similar to that of *Endopleura* (Pesce 2009).

Conservation status. Least Concern (LC).

Illustrations. Ducke (1922) fig. 10b, Ducke (1930) Figs. 14–35a, 35b; Cuatrecasas (1961), Fig. 16 g–k; Cavalcante (2010): 246.

Additional specimens examined. BRAZIL. AMAZONAS: Manaus, Estrada do Aleixo, −3.4167, −60.0333, 23 Jul 1947 (fl), *Ducke 2108* (IAC, IAN, INPA, K, MG, NY, P, RB, U, US); Reserva Ducke, Manaus, −2.8833, −59.9667, 24 Mar 1994, *Vincentini 447* (INPA, K), same loc. 3 Apr 1996 (fr), *Sothers 847* (INPA, K, MG, MO, NY, RB); Manaus, Distrito Agropecuaria, Fazenda Porto Alegre, Reserva 3402, 2°53′S, 59°54′W, 24 Jan 1989 (fr), *Pacheco* et al. *143* (INPA, K, NY, US), Fazenda Esteio, −2.3833, −59.85, 26 Jul 1989 (fr), *Sette Silva 1301.2810.2* (INPA, K, NY, US); Reserva Xixuaú-Xiparaná, Igarapé Santa Rosa, Mun. Rorainópolis, −0.7707, −61.5463, 23 Aug 2010 (fl), *Marinho 5* (INPA, MIRR); Carauari, São Raimundo, −4.8827, −66.8958, 11 Nov 2009, *Hawes 1085* (EAFM); Rio Cuieiras, −3.1019, −60.025, 27 Apr 1975 (fl), *Vilhena 50* (INPA); Cachoeira do Mangabal, Rio Tapajós, −5.3, −56.9166, 13 Feb 1913 (fr), *Ducke MG 16764* (=RB1260), (syntypes BM 000796028, BM 000796029, MG, RB 00539051, US 101212). **PARÁ:** Óbidos, −1.9, −55.5166, 11 July 1916 (fl), *Ducke MG 16325* (BM, F, MG, RB10815, NY, S, U, US); Rio Trombetas, Cachoeira Porteira, −4, −53, 5 Jun 1978 (fr), *Silva 4722* (F, MO); Estrada Sumaúma, Oriximinã, −1.7655, −55.8661 29 Sep 1985, *Menandro 47* (CEPEC, CVRD, RB, UFRGS); Porto Trombetas, Floresta Nacional Saracá-Taquera, −2.2881, −56.8375, 13 Dec 2009, *Junqueira & Ramos 1087* (EAFM); Reserva Florestal Curuá-Una, −2.4430, −54.7083, 14 Sep 1967, *Silva INPA 27679* (INPA). **RONDÔNIA**: Aug 1952, *Silva 416* (IAN); left margin of Madeira River, Porto Velho, −9.24, −64.55, 80 m, 26 Jul 1989 (fr), *Pereira-Silva 15,977* (CEN, HUEFS, INPA, RB, RON, UNIR); BR364, ramal dos Britos, toward Rio Branco, −9.4808, −64.7616, Jul 1989, *Pereira-Silva 16,041* (CEN, HUEFS, INPA, RB, RON, UNIR). **MATO GROSSO**: Nova Canaã da Norte, Colider Dam, −10.8244, −55.7016, 2 May 2015 (fr), *Lautert* et al. *s.n.* (HCF, MBM, TANG, UNEMAT).

10.2 Endopleura

Endopleura Cuatr., Contr. U.S. Natl. Herb. 35(2): 80–83. 1961. Type. *Endopleura uchi* (Huber) Cuatrecasas.

Large trees, the young branches glabrous. Leaves alternate, coriaceous, petiolate, lamina margins serrate. Inflorescence axillary, cymose-paniculate; bracts and bracteoles persistent. Sepals, orbicular, imbricate, connate at base. Petals 5, free, aestivation contorted or cochlear. Stamens 20–25, biseriate, filaments thick, complanate, papillose, connate at base, unequal, the 5 larger ones alternating with the sepals and 5 with the petals; and 10 short ones alternating with the longer ones; anthers dorsifixed, with thick elongate, acute connective, thecae 4, unilocular, disassociated, subglobose, 2 attached at base the other 2 lateral above base, sometimes 2 thecae or rarely 4 sterile in shorter stamens, dehiscence irregular. Disc of 10 ovate-triangular thick scales, united at base. Ovary glabrous, suborbicular, 5-locular, the locules uniovulate; carpels opposite the sepals; style short, stigma capitate. Drupe large, ellipsoid, exocarp fibrous-farinaceous, coriaceous when dry; endocarp woody, not resinous, with 5 apical foramina and 5 strongly elevated ribs divided into two except at the end. Seeds oblong.

Distribution. A monotypic genus of Amazonia widely distributed in Guyana and in Brazil in central and southern Amazonia, and south into Bolivia.

Differs from *Sacoglottis* by the anthers with 4 unilocular thecae two of which are basal and the other two located above and laterally, and also in the deeply sulcate endocarp with five prominent longitudinal bifid winged ribs.

1. **Endopleura uchi** (Huber) Cuatr., Contr. U.S. Natl. Herb. 35(2): 81.1961; Cuatrecasas & Huber, Fl. Venez. Guayana (Steyermark et al.) 5: 625. 1999. *Sacoglottis uchi* Huber, Bol. Hist. Nat. Mus. Paraense 2: 489. 1898. Type. Brazil, Pará, Belém, cultivated, Oct 1897, *J. E. Huber 1260* (holotype, MG 001260; isotypes, BM, G 00368452, MG 000239, P 01903246, RB 00123602, US 00101214). Fig. 10.3.

Large trees, the young branches glabrous. Leaves with petioles 1–3.5 cm long, sulcate and slightly winged at margin, glabrous or pubescent; laminas elliptic obong or elliptic-lanceolate, chartaceous or thin-coriaceous, confluent onto petioles, glabrous or slightly pubescent near to base beneath, 8–22(−33) × 2.3–8.5 cm, abruptly and obtusely cuneate at base, acuminate or cuspidate at apex, the acumen 8–15 mm long, often curved, margins crenate and slightly thickened, setae long leaving large scars; midrib prominulous above, prominent beneath; primary veins 12–14 pairs, prominulous on both surfaces; venation conspicuously reticulate, prominulous. Inflorescence axillary, cymose-paniculate, much shorter than the leaves, with trichotomous or dichotomous branching, peduncles 1–3 cm long, hirtellous or puberulous, branches puberulous; bracts persistent, ovate, clasping, 2 mm long; bracteoles persistent, acute, puberulous, minute, 0.5 mm long; pedicels hirtellous pubescent, 0.2–0.5 mm long, articulate with 0–1 mm long peduncles beneath. Sepals 0.7 mm long, orbicular, hirsute-pubescent, connate at base. Petals greenish, linear-oblong, 3–3.5 × 1–1.4 mm, subacute or subobtuse, glabrous within, hirtellous pubescent on exterior. Stamens 20–25; filaments 1.5–2 mm long, angulate, densely papillose, connate toward base, unequal in length, the longest with larger anthers and 5 opposite the sepals, 5 opposite petals; anthers 0.9 mm long, globose-elliptic, 0.2 mm long, with 2 basal and 2 inferior-sided thecae, connective, lanceolate, acute, ca 0.7 mm long, sometimes two thecae sterile and rarely all 4 on shorter anthers. Disc scales 10, 0.6 mm long, triangular, united at base, surrounding the ovary. Ovary subglobose, glabrous, 0.9–1 mm high, 5-locular, the locules uniovulate, opposite the sepals; style thick, longer than the ovary, stigma capitate, 5-lobed. Drupe oblong-ellipsoid, 4–6 × 2–3.8 cm, rounded at both ends; exocarp 1 mm thick, coriaceous when dry, almost smooth, reddish–brown; mesocarp 2–3 mm thick, fleshy, resinous; endocarp woody, deeply 5-grooved, with 5 projecting ribs divided into two except at one end, giving a 10-radiate shaped section, with 5 short opercula located at the bottom of the furrow; seeds usually 2–3, oblong, 3 cm long, 7 mm thick.

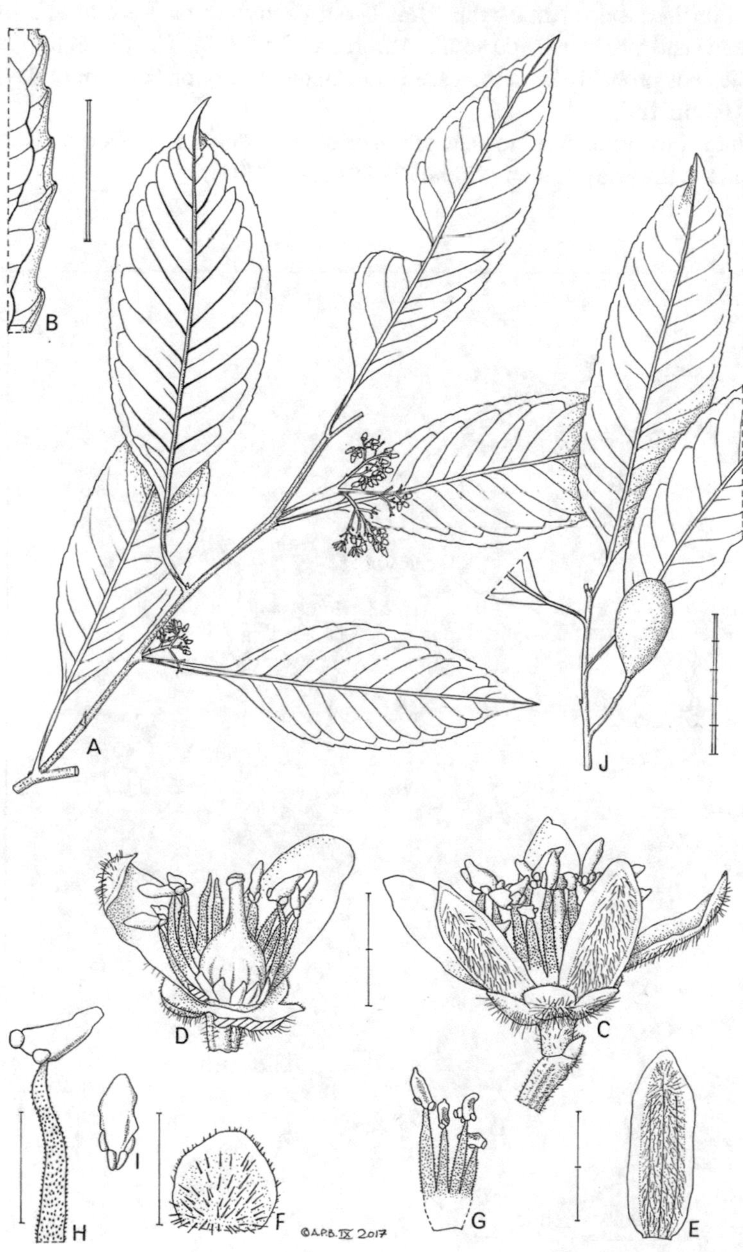

Endopleura uchi del. Andrew Brown September 2017

Fig. 10.3 *Endopleura uchi*. (**a**) Habit. (**b**) Abaxial leaf margin. (**c**) Flower. (**d**) Flower, 3 petals and 2 sepals removed. (**e**) Petal, exterior surface. (**f**) Sepal. (**g**) Part of stamina tube. (**h**) Stamen showing basal and median thecae, side view. (**i**) Outer face of anther showing thecae. (**j**) Fruiting shoot. Single bar =1 mm, graduated single bar = 2 mm, double bar = 1 cm, graduated double bar = 5 cm. (From *Ducke 241*)

Distribution and habitat. Forests on *terra firme* but often near to rivers, especially common in eastern Amazonian Brazil, but extending far beyond and into the Guianas and Venezuela and south to Bolivia (Fig. 10.4). The distribution of this species has probably been increased anthropogenically on account of the much-used edible fruit.

Phenology. Flowering April to October, mainly in the dry season, fruiting December to June in the rainy season (Pires-O'Brien et al. 1994).

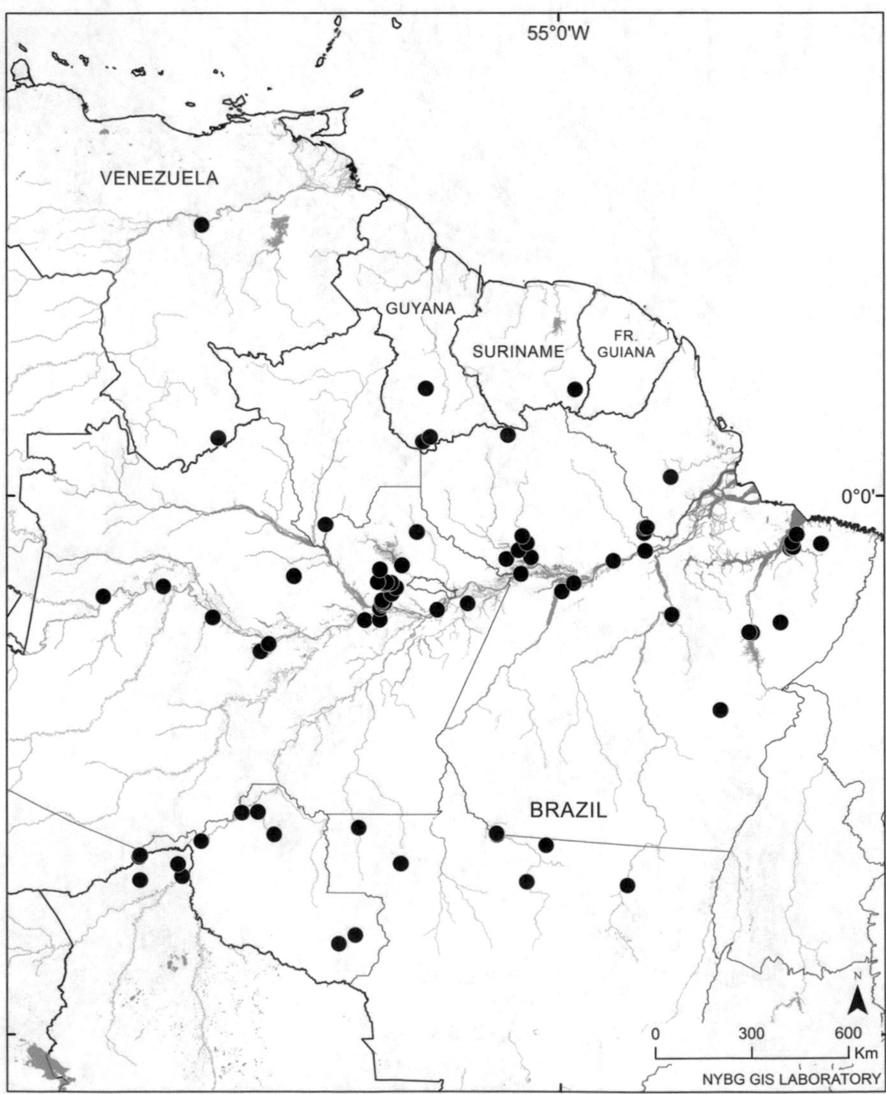

Fig. 10.4 Distribution of *Endopluera uchi*

Local names and uses. According to Shanley and Rosa (2004) and Shanley and Medina (2005), this species is much cultivated in home gardens around Belém and in eastern Amazonia for the edible fruit, and they also report that the fruit is much sought after by many animals including squirrels, agoutis, and armadillos. The fruit yields 8–18% of an edible oil that is used locally for cooking and is similar to olive oil (Pinto 1956; Pesce 2009). The oil is unsaturated and oleic acid rich also with linoleic and linolenic acids. The fruit is also used in ice cream, sweets, and local liquors. The wood is straight-grained and heavy and is used in furniture, poles, and stakes. The fruit pulp contains high amounts of dietary fibers and phytosterols, as well as notable contents of vitamins C and D (Marx et al. 2002). Da Silva et al. (2009) report antimicrobial activity from bergenin isolated from the bark of this species. Traditionally, this species has been used medicinally for the treatment of inflammations and female disorders. Silva and Teixeira (2015) found five compounds in the bark of this species using HPLC-DAD. Bergenin was the major one. Infusions contained the greatest richness in these metabolites. The anti-oxidants, acetylcholinesterase, butyrylcholinesterase, a-glucosidase, and antibacterial activity were tested by in vitro assays. Antibacterial capacity of both extracts was investigated against Gram-positive and Gram-negative bacteria, and it was more effective against the first one. Brazil; *Uchi, uchi-pucú uxi;* Bolivia: *chuendrillo.*

Conservation status. Least concern (LC).

Illustrations. Cuatrecasas (1961) figs. 16 a–f, 18 a; Cuatrecasas and Huber (1999) fig. 534; Cavalcante (2010): 244, fruit; Shanley and Medina (2005): 147.

Selected specimens examined. VENEZUELA. AMAZONAS: Mun. Río Negro, 1°37′N, 64°33′W (fl) *Valera 126* (NY). **BOLÍVAR**: Guayapo, Bajo Caura, 120 m, 7 May 1939 (fr), *Williams 12,065* (K). **GUYANA.** Essequibo River, Gunn's Landing 1.65, −58.63, 26 Sep 1977 (fl), *Jansen-Jacobs 1954* (F, INPA, K, NY, US); U-Takutu-U, Essequibo region, 1.5 km S of Kamoa River, 1°31′N, 58°49′W, 13 Nov 1996 imm fr), *Clarke 3175* (K, MO, NY, US). **SURINAME**. Ulemari River, 99 km above confluence with River Litani, 2°58′16″N, 54°33′14″W, 14 Apr 1998 (fr), *Evans & Peckham 2866* (RB); 3 km S of Kwamalasamutu village, Sipalawini, 2325, −56.7889, 3 Mar 2005, (US). **BRAZIL. AMAZONAS**: Manaus, Pensador, 10 Aug 1943 (fl), *Ducke 241* (A, F, IAN, K, MG, MO, NY, S, US); Rio Tonantins, 27 Oct 1949 (fl), *Fróes 25,565* (IAC, IAN, P); Reserva Ducke, Manaus-Itacoatiara Km 25, 2 Sep 1966 (fl), *Prance* et al *2179* (INPA, K, NY, US); Distrito Agropecuário, Fazenda Esteio, 2.38° S, 59.85° W, 5 Oct 1989, *Sette-Silva 1301.3307.2* (INPA, NY); behind Fonte Boa, −2.50°, −66.10°, 22 Aug 1973 (fr), *Lleras P17474* (F, INPA, MO, NY, US); Rio Preto da Eva, Manaus-Itacoatiara Km 134, 6 Jul 1975 (st), *Monteiro INPA50057* (INPA); Rio Juruá, Seringal Santa Rosa, Lago do Curapé, Saracura, 22 Aug 1975 (fl), *Coelho INPA52396* (INPA); Coari, Lago Coari, 22 Apr 1970 (imm fr), *Lima & Lima 350* (IFAM, INPA, MO). **PARÁ**: grounds of Museu Goeldi, Belém, 1 Jul 1908 (fl), *Baker 58* (A, GH, K, MO, NY, P, U, US); Rio Cumināmirim, Castanhal das Pedras, 13 Oct 1913, *Ducke RB 14979* (MG); Rio

Xingú, Estrada da Volta, 20 Dec 1916, *Ducke MG16641* (MG); Estrada PTR: Mina, Km 7, Oriximiná, 2 Nov 1985, *Menandro 53* (CEPEC, CVRD, RB); Belterra, 7 Jul 1947 (fl), *Black 47–1001* (IAC, IAN, INPA, NY, P, U, US, VEN, WIS): Curuá-Una dam, Belterra, −2.63°, −54.93°, 26 Sep 1977 (fl), *Pessoal do L. P. F. Brasília, 1040* (INPA); Gleba Monte Dourado, Almeirim, −0.86, −52.53, 31 Jun 1987 (fl), *Pires & Silva 1763* (HAMAB, IEPA, INPA, K, MG); Serra Sacazinho, 40 km N of Mineração Rio Norte, −1.75, −56.5, 10 Nov 1987 (fr), *Cid Ferreira 9510* (F, HAMAB, INPA, MO, NY, RB, US); Serra Carajás, 2–10 km SE of Rio Itacaiúnas, −5.92, −50.48, 14 Jun 1982 (fl), *Sperling 6154* (INPA, MO, NY); Belém-Brasília Km 94, 5 May 1959 (fl), *Kuhlmann & Jimbo 197* (CEPEC, K, MO, NY, P). **AMAPÁ**: Road Cupixi-Vila Nova, 8 km SSW of Cupuxi, Macapá, 6 Jan 1985 (fr), *Rabelo 3224* (F, HAMAB, INPA, MO, NY, US); Monte Dourado, Rio Jarí, −0.86, −52.53, Apr 1970 (st), *Moore 20* (NY). **ACRE:** Acrelândia, Assentamento Extrativista Porto Dias**,** 108 km E of Rio Branco on BR364, −9.99°, −66.78°, 5 Oct 2003 (fr), *Acevedo-Rodriguez 13,720* (MO, NY): Rio Abunã, Km 108, BR364, Palhau, 9°57′20” S, 66°45′50″W, 5 Oct 2003 (fr), *Daly* et al. *12,193* (NY, RB). **RONDÔNIA**: 10 km above Mutumparaná, road to São Lorenço mines, 25 Nov 1968 (fr), *Prance* et al. *8869* (F, INPA, MO, NY, US); Reserva Usina Hidroelectrica Samuel, 20 Jun 1986 (fr), *Cid Ferreira 7536* (F, HUNEB, INPA, MO, NY, US); PN Mapinguari, Rio Madeira, Porto Velho, 18 Aug 2012 (imm fr), *Pereira-Silva 16,307* (CEN, INPA, K, NY, RB, UNIR); Chupinguaia, 12°10′33” S, 60°44′02″W, 6 Dec 2013 (fr), *Bigio 1238* (INPA, RB, UNIR). **MATO GROSSO**: Aripuanã, 27 Apr 1977, *Gomes 1435* (INPA); Núcleo de Aripuanã 4–1-1, 3 Jun 1977, *Lima 1766* (INPA); Mun. Novo Mundo, Parque Estadual Cristalino, 9°39′S, 55°23′W, 295 m, 1 Feb 2008 (fr), *Sasaki 2100* (K, USP); Nova Canaã do Norte, 268 m, 8,789,225, 636,423, 5 Oct 2003 (fr), *Lautert 407* (MBM, RB, UFMT); Platô da Serra Sacazinho, 40 km S of Mineração Rio Norte, −1.75, −58.75, 10 Nov 1987 (fl), *Cid Ferreira 9510* (F, INPA, MO, NY, RB, US). **BOLIVIA. BENI:** Cachuela Esperanza, Río Beni, 21 Jul 1921 (fr), *Meyer 107* (NY). **PANDO**: Abuna, Cachuela El Carmen, N of Río Negro, 10°12′S, 65°42′W, 150 m, 20 Oct 1999 (fr), *Toledo* et al. *1148* (MO); Abuna, −10.65, −66.7666, 4 Jul 1992, *Gentry* et al. *77,753* (LPB, MO).

10.3 Hylocarpa

Hylocarpa Cuatr., Contr. U.S. Natl. Herb. 35(2): 84–87. 1961. Type: *Hylocarpa heterocarpa* (Ducke) Cuatrecasas.

Trees. Leaves alternate, petiolate, coriaceous, petiolate, lamina margins crenate. Inflorescences axillary, dichotomously branched, paniculate; bracts deciduous. Sepals 5, suborbicular, imbricate, connate at base. Petals 5, free, thick, aestivation cochlear or contorted. Stamens 20, glabrous, filaments biseriate, thick, papillose, connate at base; anthers club or hammer-shaped, only 15 or fewer fertile, the rest sterile, dorsifixed, thecae 2, unilocular, subglobose, basal, dehiscing by longitudinal slits. Disc of 10 linear, free, scales. Ovary ovoid, strigose, 5-locular, the locules uniovulate; carpels opposite the sepals; style thick, short, stigma capitate, 5-lobed. Drupe large, subfusiform, exocarp thick, farinaceous; endocarp woody, hard, not resinous, prominently 5-costate, the elevated furrow alternating with deep furrows with a long, linear protruding operculum; seeds oblong; usually only 1 or 2 developing.

Distribution. A monotypic genus of upper Amazonia.

1. **Hylocarpa heterocarpa** (Ducke) Cuatr., Contr. U.S. Natl. Herb. 35(2): 85. 1961; *Sacoglottis heterocarpa* Ducke, Arq. Inst. Biol.Veg. 4: 27. 1938. Type. Brazil. Amazonas: Rio Curicuriari, near cataract Cajú, 18 Nov 1934 (fl), *Ducke RB 30137* (**lectotype designated here**, RB 00539062; isolectotypes G 00368467, K 000407345, WIS, P 01903249, U 0002484, US 101225). Figs. 9.4B and 10.5.

Large trees up to 30 m, the young branches glabrous. Leaves with petioles 2–3.5 cm long, striate, sulcate above, glabrous; laminas obovate-elliptic or subobovate, 11–17 × 7.5–10 cm, thick-coriaceous, glabrous, cuneate at base and decurrent onto petiole, rounded or obtuse at apex, margins slightly crenate, almost entire; midrib plane and flattened above, prominulous and striate beneath; primary veins 14–18 pairs, prominulous on both surfaces, reticulate venation prominulous on both surfaces. Inflorescences axillary, dichotomous, much shorter than the leaves, branches complanate, whitish-pubescent; bracts deciduous, ovate-triangular, 0.5 mm long; pedicels 0.1–0.2 mm long, glabrous, articulate with a short hirtellous peduncle beneath. Sepals orbicular, connate at base, glabrous except for ciliate margins, 0.8–1 mm long. Petals 5, free, white, linear-oblong, 3–5 × 1.5–1.8 mm, obtuse, glabrous. Stamens ca 30, biseriate, connate at base, filaments 1.5–1.7 mm long, thick, broadest at middle, densely papillose; anthers dorsifixed, about 12 fertile, connective fleshy, oblong-clavate to hammer-shaped, thecae 2, globose, 0.7–0.8 mm, separated at base; sterile anthers oblong, thick, obtuse at apex, 0.7–0.9 mm long. Disc of 10 linear, thick scales, ca 3 mm high. Ovary ovoid, slightly sulcate, strigose, 0.8 mm long, 5-locular, the locules uniovulate, opposite the sepals; style short, thick, stigma 5-lobed. Drupe elliptic-fusiform, 9.5 × 4.5 cm, attenuate toward base, abruptly rounded-constricted at base, narrowed and apiculate at apex; endocarp smooth, 2–3 mm thick; endocarp woody, broadly fusiform, truncate at base, narrow at apex, deeply 5-sulcate, 5-costate, the ribs thick, prominent and robust, with long germination opercula in the bottom of each furrow; seeds oblong, usually only 1 or 2 developing.

Distribution and habitat. Confined to the caatinga forest of the upper Rio Negro, Brazil (Fig. 10.6).

Phenology. Collected flowering in October and November and fruit in February.

Local names and uses. Brazil: *Cumaté da caatinga, cumaté-rana, fruta-de-xiri.*

Conservation status. Of very restricted distribution, but in a little threatened area. Least Concern (LC).

Illustrations. Cuatrecasas (1961), figs. 16 l–p, 18c, 19a–c.

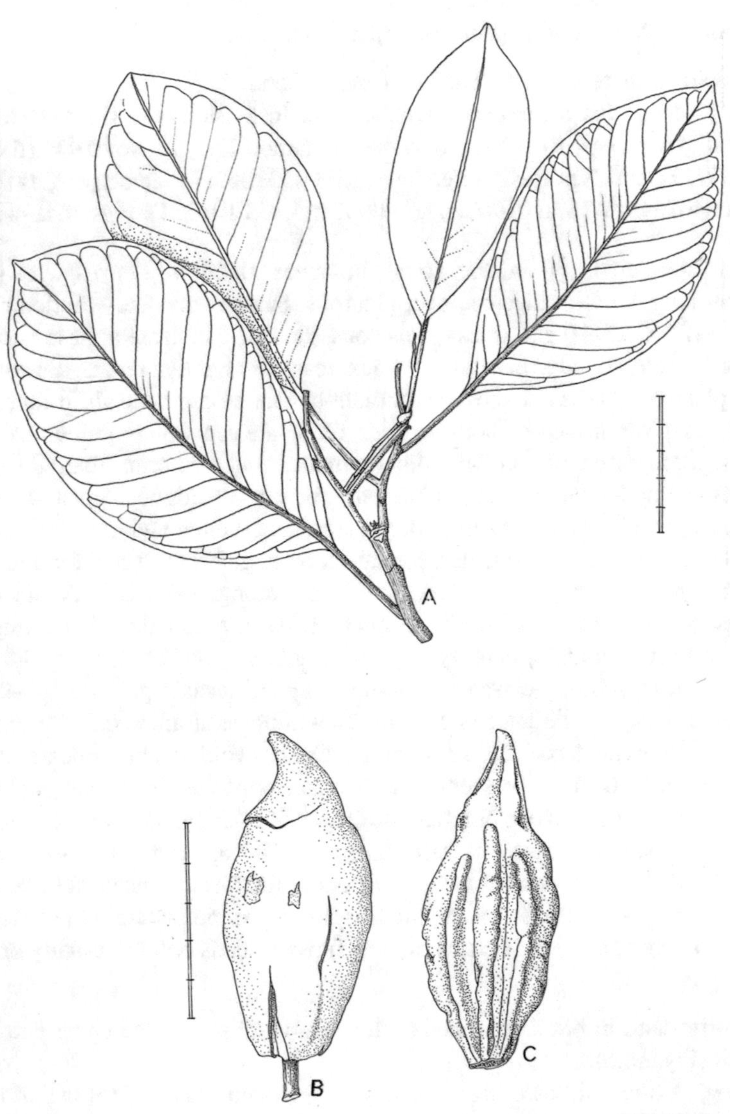

Fig. 10.5 *Hylocarpa heterocarpa*. A. Habit. B. Fruit. C. Seed. Drawn by Andrew Brown from Ducke 1937. Graduated bar = 5 cm

Fig. 10.6 Distribution of *Hylocarpa heterocarpa*

Additional specimens examined. BRAZIL. AMAZONAS: Rio Curicuriari, near cataract Cajú, 18 Nov 1936 (fl), *Ducke 265* (K 000407345, NY 00388408, NY 00388409, WIS), same data 21 Feb 1936 (fr), *Ducke RB 30137* (K, RB, U, US); Rio Içana, Tunuí, 23 Oct. 1947 (fl), *Pires 708* (IAN); Itacoatiara-Mirim, Mun. São Gabriel da Cachoeira, −0.166, −67.011, 28 Mar 2013, *Cardoso 3311* (HUEFS, INPA), same loc, Jul 2007, *Stropp & Assunção s.n.* (EAFM); Igarapé Mindú, São Gabriel da Cachoeira, −0.130, −67.089, Jul 2007, *Stropp & Assunção 629* (EAFM).

The Ducke-type specimen RB 30137 has flowers collected on 18 Nov 1936 and fruits under the same number collected on 21 Feb 1936, and the collection Ducke 265 has the same information and the date 18 Nov 1936 was also considered as an isotype by Cuatrecasas (1961). I have chosen the RB 30137 specimen collected in November as the lectotype. The distribution and ecology of this species was discussed by Cardoso et al. (2015). It is confined to the white sand habitats of the upper Rio Negro of Brazil where it seems to be quite rare.

10.4 Humiria

Humiria St. Hilaire, Exp. Fam. 2: 374. 1805; Persoon, Syn. Pl. 2: 70. 1807; Urb., Fl. Bras. (Martius) 12 (2): 437.1877, orth. cons. (ICBN App IIIA) Type: *Humiria balsamifera* (Aubl.) St. Hil.

Houmiri Aubl., Hist. Pl. Guiane 1: 546. 1775.
Werniseckia Scop., Introd. Hist. Nat. 273. 1777.
Myrodendrum Schreb., Gen. Pl. Pl. ed., 8, 1: 358.1789.
Houmiria Juss., Gen. Pl. 435. 1789.
Houmirium Rich. ex Mart., Nov. Gen. Sp. 2: 142.1827.
Myrodendron Spreng., Syst. Veg. 2: 600. 1840.
Verniseckia Steud., Nomencl. Bot. (ed. 2) 2: 752. 1841.
Wernischeckia Scop. ex Post & Kuntze, Lex. gen. phan. 288.1904.

Shrubs or small- to medium-sized trees. Leaves alternate, sessile or petiolate, chartaceous to rigid coriaceous, lamina margins usually crenulate, with dotted glands near margin of lower side. Inflorescences axillary or subterminal, paniculate and corymbiform with alternate branching or dichotomous; bracts persistent. Sepals 5, suborbicular or ovate, imbricate, connate into a cupular calyx. Petals 5, free, thick-membraneous, aestivation cochlear or quincuncial. Stamens 20, uniseriate, filaments connate into a tube for almost lower half, complanate, densely papillose or muricate, of three types, 5 longer and antesepalous, 5 intermediate and antepetalous, with 10 shorter ones adjacent to antesepalous ones; anthers ovoid-lanceolate, dorsifixed above the base, connective thickly ligniform or lanceolate, much longer than the 2 subglobose, hairy thecae inserted sublaterally on the middle at the base, opening on the ventral side by a dorsal hinge opening outward. Disc annular, surrounding ovary, formed of 20 linear, thick more or less united scales. Ovary sparsely pilose at apex or glabrous, 5-locular, each locule with 2 ovules; carpels opposite the petals; style columnar, erect, as long or longer than filaments, hirsute, stigma globose, stellate. Drupe less than 1.6 cm long, ovoid, ellipsoid or oblong; exocarp thin; mesocarp fleshy, sweet, aromatic, edible; endocarp woody, ellipsoid or ovoid, finely ten striate (rarely 8), the striae equidistant, marking 5 longitudinal germinal valves, alternating with 5 small foramina at the apex, 1–4 seeds fertile.

Distribution. Colombia, Venezuela and Peru to eastern Brazil.

Key to Species of Humiria

1. Leaves small or linear, 0.3–1.5 cm wide, margins entire; sepals acute or orbicular.
   2. Leaves linear 2.5–10 × 0.3–0.8 mm; plant entirely glabrous, sepals orbicular, obtuse..**5. H. wurdackii**
   2. Leaves small, 1.5–4 × 0.5–1.5 cm; minutely pilose, slightly velutinous on both surfaces, young branches minutely pilose, sepals subacute ....**3. H. fruticosa**

1. Leaves larger (or if small obovate or oblong-ovate), glabrous or only midrib pubescent beneath (rarely sparsely hirtellous beneath); margins usually obscurely crenulate; sepals orbicular or obtuse.
   3. Leaves broad, 7–16 × 4–9.5 cm, lamina thick-coriaceous, rigid, oblong-ovate or obovate-elliptic; margins entire; petiole stout, broadly winged, folded, clasping; drupe ellipsoid, 10–12 × 7–9 mm; plant entirely glabrous...2. **H. crassifolia**
   3. Leaves usually narrower, 4–12 × 2–6 cm, lamina subcoriaceous to coriaceous, sessile and clasping or petiolate, margins slightly crenulate; petiole flat, narrower, not clasping; drupe oblong-ellipsoid or oblong, 10–14 × 4–8 mm; plant glabrous or slightly pubescent.
      4. Leaves 3–18 × 2–7 cm, Colombia, Venezuela, the Guianas and Amazonia..1. **H. balsamifera**
      4. Leaves 1–5–4.5 (−5) × 0.8–2.7 cm, campo rupestre and restingas of S Brazil..4. **H. parvifolia**

1. **Humiria balsamifera** (Aubl.) St. Hil., Exp. Fam. 2: 374.1805; Urb., Fl. Bras. (Martius) 12 (2): 440.1877; Cuatr., Contr. U.S. Natl. Herb. 35(2): 89. 1961; Cuatrecasas & Huber, Fl. Venez. Guayana (Steyermark et al.) 5: 628–629. 1999; *Houmiri balsamifera* Aubl., Hist. Pl. Guiane 1: 564. 1775; *Myrodendron balsamiferum* (Aubl.) Raeuschel, Nomencl. Bot. 156. 1797; *Humirium balsamiferum* (Aubl.) Benth., Hooker's J. Bot. Kew Gard. Misc. 65: 102. 1853. Type. French Guiana. *J. B. C. F. Aublet s.n.* (lectotype, **here designated,** BM 000795994, possible isotype GH 00043820).

Small to large trees or low shrubs, the young branches hirtellous, puberulous, or glabrous. Leaves sessile or with petioles 0.2–2.5 cm, sometimes slightly winged; leaves chartaceous, coriaceous, or rigid-coriaceous, usually glabrous on both surfaces, midrib sometimes puberulous; laminas of varied shape, elliptic, olong-obovate or oblong, 3–18 × 2–7 cm, often broad and clasping at base, or obtuse to cuneate, sometimes confluent onto petiole; apex rounded, truncate or obtuse, rarely mucronulate, often emarginate but mucronulate in depression; midrib prominulous and inconspicuous above, prominent beneath, glabrous or hirtellous, margins entire with marginal abaxial glands; primary veins inconspicuous above, slightly prominulous and conspicuous beneath. Inflorescences axillary and terminal, cymose-paniculate, usually corymbiform, 3–11 mm long, often densely clustered; peduncle 1–6 cm long, winged in some varieties or angulate, branches more or less winged, glabrous or hirtellous; bracts persistent, triangular, 0.5–3 mm, clasping; pedicels 0.5–2 mm long, thick, usually glabrous, articulate with rigid, puberulous peduncles 1–2 mm below. Calyx cupular, 1–2 mm high. Sepals, thick, orbicular, imbricate, connate toward base, glabrous except for ciliolate margins, rarely hirtellous. Petals white or greenish white, usually glabrous, rarely puberulous on exterior, linear-lanceolate, acute at apex, 4.5–7 mm long, 1–1.6 mm broad. Stamens 20, filaments erect, 4–5 mm, in two alternating lengths, connate for lower half, free portion papillose; 5 longer and antesepalous, 5 intermediate and antepetalous, with 10 shorter ones adjacent to antesepalous ones; anthers ovate lanceolate, 0.8–1 mm long, thecae basal, globose, pilose. Disc annular, of 20 linear scales united at base. Ovary ovoid, glabrous except for a few hairs near apex, 5-locular the locules biovulate; style erect, thick, hirtellous except for glabrous apex, stigma 5-lobed, stellate-capitate. Drupe oblong-ellipsoid, 10–14 × 5–8 mm, exocarp fleshy, smooth, glabrous; endocarp woody, ellipsoid-oblong, obtuse or rounded at base, acute at apex, with 5 conspicuous foramina around apex, with 10 thin and curved longitudinal furrows and 5 irregular cells.

Distribution and habitat. A widespread and variable species with many varieties, found predominantly on white sand and rocky terrain (Figs. 10.7 and 10.8). The distribution is given under each variety. The wide dispersal from one isolated area of white sand to another is probably because the relatively small fruits are dispersed by bats and birds (Macedo & Prance, 1978) and also by tamarin monkeys according to Bhikhi et al. (2016).

This species was excessively divided into varieties and forms by Cuatrecasas (1961). It is morphologically very variable and is an ochlospecies in the sense of White (1962). I have recognized only the most distinct subspecific taxa of Cuatrecasas and reinstated *Humiria parvifolia* A. Juss. to specific rank. These varieties are probably interfertile and I was tempted to lump them all together. However, they do indicate the extensive morphological variation in this species, and there is some geographical separation of varieties. Some of the varieties of Cuatrecasas can be found growing on the same tree! A detailed molecular study of this species would make a good doctoral thesis. This was suggested by Holanda et al. (2015a) who studied the phenotypic variation of two varieties of this species. They found that the sharply discontinuous characters were not explained by either by phenological separation or pollinator preferences, suggesting that pre-zygotic reproductive mechanisms do not contribute to the demonstratable phenotypic differences.

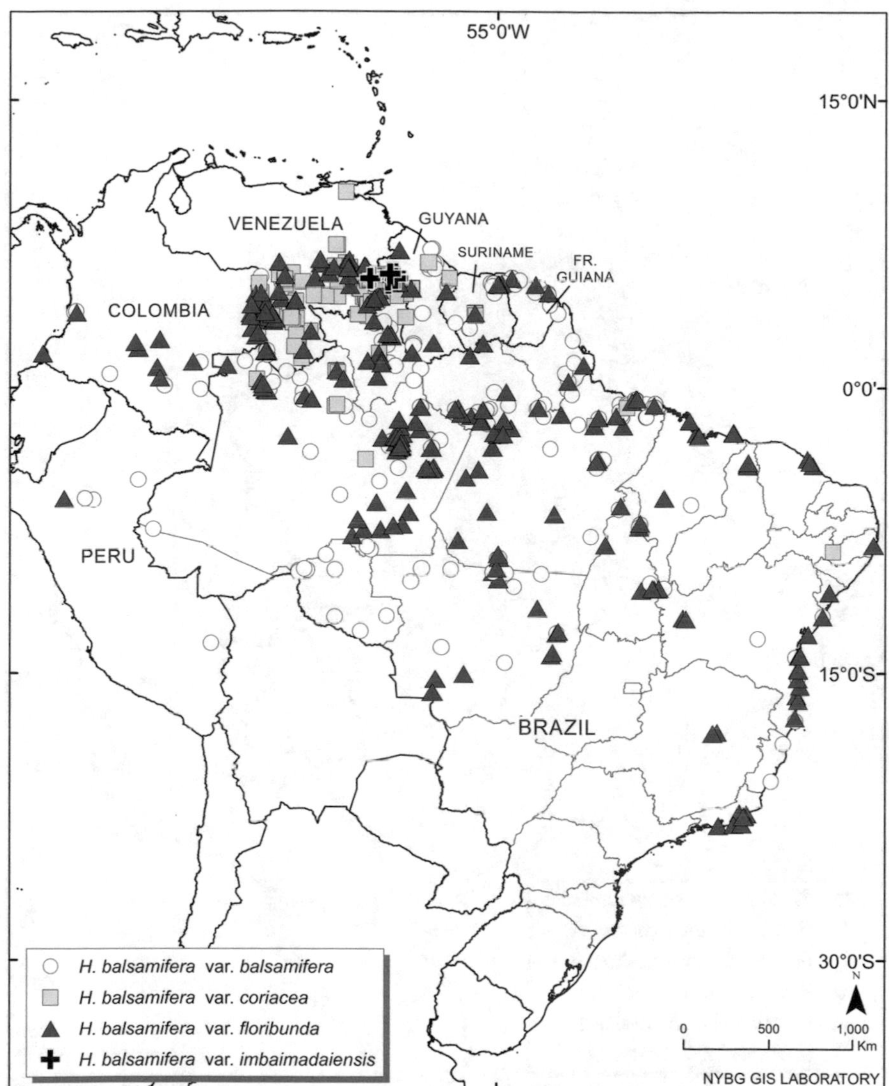

Fig. 10.7 Distribution of four varieties of *Humiria balsamifera*

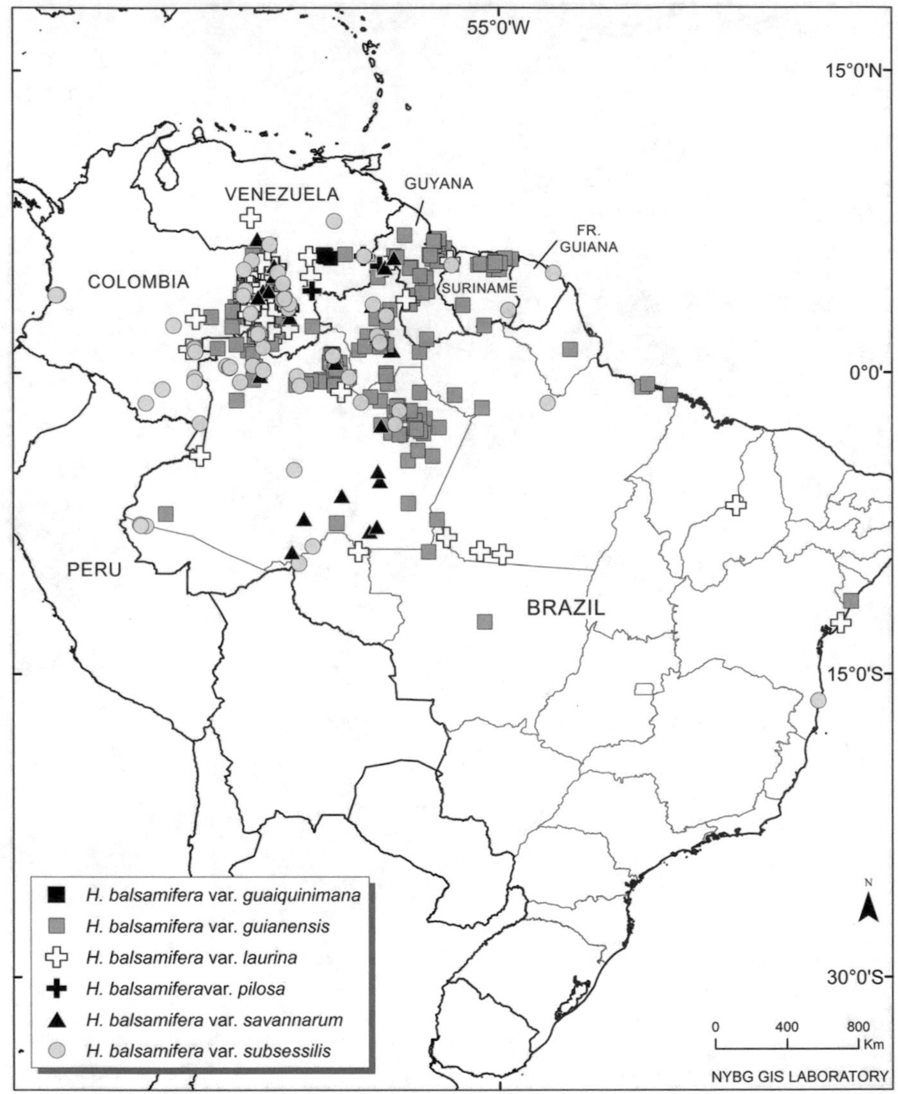

Fig. 10.8 Distribution of six varieties of *Humiria balsamifera*

Key to Varieties of Humiria balsamifera

1. Leaves sessile or subsessile, usually broadly attenuate at base.

  2. Leaves elliptic, rounded at both ends, clasping at base, rigid-coriaceous… ……………………………………………………………………**1e.** var. **imbaimadaiensis**

  2. Leaves at least slightly attenuate toward base, chartaceous, flexible-coriaceous.

3. Leaves broad toward apex, obovate or obovate-elliptic, rounded or very obtuse at apex, 6–14 × 2.5–6 cm; young branches flattened and slightly winged……………………………………………1a. var. **balsamifera**
3. Leaves narrowed toward obtuse apex, branches angulate or subterete.
    4. Leaves linear oblong, or sublanceolate, attenuate-cuneate at base, acute to acuminate at apex, 4–10 × 1.5–3.5 cm…………………………………………………………1i. var. **savannarum**
    4. Leaves elliptic-oblong or sublanceolate-elliptic, narrowed toward acute apex, obtuse or auriculate-clasping at base, 3.5–10 × 1.5–4 cm ……………………………………………1j. var. **subsessilis**

1. Leaves with petiole or winged petiolar lamina base, attenuate at base.
    5. Leaf laminas contracted into a long, winged petiole.
        6. Leaf laminas broadly obovate or oblong-ovate, elliptic or suborbicular, 3–12 × 2–6 cm; midrib pubescent beneath; petioles 0.5–2.5 cm……………………………………………………………1d. var. **guianensis**
        6. Leaf laminas elliptic-oblong, 3.5–7 × 1.5–3 cm, glabrous; petioles 0.5–1.5 cm long……………………………………………1f. var. **laurina**
    5. Leaf laminas with very short but distinct petiole 1–7 mm long.
        7. Leaf laminas obovate or obovate-elliptic, long attenuate, cuneate at base, young branches hirtellous or glabrous.
            8. Leaves thick and rigid coriaceous, high altitudes of Guayana Highland……………………………………………1b. var. **coriacea**
            8. Leaves chartaceous or flexible coriaceous, lowlands……………………………………………………………1c. var. **floribunda**
        7. Leaf laminas usually elliptic or orbicular, shortly attenuate at base, the veins abaxially prominulous, young branches always hirtellous.
            9. Leaf laminas glabrous, 4–7 × 3–5.4 cm, orbicular or suborbicular-elliptic, rounded at apex, cuneate at base into a short thick petiole 1–2 mm long margin entire, almost eglandular…………………………………………………………1h. var. **guaiquinimana**
            9. Leaf laminas hirtellous at least on the midrib abaxially, oblong-elliptic, narrowed and obtuse at apex, slightly narrowed at base into a 2–3 mm long petiole, margin slightly crenulate, with few glands…………………………………………………………1g. var. **pilosa**

1a. **Humiria balsamifera** var. **balsamifera.** *Humiria balsamifera* var. *balsamifera* forma *attenuata* Cuatr., Contr. U.S. Natl. Herb. 35(2): 97. 1961. Type. Colombia. Amazonas-Vaupés: Río Apaporis, Raudal Yayacopi, 18 Feb 1952, *R. E. Schultes & I. Cabrera 15,511* (holotype, US 0010210; isotypes, GH 00055528, U 0002481, WIS0255734).

Small to medium-sized trees, or more rarely a shrub, the young branchlets winged, usually glabrous rarely pubescent. Leaves sessile, obovate to elliptic, slightly attenuate toward apex, rounded or obtuse at base, 6–14 × 2.5–6 mm; margins entire. Petals glabrous or pubescent. Drupe oblong to ellipsoid, 10–14 × 4–8 mm, apex acute, exocarp fleshy; endocarp with 5 foramina.

Distribution and habitat. Sandy and rocky terrain throughout Amazonia, the Guianas into central and eastern Brazil (Fig. 10.7).

Phenology. Flowering and fruiting around the year.

Local names and uses. Brazil: *Umari*; Colombia: *Wa-toó-moo-ko* (Yukuna); Guyana: *Tauaranzu, tauoniro, twaranru, tawasansu, umir* (Wapisiana); Suriname: *Bastaard, bolletri, blakaberie, tawararo (*Arawak), *meerie* (Carib), *tawalāenra, tawalängro.* French Guiana: *Bonga-bita* (Paramaka). This species and most of its varieties are a source of the highly aromatic *umirí* resin which exudes from the bark. It is used medicinally as a stimulant, diuretic, and expectorant (Schultes 1979). Silva et al. (2004) showed that extracts from the leaves of this species exhibited some anti-malarial activity, and Johnson and Colquhoun (1996) reported four different medicinal uses by the Kurupukari of Guyana.

Conservation status. Least Concern (LC).

Illustrations. Cuatrecasas & Huber (1999) Fig. 536; Bhikhi et al. (2016): 165.

Selected specimens examined. COLOMBIA. CAQUETÁ: Sierra Chiribiquete, S Camp, 300–350 m, 0°55′N, 72°45′W, 8 Dec 1990 (fr), *Franco* et al. *3255* (COAH). **GUAVIARE**: NE of PNN Chiribiquete camp, 1°5′48″N, 72°44′17″W, 22 Aug 1992 (fl), *Barbosa César 7700* (COAH). **GUAINÍA**: Puerto Colombia, San José, mid Río Guainía, Caño Asamase, 200 m, 2°41′52″N, 68°1′54″W, 22 Oct 2009 (fl), *Cárdenas* et al. *24,524* (COAH). **META**: Mun. Puerto Rico, Vereda Miravalle, 2°52′39″N, 73°22′13″W. 13 Nov 2007 (fr), *Castro 4271* (COAH). **VAUPÉS**: Río Apaporis raudal Yayacopí, 250 m, 18 Feb 1952, *Schultes & Cabrera 16,893* (BM, US). **VENEZUELA. BOLÍVAR**: Region of Río Icaburú, 450–850 m, 24 Dec 1955, *Bernardi 2601* (NY). **GUYANA**. Mazaruni Station, 10 Jan 1942 (fr), *Fanshawe 715* (K, NY, U), 12 May 1933 (fl), *Tutin 83* (BM, K, U, US); Demerara River, May 1887, *Jenman 3912* (K, NY); Rupununi River, Quimatta, Oct 1889 (fl), *Jenman 5672* (K); Moraballi Creek, Essequibo River, 8 Oct 1929 (fl), *Sandwith 399* (NY, P, RB, U, US); Rupunini Basin, Isherton, *Smith 2423* (A, F, K, MO, NY, S, US, WIS); Atkinson Field, 2 May 1954 (fl), *Irwin 246* (NY, US). **SURINAME**. Brownsberg Reserve, 20 Oct, 1924, *Boschwezen 6670* (U); Zanderij I, 28 Aug 1954 (fr), *Lindeman 6541* (U); Jodensavanne-Mapane Creek, 17 Dec 1954, *Lindeman 6880* (U); Mongoe, Grote Zwiebelzwamp, 29 Sep 1948 (fr), *Lanjouw & Lindeman 573* (U); Zanderij I, 1 Oct 1942, *Stahel 90* (A, IAN, NY, U, WIS); 2 km N of Lucie River, 2 km W of Oost R., 225 m, 6 Sep 1963 (fl), *Irwin* et al. *55,420* (K, NY). **FRENCH GUIANA**. St. Jean, 24 May 1914, *Benoist 1239* (P); Acarouany road, Km 5, 5 Nov 1956, *BAFOG 7587* (U); Cayenne, *Martin s.n.* (BM, P, US); Lieu-dit Maya, route de la Carapa, Macouria, 4°56′59″N, 52°26′30″W, 27 Aug 2007 (fl), *Tostain 815* (CAY, K). **PERU**. **LORETO**: Mishuyaco, near Iquitos, May 1930 (fl), *Klug 1315* (F, US); Balsapuerto, Jan 1933 (fl), *Klug 2846* (A, BM, F, GH, MO, NY, S, US); Balsapuerto, 28–30 Aug 1929, *Killip & Smith 28,681* (F, NY, US). **BRAZIL. AMAZONAS**: Manaus, Cachoeirinha, (fl), *Ducke RB 23424* (US); Rio Vaupés, Taraquá, 10 Nov 1947 (fl fr), *Pires 989* (IAN, INPA, NY, P, US); Tefé, 25 Aug 1947 (fl), *Black 47–1276* (IAN, NY, U, US, VEN) Rio Negro, Preto Campina, 5 Nov 1957 (fl), *Fróes 22,738* (IAN, NY, US, VEN); Yanomami Village Watorikëtheri, old BR210, 62°49′W, 1°31′N, Jul 1994 (fl), *Milliken 1984* (FEMACT, INPA, K, NY); Rio Cuieiras, 2 km below Rio Brancinho, 14 Sep 1973 (fl), *Prance* et al. *17,904* (INPA, K, MO, NY, UFMT, US). **RORAIMA:** Mucajaí airport, Samauma trail, 25 Aug 1951 (fl), *Black & Magalhães 51–12,954* (IAN, P); between Fazendas Bom Intento & Capela, 31 Aug 1951 (fl), *Black 51–13,231* (IAN); Igarapé Caranã, 20 Aug 1951 (fl), *Black 51–12,776* (IAN). **PARÁ**: Belém, Lago da Agua Preta, 29 Oct 1914, *Ducke RB 15514* (MG); Belém, *Moss 57* (NY, US); Igarapé Gameleirinha, Araguaia region, 17 Jun 1953 (fl), *Fróes 29,854* (IAN); Marajó, Jutuba, 21 Jul 1902 (fl), *Huber 2785* (BM, MG); Vigia, Campinha da Palha, 29 Sep 1948 (fr), *Black 48–3249* (IAN, P); Óbidos, 12 Jul 1905 (fl), *Ducke MG7213* (BM, MG); Rio Guamá, above Ourém, Jun 1953 (fl), *Pires & Silva 4629* (IAN, NY); 10 km E of Portel, 12 Sep 1965 (fl), *Prance* et al. *1292* (IAN, K, NY, US). **AMAPÁ**: Oiapoque airstrip, 3 Oct 1949 (fl), *Black 49–8313* (IAN), 7 Dec 1954, *Cowan 38,700* (NY); Road Macapá-Clevelandia, Km 130, 27 Aug 1955 (fl), *Black 55–18,577* (IAN). **MARANHÃO**: Carolina, 11 Aug 1955, *Macedo 4034* (IAN, US); Riberão Urupuchete, 17 km N of Carolina, 7°11′S, 47°25′W, 12 Jul 1993 (fl), *Ratter* et al. *6801* (INPA, K, MO, UB); 5 km NE of Carolina, Rio Beirão, 6 Aug 1964 (fl), *Prance & Silva 58,585* (K, NY, UB, US). **RONDÔNIA**: Costa Marques, W of Cautarinho, N of Hwy BR429, 200 m, −12.01677, −63.43889, 25 Mar 1987 (fl), *Nee 34,494* (INPA, NY, MO); Porto Velho, Ramal acesso Garimpo, 103 m, −9.6, −65.03111, 8 May 2013, *Bigió 865* (CEN, INPA, NY, RB). **MATO GROSSO**: Joara, Salto Apiacás, 26 May 1988 (fl), *Macedo & Assumpção 1916* (INPA); Apiacás, Rio Juruena, −9.48611, −58.92222, Nov 2007 (fl), *Sobral 11,328* (RB, UFMG).

1b. **Humiria balsamifera** var. **coriacea** Cuatr., Contr. U.S. Natl. Herb. 35(2): 110. 1961; Cuatrecasas & Huber, Fl. Venez. Guayana (Steyermark et al.) 5: 628. 1999. Type. Venezuela. Bolívar: Mureyena Falls, 800 m, 10 Oct 1946, *F. Cardona 1823* (holotype, US 00101209; isotypes, NY 00388385, VEN 25809).

Humiria balsamifera var. *iluana* Cuatr., Contr. U.S. Natl. Herb. 35(2): 115. 1961. Type. Venezuela. Bolívar: Ilu-tepuí, 1400 m, 13 Mar 1952, *B. Maguire 33,388* (holotype, US 00101207 isotypes, NY 00388386, VEN 44533).

Humiria balsamifera var. *stenocarpa* Cuatr., Contr. U.S. Natl. Herb. 35(2): 114. 1961. Type. Brazil. Roraima: Serra Tepequém, 1000–2000 m, 29 Nov 1954, *B. Maguire & C. Maguire 40,105* (holotype, US 00101203 isotype, NY 00388388, RB 00539053, S-R7852).

Shrubs or small trees, the young branches hirtellous or puberulous. Leaves with short petioles 2–6 mm long, laminas rigid-coriaceous, 3–7 × 1.7–4 cm, obovate or obovate-elliptic, rounded or obtuse, emarginate and mucronulate at apex, cuneate at base, margin entire. Petals glabrous or hirtellous. Drupe oblong-ellipsoid, 10–14 mm long, 3–8 mm wide, exocarp fleshy, endocarp hard, 10 × 8 mm, 10-striate, apex acute, with 5 foramina.

Distribution and habitat. Highlands of Venezuelan Guyana, the Guianas and adjacent Brazil. Mainly in rocky places above 600 m (Fig. 10.7).
Phenology. Flowering and fruiting around the year with more flowers from September to January and fruits from August to December.

Local names and uses. *Amuri-yek* (Arekuna). Fruit edible.
Conservation status. Least concern (LC).
Illustrations. Cuatrecasas (1961), figs. 21 o–r; 22 i-l, 23 e–g).

Selected Specimens studied. VENEZUELA. AMAZONAS: Cerro Duida, Río Cunucunuma, 500 m, 19 Nov 1950 (fl), *Maguire* et al. *29,541* (NY, US); Cerro Duida, Savanna Hills, 300 m, (fr), *Steyermark 58,288* (NY, US), *Tate 733* (NY, US); Cerro Sipapo, 200 m, 12 Dec 1948 (fl), *Maguire & Politi 27,695* (NY, US); Cerro Yapacana, 1200 m, 2 Jan 1951 (fl), *Maguire* et al. *30,622* (NY, US); Río Ventuari, Serrania Parú, Caño Asisa, 2000 m, 7 Feb 1951 (fl), *Cowan & Wurdack 31,301* (NY, US); Cerro Ualipano, Río Paruicito, 1760 m, Feb 1962, *Cardona 2942* (MO); Dept. Río Negro, Cerro Aratitiyope, 990–1670 m, 2°10′N, 65°34′W, 28 Feb 1984 (fr), *Steyermark 130,091* (MO, US); 10–12 km NNW of Yutaje, W of Río Corocoro, 1225 m, 3°42′N, 66°09′W, 8 Mar 1987, *Liesner & Holst 21,758* (M0); Cerro Marahuaca, Río Yameduaiza, 3°38′N, 65°28′W, 17–18 Feb 1985, *Liesner 17,619* (MO). **BOLÍVAR**: Gran Sabana, Río Cuquenán, N of Santa Elena, S of Mt. Roraima, 1065 m, (fl), *Steyermark 59,186* (NY, US, VEN); 16 km N of Santa Elena, 750 m, 4°20′N, 61°45′W, *Croat 54,148* (MO, US); Río Caroní, Salto Eutouamini, 720 m, Oct 1968 (fl), *Cardona 1768* (US, paratype); Sabanas de Icaburú, El Caribe, 450–850 m, 22 Dec 1955 (fl), *Bernardi 2603* (VEN); Auyantepuí, Guayaraca, 1100 m, Apr 1956 *Vareschi & Foldats 4563* (VEN); Cerro Bolívar, W peak, 750 m, 2 Dec 1951 (fl), *Maguire 32,686* (NY, US); 15 km SW of Karavin Tepuí, 950 m, 5°19′N, 61°03′W, 29 Apr 1988 (fr), *Liesner 24,038* (MO) Cerro Avismo. 800 m, 4°40′N, 61°22′W, 2 Apr 1988 (fr), *Sastre 8425* (MO); Distr. Piar, midpoint of Auyan-tepuí, 5°42′N, 62°37′W, 11 Dec 1984 (fl), *Huber 9882* (MO); Cerro El Venado, 20 km E of Canaima, 1300 m, 6°17′N, 62°0.41′W, 31 Aug 1983 (fr), *Prance & Huber 28,414* (MO); Auyan-tepuí, E section, 1800 m, 5°56′N, 62°34′W, 28 Aug 1983 (fl), *Prance & Huber 28,298* (K, MO); Sabana de Arekuna, E margin Río Caroni, 6°31′N, 62°53′W, 29 Aug 1983 (fl), *Prance & Huber 28,332* (K, MO, US); Cerro Marutani, Río Paragua, 1200 m, 3°50′N, 62° 15′W, 11–14 Jan 1981 (fl), *Steyermark 123,896* (MO); Chimantá, Anurí-tepuí, 1850 m, 5°10′N, 62°07′W, 2–5 Feb 1983 (fl), *Steyermark* et al. *128,612* (MO). **SUCRE**: Lago de Guanoco, Aug 1955 (fl), *Lasser & Vareschi 3888* (VEN). **GUYANA**. Kaieteur Plateau, 8 May 1944 (fl), *Jenman 1023* (K; NY, U, US, VEN), Kaieteur Falls, 11 Feb 1962 (fr), *Cowan & Soderstrom 1820* (K, US); Karowtipu, 2 Oct 1960 (fl), *Tillett & Tillett 45,600* (K, NY, US); Pakaraima Mts, Mt. Aymatol, 5°55′N, 61°00′W, 1150 m, 15 Oct 1981, (fl fr), *Mennega* et al. *5649* (K, MO, U). **SURINAME**. Tafelberg, 300 m, 4 Aug 1944 (fl), *Maguire 24,223* (NY, US, U, VEN), Tafelberg Savanna No II, 12 Nov 1944 (fl), *Maguire 24,707* (K, NY, U, US). **BRAZIL. AMAZONAS:** Mun. São Gabriel da Cachoeira, Rio Cubate, tributary Rio Içana, 2 Nov 1987 (fl), *Farney 1882* (IEPA, INPA, K, MO, NY); Serra Aracá, 1200 m, 11 Feb 1984 (fr), *Prance* et al. *28,974* (INPA, K, MO, NY). **RORAIMA**: Serra Tepequém, 1000–1200 m, 4 Dec 1954 (fl),

Maguire & Maguire 40,159 (NY, US), 16 Feb 1967 (fr), 1200 m, *Prance* et al. *4404* (INPA, K, RB, NY, US), Serra Tepequém, 1800 m, 3°45′N, 61°50′W, 6 Jan 1986 (fl), *Sette Silva 501* (INPA, K, MO).

This variety is quite distinct by the rigidly coriaceous leaves that are much more rigid than any of the other varieties; it is also generally an high-altitude variety.

1c. **Humiria balsamifera** var. **floribunda** (Mart.) Cuatr., Contr. U.S. Natl. Herb. 35(2): 99. 1961; Cuatrecasas & Huber, Fl. Venez. Guayana (Steyermark et al.)5: 628. 1999 *Humirium floribundum* Mart. Nov. Gen. sp. pl. 2: 143.1827; *Humiria floribunda* (Mart.) Urb., Fl. Bras. (Martius)12 (2): 437.1877. Type. Brazil. Pará: Rio Xingu, *C. F. P. von Martius s.n.* (holotype M; isotype, FI-W, K 000407344).

Humirium multiflorum Pritz., Icon. Bot. Index 560. 1866. Nom illegit. This name is the result of an erroneous quotation by Pritzel and has never been validly published.

Humirium ellipticum Klotsch ex Urb., Fl. Bras. (Martius) 12(2): 437. 1877, in synonymy.

Shrubs or small- to medium-sized trees up to 15 m tall, the young branches glabrous or rarely puberulous, angulate or subterete, rarely narrowly winged. Leaves with short petioles 1–3 mm long, laminas subcoriaceous to coriaceous, elliptic, obovate-elliptic to oblong, 5–12 × 2–6 cm, obtuse, emarginate, rounded or truncate at apex, margins usually minutely crenate. Petals 3 mm long. Drupe elliptic 8–9 × 4–6 mm, pointed at apex; exocarp fleshy; endocarp striate, with 5 foramina.

Distribution and habitat. Widespread in lowland savannnas and campinas of Colombia and Amazonian Venezuela, Brazil, and the Guianas and south into eastern Brazil (Fig. 10.7).

Phenology. Flowering and fruiting around the year, often with flowers and fruit simultaneously.

Local names and uses. Brazil: *Umiri, mirim doce, jua preta*. Suriname: *Blakaberie*.

Conservation status. Least concern (LC).

Illustrations. Cuatrecasas (1961): figs. 21 d–f, 23 a–d).

Selected specimens examined. COLOMBIA. VALLE: Buenaventura, 100 m, 3°55′N, 77°00′W, 12 Nov 1986 (fl), *Monsalve 1336* (CUVC, MO). **NARIÑO:** Tumaco, 9 Jun 1955 (fl), *Romero Castañeda 5129* (COL). **META:** Reserva La Macarena, 400 m, 2°04′N, 73°45′W, 2 Aug 1988 (fl), *Callejas & Marulanda 6974* (HUA, MO). **CAQUETÁ:** Sierra Chiribiquete, 550 m, 1°04′N, 72°46′W, 22 Nov 1992 (fl), *Franco* et al. *4261* (COL, MO). **GUAVIARE**: Mun. San José del Guaiviare, ésSerrania de la Lindosa, 2°29′N, 72°41′W, 20 Jul 1997 (fl), *Duque 3244* (COAH). **GUIANÍA**: Mun. Puerto Inírida, Cao Vitina, 220 m, 3°48'213′N, 67°49′194″, 150 m, 8 Sep 2014 (fr), *Nórida & Rodríguez 1648* (COAH). **VAUPÉS**: Mun. Mitú, Wacurabá, Caño Cuduyañí, 400 m, 1°21′1″N, 70°53′33″W, 26 Sep 2007 (fl), *Cárdenas* et al. *21,199* (COAH). **VENEZUELA. AMAZONAS**: Base of Cerro Sipapo, 125 m, 28 Dec 1948 (fl), *Maguire & Politi 27,974* (NY, US); Base of Serrania Yutaje, Río Manapiare, 200 m, 29 Jan 1953 (fr), *Maguire & Maguire 35,040* (NY, US); Base of Cerro Moriche, 150 m, 17 Jan 1951 (fr), *Maguire* et al.

30,987 (K, NY, US); Atabapo, 120 m, 3°29′N, 66°41′W, (fl), *Huber 3089* (K, US); Río Mavaca, 200–230 m, 1°58′N, 65°06′W, 3 Feb 1991 (fr), *Stergios & Yánez 15,106* (MO); San Carlos de Río Negro, 100 m, 1°56′N, 67°04′W, 23 Nov 1984 (fl), *Croat 59,251* (MO, US); Dept. Atabapo, Cerro Moriche, 0 m, 4°40′N, 66°17′W, Oct 1989 (fl), *Delgado 931* (MO); 20 km from confluence of Ríos Negro and Casiquiare, 119 m, 1°56′N, 67°00′W, 21 Mar 1981 (fr), *Delascio* et al. *9361* (VEN). **BOLÍVAR**: Distr. Piar, 950 m, 6°20′N, 62°42′W, 6 May 1987 (fl), *Huber 12,103* (K, US); Mun. Gran Sabana, 950 m, 5°57′N, 61°46′W (fr), *Huber 12,539* (K, US); 96 km N of Santa Elena de Uairén, 500–1000 m, 5°05′N, 61°07′W, 24 Jul 1982 (fr), *Croat 54,008* (MO); Mun Sucre, 70 km SE Hato La Vergaseña, 6°28′N, 64°12′W, 750 m, Jun 1989 (fl), *Fernández & Yánez 451* (MO, US); Mun. Cedeno, Río Túriba, 45 km E of Pijiguas, 800 m, 6°34′N, 66°23′W, Aug 1989 (fr), *Fernández & Sanjoa 5887* (MO); Piar, Río Acanán, 240 m, 6°42′N, 64°14′W, 29 Apr 1986 (fl), *Liesner & Holst 20,487* (MO); Distr. Piar, Río Cucuritai, facing Auyan-tepuí, 420 m, 5°51′N, 62°43′W, 31 Mar 1984 (fl), *Huber 9322* (MO), Cerro Kurún-tepuí, 17 km E of Canaima, 6°14′N, 62°42′W, 21 Aug 1983 (fl), *Huber* et al. *8212* (MO); Dist Roscio, 13 km NE of Santa Elena de Uairén, 900 m, 4°45′N, 61°05′W, 2 Dec 1982 (fl), *Steyermark & Liesner 127,537* (MO). **GUYANA**. Berbice-Rupununi cattle trail, 29 May 1919, *Abraham 152* (NY); Upper Mazaruni River Karowtipu Mt. 5°45′N, 60°35′W, 1000 m, 23 Apr 1987 (fl, fr), *Boom & Gopaul 7669* (MO); Berbice-Corantyne, Baba Grant Sawmill, 5°02′N, 57°42′W, 18 Apr 1990 (fl) *McDowell 2325* (MO); Wakadanawa savannah, 290 m, 7 Sep 1997 (fl), *Jansen-Jacobs 5399* (K, MO, US). **SURINAME**. Jodensavanne-Mapane Creek, 10 Jul 1953 (fl), *Lindeman 4201* (U); Blakawatra 17 Dec 1954 (fr), *Lindeman 6862* (U); Moengo, Grote Zwiebelzwamp, 25 Oct 1948 (fr), *Lanjouw & Lindeman 968* (U). **FRENCH GUIANA**. Cayenne west, 26 Oct 1954, *Cowan & Maguire 38,034* (NY); Saint Elie Savanne, 5°23′N, 53°00′W, 28 Nov 1995 (fl), *Loubry 2211* (MO). **PERU. SAN MARTÍN**: Rioja, 980 m, 5°51′S, 77°40′W, 980 m, 1 Jun 1986 (fl), *Knapp* et al. *7457* (MO). **BRAZIL. AMAZONAS**: Manaus, Igarapé do Buião, 31 Aug 1954 (fl), *INPA 1773* (INPA, MG, US); Manaus, Cachoeira Tarumã, 25 Apr 1932, *Ducke 87* (A, WIS); Cachoeira Mundú, 2 Sep 1942 (fl), *Ducke 541* (IAN, K, MG, NY, S, US); Rio Negro, Airy-Pirá, Rio Aiary, 5 Nov 1945 (fl), *Fróes 21,346* (IAN, NY, US); Rio Negro near Manaus, Oct 1851 (fl, fr), *Spruce 1499* (BM, GH, F, FI-W, K, M, NY, OXF, P, S, US); Manaus, Aleixo road, Aug-Sep 1936 (fl), *Krukoff 7928* (A, BM, K, NY, S, U); Rio Negro, Praia Grande 27 Mar 1978 (fl), *Revilla 4049* (HUFU, INPA, MO); Manaus- Caracaraí road Km 45, 10 Jun 1971 (fl), *Prance* et al. *13,340* (INPA, K, MO, NY, US); Manaus, May 1902, *Ule 6142* (MG, K). **RORAIMA**: Serra da Malacacheta, *Kuhlmann RB3509* (K, US); near Boa Vista, Oct 1908, *Ule 7625* (K, MG, UC, US). **PARÁ**: Caripi, Aug 1849 (fl), *Spruce s.n.* (FI-W, K); Santarém, Jun 1850, *Spruce s.n.* (BM, GH, P); Casa Santa Izabel near Santarém, 28 Oct 1950, *Black & Ledoux 50–10,371* (IAN); Caripi, Aug 1849 (fl), *Spruce 164* (K, P); Gurupá, Igarapé Jucupí 18 Aug 1954 (fl), *Pires & Silva 4702* (IAN, US); Belterra, Rio Tapajós, 24 Oct 1947 (fl), *Black 47–1756* (IAN, NY, U, US, VEN); Oriximinã, 70 m, 8 Jun 1980 (fl), *Martinelli 6881* (INPA, MO, RB); Mun. Oriximinã, Campos de Ariramba, 8 Jun 1980 (fl), *Martinelli 6860* (INPA, MO, NY, RB); Road Tucuruí-Goianésia, 19 Aug 1980 (fl), *Rodrigues 10,228* (INPA, MO, NY); Cururú, 10 May

1977 (fl), *Rosa & Santos 1892* (IAN, IEPA, MO, NY, RB); Cachimbo Airport, 23 Jun 1974 (fl), *Rodrigues 9428* (INPA, MO). **AMAPÁ**: Rio Araguari, 22 Jul 1951 (fl*), Fróes & Black 27,572* (IAN). **MARANHÃO**: Ilha de São Luiz, 15 Jan 1951 (fl), *Fróes 26,812* (IAN); Alacántara, 28 Sep 1903, *Ducke 440* (BM, MG); Mun. Carolina, Pedra Caida, 7°08′S, 47°25′W, 13 Apr 1983 (fl bud), *Silva* et al. *1060* (IEPA, INPA, K, MO, NY). **MATO GROSSO**: Mun. Xavantina 4 km N of base camp, 12°54′S, 51°52′W, 23 Apr 1968 (fl), *Ratter 1139* (K, MO); 86 km N of Xavantina, 550 m, 31 May 1966 (fl), *Irwin* et al. *16,311* (ASU, K, MO, NY, UB, US). **TOCANTINS**: TO25, Mateiros-Ponte Alta do Tocantins. 10°39′05"S, 47°20′46"W, 10 Nov 2015 (fr), *Goes* et al. *193* (IBGE, RB); Margin of Rio Novo, −10.55, −46.13333, 8 May 2001, *Proença* et al. *2514* (fl) (CEN, MBM, UB, UFG). **MINAS GERAIS**: Parque Estadual do Rio Preto, São Gonçalo do Rio Preto, −18.1166, _43.3333, 13 Jun 1999 (fl), *Lombardi 3048* (SPF). **CEARÁ**: Fortaleza, 10 Sep 1907, *Huber 96* (BM, MG, US). **BAHIA**: *Salzman s.n.,* no locality, s.d. (MO, P): Mun. Ilhéus, km 10 Pontal-Olivença, 10 Feb 1985 (fr), *Gentry* et al. *50,012* (CEPEC, MO). **RIO DE JANEIRO**: Restinga de Cabo Frio 20 Jun 1877 (fl), *Glaziou 10,437* (P); vic. Rio de Janeiro, *Weddell 526* (P).

1d. **Humiria balsamifera** var. **guianensis** (Benth.) Cuatr., Contr. U.S. Natl. Herb. 35 (2): 103. 1961; Cuatrecasas & Huber, Fl. Venez. Guayana (Steyermark et al.) 5: 628. 1999; *Humirium guianense* Benth., London J. Bot. 2; 374. 1843; *Humiria floribunda* var. *guianensis* (Benth.) Urb., Fl. Bras. (Martius) 12 (2): 439. 1877. Type. Guyana, *R. Schomburgk 270* (holotype, K 000407340; isotypes, BM 000796008, F V0060777, F V0060778, G 00368459, G 00368460, IAN 40224, L 1162529, L 2120708, NY 00388394, OXF, P 01903268, TCD 0003828, U 0002479, US 000101199).

Myrodendron petiolatum Mart. ex Urb., Mart. Fl. Bras. 12 (2): 4339 1877, in synonymy.

Humirium surinamense Miq., Stirp. surinam. select. 86, pl 24. 1850. Type. Suriname. "In savana prope plantations Berlyn, l'Inequtitude. M. Septembri florentem" = Plantage Berlin, *H. C. Focke 1018* (holotype, U 0002478).

Humiria cassiquiari Sussenguth & Bergdolt, Feddes Repert. Spec. Nov. Regni. Veg. 39: 16. 1935. Type. Venezuela. Amazonas: Río Cassiquiare, Laja de Caraça, 4 m, 5 Oct 1928 (fl), *P. von Luetzelburg 22,627* (holotype, M, photos of holotype NY, US).

Shrubs or small- to medium-sized trees, the young branches usually pubescent or puberulous. Leaves obovate, ovate or elliptic, 3–12 × 2–6 cm, subcoriaceous, emarginate, or obtuse at apex, contracted at base into a long petiole 0.5–2.5 cm long, margins conspicuously crenulate. Petals glabrous with a glandular tip. Drupe 7–8 × 4–5 mm, exocarp very fleshy, endocarp striate, with 5 foramina.

Distribution and habitat. Amazonian Colombia, Venezuela, Guyana, Suriname, and Amazonian and NE Brazil (Fig. 10.8).

Phenology. Flowering around the year, flowers and fruits often simultaneous.

Local names and uses. Guyana: *muri*; Suriname: *Tauraroe* (Arawak), *blakkaberie*. Brazil: *Umiri*. Fruit edible.
Conservation status. Least concern (LC).
Illustrations. Cuatrecasas (1961) figs. 21 h-I, 22 g-h,23 l-m.

Selected specimens examined. COLOMBIA. CAQUETÁ: Mun. Solano, Huitorá, Río Caquetá, 0°12′4″N, 74°42′28″W, 200 m, 28 Dec 2017 (fr) *Cárdenas* et al. *49,375* (COAH). **GUAINÍA:** Raudal Guacamayo, Río Inírida, 4 Feb 1953 (fr), *Fernández 2148* (COL, US); Río Negro at confluence of Ríos Guianía & Cassiquiare, Caño Ducuruapo, *Schultes & López 9363* (F, IAN); Río Negro, San Felipe, 200 m, 13–25 Nov 1952, *Humbert 27,440* (US). Río Atabapo, Cacahual, 130 m, 19 Nov 1953 (fl), *Maguire* et al. *36,295* (IAC, NY, US); Rio Guainía, between Moroa & Tonina, 175 m, 2°45′N, 67°40′W, 2 Jul 1991 (imm fr), *Berry & Sánchez 5032* (K, MO); Guainía, 120 m, 1°47′N, 67°16′W, *Gentry & Stein 46,400* (MO); Caño Colorado, Aeropuerto La Esperanza, 2°16′N, 68°22′W, 150 m, 15 Apr 1993 (fr), *Madriñán & Barbosa 1046* (K). **VENEZUELA. AMAZONAS**: Río Guainía, Caño San Miguel 2 km above Limoncito, *Maguire* et al. *41,917* (NY, US); S of Maroa, 28 Nov 1953 (fr), *Maguire* et al. *36,456* (NY, US); Orinoco, Cerro Yapacana, 125 m, 1 Jan 1951 (fl), *Maguire* et al. *30,543* (IAC, NY, US); Yavita, 22 Jan 1942 (fr), *Williams 13,903* (US); Pacimoni, Feb 1854 (fl), *Spruce 3409* (K, P, S); Laja de Caraça, 9 Oct 1928, *Luetzelburg 22,575* (M); San Carlos de Río Negro, 1°56′N, 67°03′W, 1 Mar 1979 (fr), *Clark 7048* (MO); Río Caname, 1 km from Río Atabapo, 95 m, 3°41′N, 66° 27′W, Nov 1989 (fl), *Yánez 47* (MO); Bana, S of Yavita, 110 m, 2°55′N, 67°26′W, 25 Nov 1995 (fl), *Berry* et al. *5626* (MO); 10 km NE of San Carlos de Río Negro, 120 m, 1°54′N, 67°00′W, 8 Nov 1995 (fl), *Liesner & Carnevali 22,880* (MO, US); Dept. Atures, 5 km N of Santa Rosa de Ucata, 80–85 m, 4°24′N, 67°46′W, 21 Jun 1992 (fl fr), *Berry* et al. *5194* (MO). **GUYANA**. Iwokrami, 75 m, 4°20′N, 58°50′W, 25 Nov 1995 (fr), *Clarke 656* (K) Kurupukari, basin of Essequibo River 3 Oct 1937 (fl), *Smith 2176* (A, F, K, NY, S, U, US); Upper Mazaruni River, Sep-Oct 1922, *De La Cruz 2210* (F, GH, MO, NY, US); Malali, Demerara, 5°37′0.12″N, 58°21′W, 30 Oct 1922, *De La Cruz 2645* (F, GH, K, MO, NY, UC, US); Kaieteur Plateau, Potaro River, 14 May 1944 (fl), *Maguire & Fanshawe 23,450* (BM, K, NY, U, US); E of Itumi, 35 miles S. of Makenzie, 17 Jan 1955 (fl fr), *Cowan 39,266* (K, NY,U, US); Apotere, Rupununi R, 13 Jun 1931 (fl), *Davis 643 (FD2055)* (K); S of Timeri, 16 Oct 1979 (fl), *Maas & Westra 3564* (K, MO); Demerara-Mahaica Region, hwy to Timeri airport, 6°30′N, 58°30′W, 22 Mar 1987 (fl), *Boom 7124* (K, MO, NY). **SURINAME**. Sectie 0, 3 Sep 1920 (fl), *Pulle 150* (U), Zanderij I, 10 Mar 1949 (fr), *Lanjouw & Lindeman 3259* (U), 9 Sep 1948 (fl), *Lindeman 258* (U); Zanderij II, 3 Jun 1944 (fl), *Maguire & Stahel 23,654* (BM, NY, U, US), *Maguire & Stahel 23,696* (A, BM, F, NY, U, US, VEN); Jodensavanne, May 1909, *Focke 1286* (U); Jodensavanne Mapane Creek, 17 Dec 1954, *Lindeman 6881* (U); Moengo, Grote Zwiebelswamp, 20 Oct 1948 (fl), *Lanjouw & Lindeman 911* (BM, F, GH, K, MO, U); Zuid R., 3.5 km E of Kayser Airstrip, 3°20′N, 56°49′W, 25 Sep 1963 (fr), *Irwin* et al. *57,565* (K, MO, NY); 0.5 km SE of Zanderij Hwy, N of Airport, 5°32′N, 55°12′W, 25 Jun 1994 (fl), *Evans & Koemar 2610* (K, MO); Km 37 road

Paranam-Afobaka, 3 Nov 1974 (fl), *Maas* et al. *2294* (K). **BRAZIL. AMAZONAS**: Rio Içana, Tunuí, 10 May 1948, *Black 48–2589* (IAN, NY, VEN); Mun. São Gabriel da Cachoeira, Vista Alegre, mouth of Rio Xié, 0°55′N, 67°20′W, 22 Oct 1987 (fl), *Farney 1765* (IEPA, INPA, K, MO, NY); Manaus, Aug 1942, (fl, fr), *Sandeman 2699* (K); Upper Rio Aracá, 30 Oct 1952, *Fróes & Addison 29,211* (IAN); 3 km S of Serra Aracá, 3 km E of Rio Javari, 0°49′N, 63°19′W, 6 Feb 1984 (fl), *Prance* et al. *28,833* (K, INPA, MO, NY, RB, US); Rio Negro, Aiary-pirá, Rio Aiary, 5 Nov 1945, *Fróes 21,342* (F, IAN, K, NY, US) Preto, Malupiry, 13 Nov 1947 (fl), *Fróes 22,760* (IAN, NY, U, US, VEN); Mun. Borba, 8 km from Borba, 22 Jun 1983 (imm fr), *Cid Ferreira 3870* (INPA, MO): Manaus-Caracaraí, Km 45, 1°40′S, 60°05′W, 16 Aug 1986 (fr), *Croat 62,194* (INPA, MO); Manaus-Itacoatiara Km 201, Rio Urubu, 19 Dec 1966 (fl), *Prance 3740* (INPA, K, NY, US). **RORAIMA**: Vista Alegre, *Kuhlmann RB2894* (K, US); Manaus-Caracaraí, Km 350, 18 Nov 1977 (bud), *Steward 78* (INPA, MO, NY). **PARÁ**: Lago de Fero, Praia, 20 Aug 1907 (fl), *Ducke MG8410* (BM, MG); Rio Mapuera NE of Tabolierinho, 12 Dec 1907, *Ducke 9123* (MG); BR22, Km 64, near Piritoro, 6 Nov 1965 (fr), *Prance & Pennington 2003* (IAN, K, NY, US).

This is probably the easiest variety of this species to identify because of the long, often slightly winged, petiole.

1e. **Humiria balsamifera** var. **imbaimadaiensis** Cuatr., Contr. U.S. Natl. Herb. 35(2): 115. 1961; Cuatrecasas & Huber, Fl. Venez. Guayana (Steyermark et al.) 5: 628. 1999. Type. Guyana. Imbaimadai savannas, Upper Mazaruni River, 21 Oct 1951, 550 m, *B. Maguire & D. B. Fanshawe 32,158* (holotype, US 00101206; isotypes, K 000407338, NY 00388387, S-R7850).

Low shrubs, the young branches glabrous. Leaves sessile, rigid-coriaceous, rounded at both ends, 2.5–5 × 1.7–3.2 cm, clasping at base, margins entire and glandular beneath. Petals sparsely pubescent on exterior. Drupe ellipsoid, 7–8 × 5.5 cm, exocarp fleshy, endocarp striate.

Distribution and habitat. Dominant shrub on muri savannas of Guyana and in adjacent Venezuela (Fig. 10.7).
Phenology. Flowering and fruiting around the year.
Local names and uses. None recorded.
Conservation status. Vulnerable (V), B1, very local distribution, but common there.
Illustrations. Cuatrecasas (1961: 95, fig 22m).

Selected specimens examined. VENEZUELA. BOLÍVAR: W summit Torónmreu, NW of Parupa, 1200 m, 14 Dec 1984 (fr), *Kral 72,051* (MO). **GUYANA**. Cuyuni-Mazaruni Region: Imbaimadai, Aug 1967 (fl), *Davis 334* (K); Imbaimadai, 5°42′N, 60°17′W, 516 m, 29 Jul 2010 (fl), *Redden 7121* (K, US); Imbaimadai, 5°44′N, 60°16′W, 13 May 2012 (fl), *Redden 7188* (K, MO, US); 5°37′N, 60°40′W, 491 m, 19 May 2009 (fl fr) *Redden* et al. *6678* (K, MO, US); 1 km W of Imbaimadai, 500 m, 16 May 1991 (fr), *Hoffmann 1624* (K); 0.5 km NW of Imbaimadai, 17 Nov 1992 (fl), *Hoffmann 3415* (K, US); Pakaraima Mts, Mt.

Membaru, 5°57′N, 60°33–34′W, 17 Nov 1992 (fl, imm fr), *Maas & Westra 4293* (K, MO, US).

Described by Cuatrecasas (1961) from a single collection, but now collected many times in the same region of Guyana and also in Venezuela. This local variety is distinct by the sessile leaves that are rounded at both ends.

1f. **Humiria balsamifera** var. **laurina** (Urb.) Cuatr., Contr. U.S. Natl. Herb. 35(2): 107. 1961; Cuatrecasas & Huber, Fl. Venez. Guayana (Steyermark et al.) 5: 629. 1999; *Humiria floribunda* var. *laurina* Urb., Fl. Bras. (Martius) 12 (2): 439. 1877; *Humirium laurinum* Klotsch ex Urb., Fl. Bras. (Martius) 12 (2): 439. 1877, in synonymy. Type. Guyana. *R. Schomburgk 560* (holotype, M; isotypes, K 000407339, NY 00388391, US 00101205).

Humiria floribunda Mart. var. *spathulata* Gleason, Bull. Torrey Bot. Club 58: 374. 1931. Type. Venezuela. Amazonas. Esmeralda, Gran Sabana, Nov 1928, *G. H. H. Tate 286* (holotype, NY 00388392).

Sprawling shrubs, 0.2–3 m tall or small tree, the young branchlets glabrous. Leaves cuneate at base, rounded, spathulate or obtuse at apex, contracted into a long petiole 0.5–1.5 cm long; laminas 3.5–7 × 1.5–3 cm. Petals puberulous or glabrous. Drupe ellipsoid, pointed at apex, round at base, 7 × 3 mm, exocarp fleshy; endocarp striate with 5 foramina.

Distribution and habitat. Abundant shrub in savannas and rocky places from SE Colombia to Maranhão, Brazil (Fig. 10.8).
Phenology. Flowering and fruiting around the year.
Local names and uses. Guyana: *Tauroniro* (Arawak). Fruit edible.
Conservation status. Least Concern (LC).
Illustrations. Cuatrecasas (1961) figs. 21j, 22f, 22n, 24d.

Selected specimens examined. COLOMBIA. VAUPÉS: Río Paraná, Pichuna, Jun 1953, *Schultes & Cabrera 19,951* (US). **GUIANÍA**. Río Inírida, Raudal Alto Mariapiri, 250 m, 3 Feb 1953, *Fernández 2084* (COL, US). **VENEZUELA. AMAZONAS**: Esmeralda, Dec 1853, *Spruce 3419* (K, P); Esmeralda, 143 m, 16 May 1942, *Williams 15,418* (US, VEN); SE base of Cerro Duida, 200 m, 22 Aug 1944, *Steyermark 57,817* (A, US, VEN), 175 m, 24 Mar 1953, *Maguire & Wurdack 34,677* (NY, US); Esmeralda Ridge, 6 Oct 1929, *Tate 209* (NY, US); Río Guainía, Caño Pimichin, above Pimichin, 140 m, 14 Apr 1953, *Maguire & Wurdack 35,579* (NY, US); 2 km above Pimichin, 10 Oct 1957 (fl), *Maguire* et al. *41,821* (NY, US); Orinoco, San Antonio, 121 m, 27 Apr 1942 (fr), *Williams 15,052* (MO, US); Santo Antonio del Orinoco, 7°40′N, 61°26′W, Apr 1990 (fl bud), *Yánez 467* (MO, US); Dept. Cassiquiare, Caño San Miguel, 2°39′N, 66°45′W, 25 Apr 1991 (fr), *Aymard 9161* (MO); 20 km SE of San Fernando de Atabapo, 3°50′N, 67°47′W, 10 Jan 1988 (fl bud, fr), *Aymard* et al. *6505* (MO, US); Cucurital de Caname, Caño Caname, 3°40′N, 67° 22′W, 30 Apr 1979 (fr), *Davidse* et al. *16,892* (MO, US). **GUYANA.** Roraima, *Schomburgk 628* (BM, K, OXF, P); Pirara, *Schomburgk 346* (K, OXF, P); Canje River, Ikuruwa Island, 2 Jan 1915 (imm fr), *Hohenkerk FD663* (K). **BRAZIL. AMAZONAS:** Rio Negro, Cemiterio do Castelo, acima de São Luis, 28 Jun 1979 (fr), *Maia 194* (INPA, MO). **PARÁ**: Rio São Manoel, Caiabi, 8 Jan 1952, *Pires 3877* (IAN, US). **MARANHÃO**: Grajahú, 13 Aug 1909, *Lisboa 2327* (BM, MG, US). **MATO GROSSO**: Tabajara, Upper Machado River, Nov–Dec 1931 (fl, fr), *Krukoff 1483* (A, BM, K, NY, P, S, U).

1g. **Humiria balsamifera** var. **pilosa** (Steyerm.) Cuatr., Contr. U.S. Natl. Herb. 35(2): 116. 1961; Cuatrecasas & Huber, Fl. Venez. Guayana (Steyermark et al.) 5: 629. 1999; *Humiria pilosa* Steyerm., Fieldiana, Bot. 28: 270. 1952. Type. Venezuela, Bolívar: Ptari-tepuí, 1615 m, 15–17 Nov 1944, *J. A. Steyermark 60,289* (holotype, F V0060781; isotype, NY00388396).

Trees to 12 m tall, the young branches hirtellous. Leaves oblong-elliptic, 3–5 × 1.5–2.5 cm, coriaceous, cuneate, narrowed to a short petiole 2–3 mm long, subcuneate at base, rounded or obtuse at apex, hirtellous beneath, margins entire. Calyx, inflorescence branches, pedicels and petals pubescent. Drupe not seen.

Distribution and habitat. Endemic to the region of Ptari-tepuí of Venezuela and nearby Brazil in low montane forest (Fig. 10.8).
Phenology. Collected in flower and fruit in November.
Local names and uses. None recorded.
Conservation status. Data deficient (DD).
Illustrations. Cuatrecasas (1961) fig. 24 f.

Additional specimens examined. VENEZUELA. BOLÍVAR: Ptari-tepuí, 1600 m, 1 Nov 1944 (fr), *Steyermark 59,621* (F, NY); Mun. Gran Sabana, Ayava, 1110 m, 5.27, −61, *Hernández 391* (US). **BRAZIL. RORAIMA**: Serra Parima, S of Auaris, 1400–1520 m, 10 Feb 1968 (fl), *Prance* et al. *9795* (INPA, MG, NY, US).

1h. **Humiria balsamifera** var. **guaiquinimana** Cuatr., Contr. U.S. Natl. Herb. 35(2): 113. 1961; Cuatrecasas & Huber, Fl. Venez. Guayana (Steyermark et al.) 5: 628. 1999. Type. Venezuela. Bolívar: Cerro Guiaquinima, Río Paragua, 1760 m, 15 Jul 1944, *F. Cardona 1112* (holotype, US 00101208; isotype, VEN 48656).

Shrubs or low trees, the young branches hirtellous glabrescent. Leaf laminas rigid-coriaceous, 4–7 × 3–5.4 cm, orbicular to suborbicular, almost sessile with short petiole ca. 2 mm long, cuneate at base, rounded, retuse or obtuse at apex, margin entire. Petals glabrous. Drupe not seen.

Distribution and habitat. Scrub and shrubby savannas of Bolívar, Venezuela (Fig. 10.8).
Phenology. Collected in flower in July and October.
Local names and uses. None recorded.
Conservation status. Data deficient (DD).
Illustrations. Cuatrecasas (1961) figs. 21 w, 24 a.

Additional specimens examined. VENEZUELA. BOLÍVAR: Río Paragua, Cerro Guaiquinima, 1760 m, Oct 1943 (fl), *Cardona 965* (F, NY, US); Cerro

Guaiquinima, Cumbre Camp, 1800 m, 25 Dec 1951, *Maguire 32,763* (NY, US), 14 Apr 1952, *Maguire 33,099* (NY, US).

1i. **Humiria balsamifera** var. **savannarum** (Gleason) Cuatr., Contr. U.S. Natl. Herb. 35(2): 108. 1961; Cuatrecasas & Huber, Fl. Venez. Guayana (Steyermark et al.) 5: 629. 1999; *Humiria savannarum* Gleason, Bull. Torrey Bot. Club 58: 378. 1931. Type. Venezuela. Amazonas: Esmeralda, 325 ft., 2 Nov 1929, *G. H. H. Tate 330* (holotype, NY 00388398).

Low shrubs. Leaves sessile, coriaceous, linear lanceolate to narrowly oblong, narrowed-cuneate at base, emarginate or minutely mucronate at apex, 4–10 × 1.5–3.5 cm. margins entire. Petals puberulous on exterior. Drupe oblong-ellipsoid, 14 × 8 mm.

Distribution and habitat. White sand savannas of Venezuela, Guyana, and adjacent Brazil (Fig. 10.8).
Phenology. Flowerings and fruiting around the year.
Local names and uses. None recorded.
Conservation status. Least concern (LC).
Illustrations. Cuatrecasas (1961) fig. 21 v.

Additional specimens examined. VENEZUELA. AMAZONAS: Dept. Atabapo, SSE of Santa Bárbara del Orinoco, 100 m, 3°45′N, 67°04′W (fr), *Huber & Tillett 2832* (INPA, K, US); Canaripó, 98 m, 4°03′N, 66°49′W, 29 May 1978 (st), *Huber 1857* (INPA, K, US); 2 km SE of San Fernando de Atabapo, 3°50′N, 65°48′W, 110 m, 10 Jan 1988 (fr), *Aymard* et al. *6317* (MO); Río Asiva, 4°30′N, 65°48′W, Oct 1989, *Delgado 805* (MO). **BOLÍVAR**: Cerro Matimarota, 100–250 m, 26 Jan 1956 (fr), *Wurdack & Monachino 41,380* (K, NY, RB, US). **GUYANA**. Roraima, 1842–3, *Schomburgk 576* (BM, K, P), Oct 1842, *Schomburgk 845* (K, P). **BRAZIL. AMAZONAS**: 3 km S of Serra Aracá, 0°49′N, 63°19′W, 16 Mar 1984 (fl), *Amaral 1705* (INPA, MO, NY, US); Mun. Manicoré, BR230, 242 km from Humaitá, 7°40′S, 61°10′W, 24 Apr 1985 (fr), *Cid-Ferreira 5810* (INPA, K, NY, RB).

1j. **Humiria balsamifera** var. **subsessilis** (Urb.) Cuatr., Contr. U.S. Natl. Herb. 35(2): 102. 1961; Cuatrecasas & Huber, Fl. Venez. Guayana (Steyermark et al.) 5: 629. 1999; *Humiria floribunda* Mart. var. *subsessilis* Urb., Fl. Bras. (Martius) 12 (2): 439. 1877; *Humirium subsessilis* Spruce ex Urb., Fl. Bras. (Martius) 12 (2): 439. 1877, in synonymy. Type. Brazil. Amazonas: Vaupés, Panure, Nov-Dec 1852, *R. Spruce 2454* (holotype, B, lost; lectotype **designated here**, K 000407341, isolectotypes, BM 000624894, C 00043824, F V0060775, G 00368461, GH 00043824, K 000407342, K 000407343, LD 1775651, MPU 022180, NY 00388399, OXF, P 019032, P 01903273, S-R10008, TCD, US 00101201).

Humiria balsamifera (Aubl.) St. Hil. forma *acuminata* Cuatr., Cíencia (México) 23: 137. 1964. Type. Suriname. Lobin Savanna, between Zanderij and Hannover, 29 Sep 1958 (st), *J. van Donselaar 109* (holotype, U 0002477).

Small trees, the young branches wingless, hirtellous, puberulous or glabrous. Leaves sessile or subsessile, elliptic-oblong or sublanceolate-elliptic, narrowed toward acute apex, obtuse or auriculate at base, 3.5–10 × 1.5–4 cm, margins crenulate and glandular-punctate beneath. Petals glabrous or puberulous. Drupe ovoid-ellipsoid, 9–11 × 5–7 mm, acute at apex; exocarp fleshy, endocarp verruculate.

Distribution and habitat. Savannas of Colombia, Venezuela, northern Brazil, and the Guianas, where it can be locally common (Fig. 10.8).
Phenology. Flowering and fruiting around the year.
Local names and uses. Colombia: *Ta-ta-wee-tee-go* (Barasana), *tatapejá* (Tatuyo) *oloroso*; Venezuela: *Niña, coporik-warei-yek, tatapejá* (tatuyo). Guyana: *Touranero*.
Conservation status. Least concern (LC).
Illustrations. Schultes (1979) fig. 1b; Cuatrecasas (1961) figs. 21 k–n, 22e, 23j–k.

Selected specimens examined. COLOMBIA. AMAZONAS: Río Igaráparaná, caserio Kuirú 15 km below La Chorrera, 1 Mar 1974 (st) *Idrobo 6869* (COAH). **CAQUETÁ:** Mun. Solano, Parque Nacional Natural Serrania de Chiribiqucte N sector, 550 m, 1°21′49″N, 72°55′14″W, 23 Feb 2017 (st), *Cárdenas 46,439* (COAH). GUAVIARE: Mun. Miraflores, headwater Rio Apaporis, 10 km W of Parque Nacional Natural Chiribiquete, 300 m, 1°17′27″, 72°38′36″W, 3 Mar 2018 (st), *Cárdenas* et al. *50,076* (COAH). **VAUPÉS**: Río Piraparaná, Caño Timiña, 6 Sep 1952 (fl), *Schultes & Cabrera 17,231* (BM, US); Río Cubiyú, Cerro Cañendá, 10 Nov 1052 (fl), *Schultes & Cabrera 18,319* (US); Río Cubiyú, 350 m, *Humbert & Schultes 27,364* (US); Alto Río Papurí, Yapú, 16 Jan 1978 (fr), *Patmore & Dufour 162* (K). **GUIANÍA**. Río Negro, Piedra Cocui, 27 Dec 1947, *Schultes & López 9510* (IAN, K, NY); San Felipe, Río Negro, 200 m, 13–25 Nov 1952, *Humbert 27,422* (P), San Felipe, mouth of Río Guainía, 250 m, 21 Nov 1948 (fr), *Araque & Barkley 18 Va021* (MO, US). **VALLE**: Bahía de Buenaventura, Quebrada de Aguadulce, 0–10 m, (fl fr), *Cuatrecasas 19,727* (COL, F, G, GH, K, S, U). **VENEZUELA. AMAZONAS**: San Fernando de Atabapo, Río Cupuení, opposite mouth Río Atabapo, 12 Nov. 1953 (fl), *Maguire* et al. *36,210* (NY, US); Between Sabana Grande & SE base Cerro Duida, 23 Sep 1944 (fl), *Steyermark 57,880* (F, US); Río Orinoco, Río Cunucunuma, N base Cerro Duida, 27 Nov 1950 (fl), *Maguire* et al. *29,769* (NY, US); Río Ventuari, Caño Arisa, 200 m, 16 Feb 1951 (fr), *Cowan & Wurdack 31,502* (K, NY, US); Esmeralda, 120 m, *Tate 331* (NY, US); Base slope of Cerro Duida, Río Cunucuma, 210–280 m, 3°44′N, 65°44′W, 24 Feb 1985 (fr), *Liesner 17,929* (MO); Yavita, 280 m, 22 Jan 1942 (fr), *Williams 13,868* (RB, US); Mt. Moriche, 5°10′N, 66°30′W, 230 m, 14 Mar 1976 (fr), *Colchester 2352* (K). **BOLÍVAR**: Soropán-tepuí, 1656–1980 m, *Steyermark 60,192* (MO, NY); Sabana La Mariposa, SE of Turiba, 15 Dec 1970 (fl), *Berti 2614* (INPA, IPA, MBM, MO, US). **GUYANA**. Corentyne-Berbice Region, 5°20′N, 57°22′W, 26 Apr 1990 (fl), *McDowell 2509* (K). PERU. Loreto: Requena, Jenaro Herrera, no date (st), *Peñuela 1600* (COAH). **BRAZIL. AMAZONAS**: Vaupés, Panure, 15 Nov 1947 (fl), *Pires 1029* (IAN, K, NY, U, US, VEN), Nov 1853, *Spruce 2457* (BM, BR, E, GH, K, MG, NY, OXF, P, S); Ipanoro, Rio Vaupés, 14–15 Nov 1947 (fl), *Schultes*

& Pires 9103A (IAN, US); Rio Xiborem, 15 m, 30 Sep 1928 (fl), *Luetzelburg 24,014* (M); Rio Curicuriari, Cachoeira Cajú, 29 Feb 1936 (fl), *Ducke RB 30128* (F, K, MO, P, U, US). **RORAIMA**: N of Casa de Maracá, 3°21′N, 61°16′W, 2 Dec 1987 (fl), *Milliken* et al. *758* (K). **ACRE:** 5 km from Cruzeiro do Sul, Estrada do Isac, −7.62, −72.92, 24 Mar 1992, *Tate 10,936* (INPA, MO, NY, UFAC, US); 5 km from Aldeia dos Poyanawa, −7.5725, −72.95528, 18 Feb 2009, *Quinet* et al. *10,160* (INPA, RB, NY).

2. **Humiria crassifolia** Mart., Nov. Gen. Sp. Pl. (Martius) 2(2): 143. 1827 (as *Humirium crassifolium*); Mart. ex Urb., Fl. Bras. (Martius) 12 (2): 441.1877; Cuatr., Contr. U.S. Natl. Herb. 35(2): 121. 1961; Cuatrecasas & Huber, Fl. Venez. Guayana (Steyermark et al.) 5: 629. 1999. Type. Colombia. Caquetá: Sierra Araracuara, *C. F. P. von Martius s.n.* (holotype, M; isotype, M).

Myriodendrum subvaginale Mart. ex Urb., Fl. Bras. (Martius) 12 (2): 441.1877, in synonymy.

Shrubs or small trees to 8 m tall, the young branches glabrous, smooth shiny. Leaves with petioles 1–2.5 cm long, stout, winged, clasping at base, often with glands on wing; laminas elliptic-ovate, 7–16 × 4–9.5 cm, rigid-coriaceous, glabrous, attenuate toward base, obtuse or rounded at apex, margins entire, some with gland spots around margin; midrib flat, broad, prominulous on both surfaces; primary veins 11–12 pairs, prominulous beneath, reticulate venation not conspicuous. Inflorescences axillary and subterminal, cymose-paniculate, dichotomous, corymbiform, shorter than the leaves; peduncle compressed, glabrous; bracts ovate or triangular, 1–2 mm long, persistent, clasping, acute; pedicels thick, glabrous, 1 mm long. Sepals rounded, connate at base, glabrous except for ciliolate margins, 1 mm long. Petals white, thick, linear-oblong, 5 mm long, 1–1.5 mm broad, obtuse at apex, puberulous on exterior. Stamens 20, filaments 4–5 mm long, papillose, connate for lower half, 10 shorter ones alternating with 10 longer ones; anthers 0.8 mm long, thecae globose, hairy, connective ovate-lanceolate. Disc annular, scales linear, acute, united at base. Ovary globose, 1 mm high, glabrous, 5-locular, the locules biovulate; style rigid, 3.5 mm long, hirtellous, stigma 5-lobed, oblong-ellipsoid, translucent, capitate. Drupe ellipsoid, 10–12 × 7–9 mm; endocarp obovoid, rounded at apex, slightly acute at base, 5-foraminate at apex.

Distribution and habitat. Rocky and sandy places in the Guayana Highland region from Colombia to Guyana and adjacent northern Brazil (Fig. 10.10).

Phenology. Flowering mainly October to January and fruiting March to May, but often also outside these times.

Local names and uses. The Tawano Indians boil the flowers and young leaves to use as a cataplasm to persistent sores and ulcers (Schultes 1979).

Conservation status. Least Concern (LC).

Illustrations. Cuatrecasas (1961) figs. 21u, 23o, qr., 25 a–b; Cuatrecasas and Huber (1999) fig. 538.

Selected specimens examined. COLOMBIA. AMAZONAS: Puerto Santander, Perdido-Araracuara, 0°36′S, 72°26′W, 160–200 m, 26 Mar 1994 (fl), *Cárdenas* et al. *4531* (COAH, MO); Río Kananarí, Cerro Isibukuri, Jan 1952 (fl), *Schultes & Cabrera 15,054* (NY, US). **CAQUETÁ**: Solano, Paujil, 10 km NE of Araracuara, 0°45′S, 72°20′W, 100–350 m, 2 Dec 1993 (fl), *Arbeláez & Sueroque 543* (COAH, MBM, MO). **GUAINÍA**: Serranía de Naquém, Maimachi, 2°12′N, 68°13′W, 900 m, 7 Apr 1993 (fr), *Barbosa & Madriñán 8364* (COL, GH, MO); Serrania de Naquén, Maimachi, 800 m, 2°13′N, 68°14′W, 1 Aug 1992 (fl), *Cortés* et al. *201* (COAH, NY). **VAUPÉS**: Pacoa, Cerro Morroco, 595 m, 0°7′46″N, 70°56′38″W, 27 Feb 2018 (fl), *Castaño* et al. *10,593* (COAH). **VENEZUELA. AMAZONAS**: Dept. Río Negro, Río Síapa, Matapire, 1°36′N, 65°41′W, 600 m, 14 Feb 1081 (fl), *Huber 6011* (K, NY, US); Cerro Aracamuni, 1°32′N, 65°49′W, 1415 m, 16–18 Oct 1987 (fr), *Delascio & Liesner 13,560* (MO); Cerro Aracamuni, Proa Camp, 1°32′N, 65°49′W, 29 Oct 1987 (fl, fr), *Liesner & Carnevali 22,633* (MO, US); Cerro Neblina above Puerto Chimo Camp, 0°50′N, 66°07′W, 14 Feb 1984 (fr), *Liesner 15,920* (MO, NY, US). **BOLÍVAR**: Piar, Sifontes, Río Uaiparú, 4°12′N, 61°49′W, 900 m, 18 Feb 1986 (fr), *Huber 11,331* (NY, US); Roscio, Cerro Chirikayen, 1580 m, 4°41′N, 61°14′W (fl), *Huber 9619* (NY, US). **GUYANA**. Kaieteur Plateau, 5 May 1944 (fr), *Maguire & Fanshawe 23,233* (BM, K, NY, RB, U, US, VEN); Kaieteur, MureMure Creek, 12 Mar 1962 (fr), *Cowan & Soderstrom 2145* (F, K, NY, US); Potaro-Siparuni Region, Kaieteur Nat. Park, 5°10′N, 59°29′W, 500 m, 14 Apr 1988 (fr), *Hahn* et al. *4544* (MO, US); Cuyuni Mazaruni Region, 0.3 km N of Utshe R, 5°45′N, 61°08′W, 975 m, 22 May 1990 (fr), *McDowell & Gopaul 2730* (K, MO, NY, US); Upper Mazaruni, Karowtipu Mt., 5°45′N, 60°35′W, 1000 m, 23 Apr 1987 (fl), *Boom & Gopaul 7644* (F, K, MO, NY); Pakaraima Mts, Imbaimadai, Karourieng River 5°40′N, 60°13′W, 719 m, (fl), *Redden 1574* (K, NY, US). **BRAZIL. AMAZONAS**: Road São Gabriel to Cachoeira Cucuí, Km 153, 1°24′S, 66°38′W, 250 m, 27 Nov 1987 (fl), *Stevenson 1059* (F, INPA, K, MO, NY); Mun. São Gabriel, road to Cucuí, 27 Nov 1987 (fl), *Lima* et al. *3307* (F, INPA, MO, NY, RB). **RORAIMA**: BR174, Km 530, Bacia do Anauá, 1.13130, –60.40063, 12 Feb 1979, *Cid Ferreira 10,130* (INPA).

The flowers of this species are noted on herbarium labels as very fragrant,

3. **Humiria fruticosa** Cuatr., Contr. U.S. Natl. Herb. 35(2): 118. 1961; Cuatrecasas & Huber, Fl. Venez. Guayana (Steyermark et al.) 5: 629. 1999. Type. Venezuela. Amazonas. Cerro Yapacana, 125 m, 20 Nov 1953 (fl), B. *Maguire, J. Wurdack & G. Bunting 36,580* (holotype, US 00101200; isotypes, F V0060776, K 000407336, NY 00388393, S-R9878, VEN 44496).

Low sprawling shrubs, suffruteces or subshrubs, 0.2–1 m tall, the young branches hirtellous-pubescent. Leaves sessile; laminas elliptic-oblong or oblong, 1.5–4 × 0.5–1.5 cm, rigid-coriaceous, with minute pointed trichomes, densely pubescent beneath, subcordate or rounded at base, clasping, slightly attenuate at apex, obtuse and mucronulate, margins entire and mostly with 2–4 small glands near to base on abaxial surface; midrib prominent on both surfaces; primary veins 9–10 pairs, slightly prominulous beneath, reticulate veins not conspicuous. Inflorescence axillary, shorter than the leaves, usually with only 3–4 flowers; peduncle 0.5–1.4 mm long, compressed, striate, minutely spreading pubescent, branches angulate, minutely pilose; bracts persistent, ovate-triangular to ovate-lanceolate, 0.6–1.5 mm long, minutely pilose; pedicels 1–2 mm long, sparsely puberulous, thickened at top. Sepals triangular-ovate, subacute, thick, 1–1.2 mm long, sparsely pilose on exterior, margins densely ciliate. Petals 6 × 1.3–1.5 mm, linear, white, attenuate, mucronulate, with sparse minute hairs near apex of exterior. Stamens 20, filaments 3–4.5 mm long, of 3 lengths, papillose, connate into a tube for lower third; anthers 1.20–1.4 mm long, connective lanceolate, acute, 0.9–1 mm long; thecae basal, ellipsoid, hairy, 0.4 mm long. Disc annular, 0.8 mm high, dentate. Ovary ovoid, slightly pilose at apex, 5-locular, locules biovulate; style 3 mm long, erect, densely hirtellous, stigma 5-lobed, capitate-lobate. Drupe globose to ovoid, 5–10 × 6–8 mm.

Distribution and habitat. White sand savannas of Amazonas, Venezuela (Fig. 10.10).
Phenology. Collected in flower from December to April.
Local names and uses. None recorded.
Conservation status. Least concern (LC).
Illustrations. Cuatrecasas (1961) figs. 24 g-h; Cuatrecasas & Huber (1999) fig. 537.

Additional specimens examined. VENEZUELA. AMAZONAS: Cerro Yapacana, NW base, 125 m, 31 Dec 1950 (fl), *Maguire* et al. *30,483* (NY, US), 1 Jan 1951 (fr), *Maguire* et al. *30,561* (NY, US); Canaripo, Río Ventuari, 4°05′N, 66°50′W, 125 m, 28 Dec 1976 (fr), *Steyermark 112,828* (MO); Dept. Cassiquiare, Caño San Miguel, 2°40′N, 66°50′W, 21 Apr 1991 (fl), *Aymard 9116* (MO); Dept. Atabapo, E of Caño Perro de Agua, 3°47′N, 67°00′W, 100 m, 20 Feb 1979 (fl), *Huber 3277* (INPA, K, NY, US); Cerro Yapacana, 125 m, 3°45′N, 66°45′W, 7 May 1970 (fl), *Steyermark & Bunting 103,255* (NY, US, VEN).

4. **Humiria parvifolia** A. Juss., Fl. Bras. merid. (A. St.-Hilaire) 2: 89.1829 (sub *Humirium parvifolium* A. Juss); *Humiria floribunda* var. *parvifolia* Urb., Fl. Bras. (Martius)12 (2): 438. 1877. Type. Brazil. Rio de Janeiro, near Cabo Frio. Sep, (fl), *A de St. Hilaire 1145* (holotype, P 01903275; isotypes MPU 012217, P 01903276). Fig. 10.9.

   Humirium montanum A Juss., Fl. Bras. merid. (A. St.-Hilaire) 2: 90.1829; *Humiria floribunda* var. *montana* Urb., Fl. Bras. (Martius) 12 (2): 438. 1877. Type. Brazil. Minas Gerais, Itambe, 1830, *A de St. Hilaire s.n.* (holotype, P; isotypes G, MPU 012218, MPU 012219; photo F).

   Humirium arenarium Guill. *in* Baill., Adansonia 1: 208. 1961. Type. Brazil. Minas Gerais, Catingas de Tocaia, Dec 1838, *J. B. A. Guillemin 205* (lectotype of Cuatrecasas (1961). p. 109, P 01903271 (specimen marked probable type by Cuatrecasas); isotypes, P 01903270, P 01903272).

   Humiria balsamifera var. *minarum* Cuatr., Contr. U.S. Natl. Herb. 35(2): 117. 1961. Type. Brazil. Minas Gerais, Serra do Rio Grande, near Diamantina, 1280 m, 12 May 1931 (fl), *Y. Mexía 5815* (holotype, US 00101204; isotypes, A,

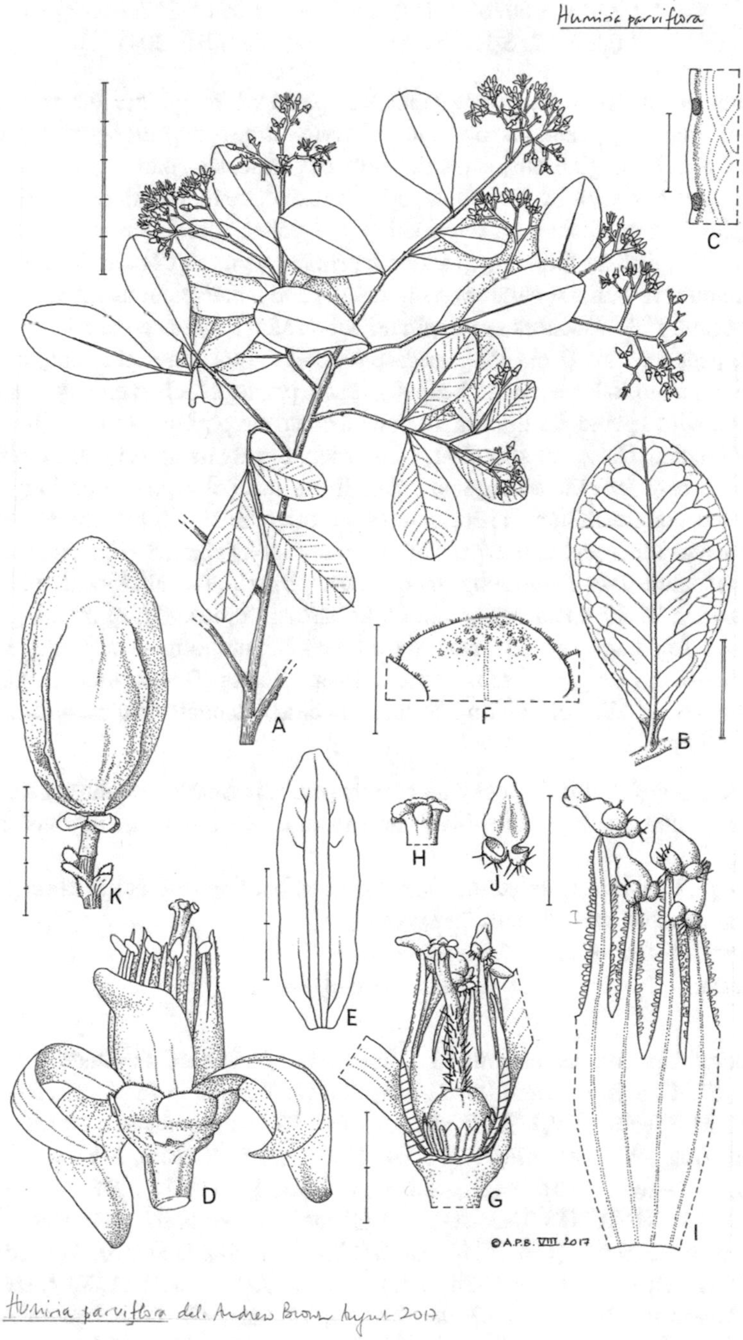

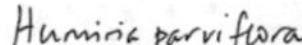

Humiria parviflora del. Andrew Brown August 2017

Fig. 10.9 *Humiria parvifolia*. (**a**) Habit. (**b**) Abaxial leaf surface. (**c**) Leaf margin showing glands. (**d**) Flower. (**e**) Petal. (**f**) Calyx lobe. (**g**) Flower with 2 petals and sepals removed. (**h**) Stigma. (**i**) Part of androecium opened flat. (**j**) Inner surface of anther. (**k**) Fruit. Single bar = 1 mm, graduated single bar = 2 mm and 5 mm, double bar = 1 cm, graduated double bar = 5 cm. (A–J from *Webster 25,089*; K from *Harley* et al. *16,664*)

BM 000611197, F V0060765, GH 00043829, K 000407337, MO 1836383, NY 0022688, R 000029772, S-R7851, U 0002480, US 00101204).

Shrubs or low trees, the young branches sparsely hirsutulous; leaves with petioles 1–3 mm long, laminas obovate, obovate-elliptic or oblong-ovate, 1.5–4.5 (−5) × 0.8–2.7 cm, glabrous, subcoriaceous or coriaceous, cuneate at base, decurrent onto petioles with base revolute, apex retuse or rounded, margins entire, with many glands around margin, terminating the veins; midrib prominulous above, prominent beneath, terminating in a gland; primary veins ca 12 pairs inconspicuous, prominulous beneath, reticulate venation plane and inconspicuous above, prominulous beneath. Inflorescences terminal and subterminal, cymose-paniculate, corymbiform, peduncle 0.5–3 cm long, pedicels 0.5–1.5 mm long, glabrous, articulate with short puberulous peduncles beneath; bracts persistent, chartaceous, rounded or triangular with a glandular tip. Glabrous with cilate margins, ca. 1 mm high. Flowers heavily scented. Calyx cupular, 1–1.5 mm high. Sepals thick, orbicular, imbricate, glabrous except for ciliate margins. Petals linear-lanceolate, 3–4 mm long, greenish–yellow, glabrous. Stamens 20, filaments erect, connate for two thirds of length, papillose and pubescent in free part, of three alternating sizes, 3–4 mm long; anthers ovate-lanceolate, 0.8–1 mm long, thecae basal, hirtellous; subglobose, pubescent. Disc annular, of 20 linear scales united at base, glabrous. Ovary ovoid, glabrous except for few hairs toward apex; style erect, hirsute, stigma capitate, 5-lobed. Drupe 0.8–1 cm long, 4–5 mm wide, oblong, exocarp fleshy, smooth, glabrous; endocarp woody, ellipsoid-oblong, rounded at base, attenuate at apex, with 5 foramina around tip.

Distribution and habitat. Rocky and sandy places in South-Central Brazil, campo rupestre and caatinga and dunes and disjunct to mountain regions of Peru (Fig. 10.10).

Phenology. Flowering around the year and often bearing fruit at the same time.

Local names and uses. Brazil: *Pau preto.*

Conservation status. Least Concern (LC).

Illustrations. Cuatrecasas (1961) figs. 22 o–p, 24e.

Selected specimens examined. PERU. CAJAMARCA: Distr. Huarango, 1583 m, 5°04′58"S, 78°43′31″W, 21 May 2007 (fr), *Perea & Flores 3176* (AMAZ, HUT, K, MO, MOL, NY, US, USM). **SAN MARTÍN:** Sepalacio near Moyabama, 1100 m, Aug 1934 (fl), *Klug 3706* (A, BM, GH, F, K, MO, NY, S, US); near Tarapoto, Aug 1856 (fl), *Spruce 4335* (BM, BR, GH, S). **BRAZIL. PERNAMBUCO:** APA de Guadelupe, −8.6597, −35.1047, 28 Jan 1992, *Sacramento 643* (PEUFR). **ALAGOAS:** Marechal Deodoro, *Andrade-Lima 72–7097* (IPA); Maceio, Apr 1838 (fl, fr), *Gardner 1263* (BM, IPA, K). **SERGIPE:** Santo Amaro de Brotas, −10.789, −37.054, 11 Jan 1978 (fl), *Fonseca ASE511* (ASE); 2 km from São José (Japaratuba), Pirambu, −10.737, −36.855, 1 Jun 1983 (fl), *Carneiro 736* (ASE); 1 km from São Cristovão, −11.014, −37.206, 13 Nov 1981 (fl), *Carneiro 176* (ASE); Serra da Miaba, Campo de Brito, −10.733, −37.493, 1 Jan 1981 (fl), *Carneiro 27* (ASE). **GOIÁS:** 1890, *Glaziou 18,180* (BR,

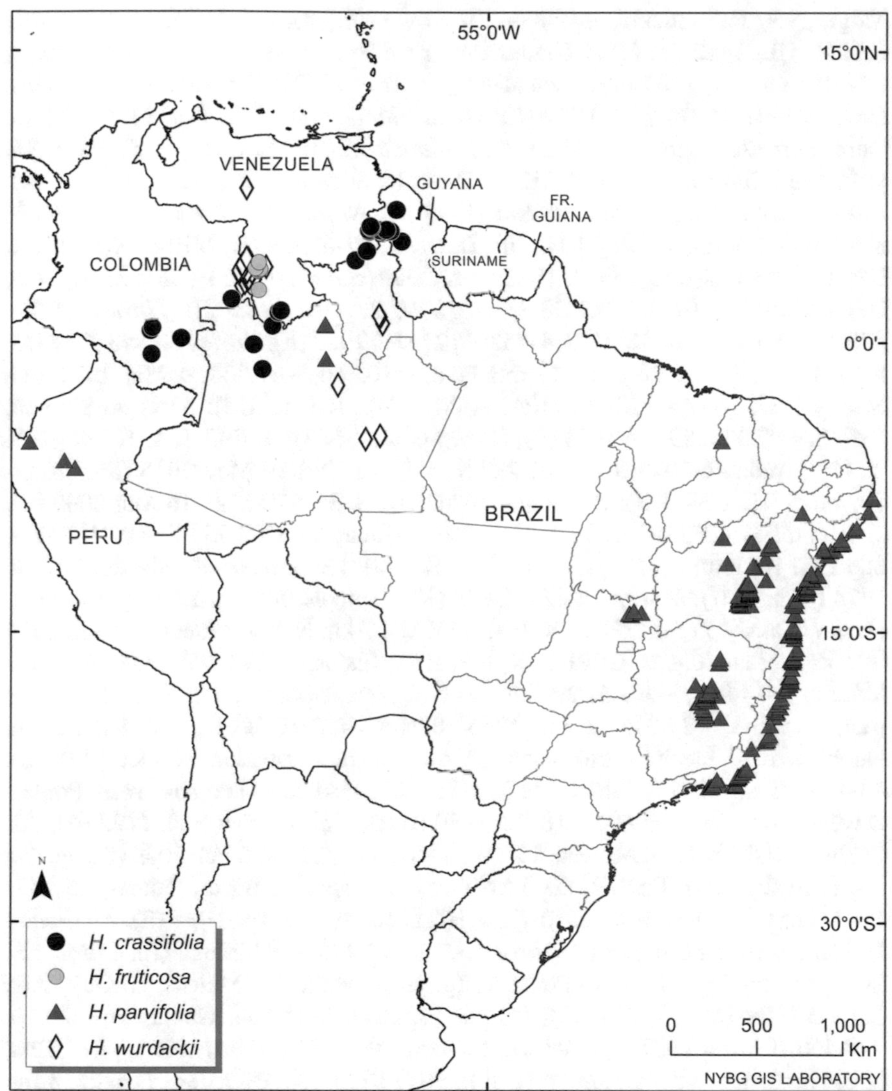

Fig. 10.10 Distribution of four species of *Humiria*

F, NY, P, RB, US); Parque Nacional Chapada dos Veadeiros, 16 Aug 1995 (fl), *Oliveira* et al. *425* (NY, RB, UFG, US). **MINAS GERAIS**: Serra do Cipó, 16 Jan 1951 (fl), *Andrade-Lima 51–821* (IPA); Presidente Juscelino, 16 km toward Diamantina, 9 Feb 1994 (fl), *Sakuragui CFCR 13981* (K); Rio Vermelho, 15 Jul 1984 (fl), *Varanda* et al. *CFCR 4497* (K); 9 km from Itacambira, 29 Nov 1984, *Oliveira* et al. *CFCR6562* (fl), (K); Km 116 road Belo Horizonte-Mato Dentro, Rio Santo Antonio, −19.169, −43.714, *Rossi s.n.* (CEPEC, MBM, USP); Km 120 road Belo Horizonte-Conceição do Mato Dentro, Santana de Riacho, −19.169, −43.714, 5 Nov 1983 (fl), *Arraes* et al. *31,879* (UEC); road Francisco de Sá-Grão

Mogul, 950 m, −16.559, −42.8897, 9 Nov 1995 (fl), *Leitão Filho* et al. *7892* (MBM, UEC); 12 km N. of Diamantina, road to Mendanha, Serra do Espinaço, −18.18, −43.55, 1300 m, 27 Jan 1969, *Irwin* et al. *22,719* (LD, NY, UB, US). **BAHIA**: −15.159747, −39.039072, 1838, *Blanchet 1005* (BM, NY); Gentio do Ouro, Serra do Açurua, −11.42, −42.5, *Blanchet 2810* (BM, BR, F, K, NY, OXF, US), 1838, *Blanchet 3570* (fl) (K); without locality, s.d. (fl), *Salzmann s.n.* (K); Rio de Contas, Pico Itoibira, 23 Mar 1999 (fl), *Nascimento 158* (HUEFS, INPA, K); Rio de Contas, 17 May 1983 (fl), *Hatschbach 46,551* (K, MBM, MO, UFPR, UPCB); Nova Viçosa, 19 Oct 1983 (fl fr), *Hatschbach* 47,031 (CEPEC, K, MO, UPCB); Rio de Contas, 13°28′S, 41°50′W, 27 Dec 1988 (fl), *Harley 25,363* (CEPEC, K); Catolés, 13°17′S, 41°52′W, 27 Dec 1988 (fl), *Harley 27,832* (CEPEC, K, SPF, US, USP); Serra do Curral Feio, −10.366, −41.333, 4 Mar 1974 (fr), *Harley* et al. *16,664* (CEPEC, IPA, K, MO, NY, RB, U, US); Serra do Sincorá, 13°0′S, 41°23′W, 29 Jan 1974 (fl), *Harley* et al.*15649* (CEPEC, IPA, K, MO, NY, U, US); Andaraí, 1012 m, 12°57′28” S, 41°18′11″W, 19 May 2015 (fr), *Sousa-Silva 628* (HUEFS, RB); Belmonte, 16°02′26” S, 38°57′51″W, 16 Aug 2000 (fl), *Sant'Ana 947* (CEPEC, HUEFS, NY, RB); Mucugê, 13°08′55” S, 41°31′2″W, 6 Sep 1981 (fl, imm fr), *Pirani* et al. *1906* (K, USP); 35 km NE of Salvador 30 Aug 1978 (fl, imm fr), *Morawetz 27–30,878* (K); Lençois, 800 m, 27 May 1980 (fl), *Harley* et al. *22,718* (CEPEC, K, NY, USP, U); 7 km NW of Mucuri, 13 Sep 1979 (fl), *Mori* et al. *10,478* (CEPEC, K, NY, RB); Ilhéus, 5.4 km SW road Olivença-Maruím, −14.983, −39.05, 29 Jan 1992 (fl), *Thomas 8932* (CEPEC, NY, RB); Una, −15.293, −39.075, 2 Jun 1966, *Belém 2408* (CEPEC, NY, UB); Mun. Jacobina, road Jacobina-Morro do Chapéu, Serra Tombador, 28 Oct 1995 (fl), *Amorim* et al. *1807* (CEPEC, MO, NY, RB); Fazenda Guanaba, road Pontal-Olivença, −14.783, −39.033, 16 Oct 1980 (fl fr), *Mattos Silva* et al. *1193* (ALCB, CEPEC, HUEFS, US); Arateca, 15°10′26” S, 39°20′22″W, 6 Feb 2009 (fl), *Jardim 224* (CEPEC, RB). **ESPÍRITO SANTO**: Conceição da Barra, Itaúnas, 15 May 1987 (imm fr), *Lima* et al. *2975* (K, RB); Linhares, 12 Apr 1934 (fl), *Kuhlmann 179* (K); between Campos and Vitoria, −21,042,996, − 41.22626 (fl), *Sellow s.n.* (K); Road linking BR101 to Ponta do Ipiranga, Mun. São Mateus, 13 Oct 1992, *Cervi 58,074* (MO, NY, UCS); Parque Estadual de Setiba, Guarapari, −20.60, −40.466, 19 Aug 1987 (fl), *Pereira & Gomes 949* (RB, VIES); Mun. Vila Velha, Lagoa do Milho, s.d, *Peixoto 340* (MO). **RIO DE JANEIRO**: Jacarepaguá, 4 Jun 1969 (fl), *Sucre 5423* (K, NY, RB); Tijuca, 23 Nov 1971 (fr), *Sucre 7937* (K, NY, RB); Mun. Maricá, 23 Nov 1986 (fl), *Farney 373* (K, NY, RB); Praia de Sernambetiba, 23°0′S, 43°20′W, 4 Apr 1952 (fl), *Smith 6406* (K, NY, US); Restinga de Itapeba, 23°0′S, 43°20′W, (fl), *Segadas-Vianna 3506* (K, US); Rio das Ostras, −22.478, −4.883, 17 Oct 2007, *Marquete 4132* (CEPEC, HUEFS, RB).

5. **Humiria wurdackii** Cuatr., Contr. U.S. Natl. Herb. 35(2): 119. 1961; Cuatrecasas & Huber, Fl. Venez. Guayana (Steyermark et al.) 5: 629. 1999. Type. Venezuela. Amazonas: Río Atabapo, 20 km above San Fernando de Atabapo, Sabana Cumare, 125 m, 3 Jun 1959 (fl), *J. J. Wurdack & L. S. Adderley 42,760* (holotype, US 00101195; isotypes, GH 00043825, K 000407335, NY 00388400, RB 00539054, US 006921448, VEN 51804).

Shrubs or small trees to 8 m tall, the young branches glabrous, spreading. Leaves with pseudopetiole 0.5–2 cm long, canaliculate; laminas linear, 2.5–10 × 0.3–0.8 cm, coriaceous, glabrous, narrowed into a pseudopetiole at base, obtusely narrowed and minutely mucronate at apex, margins entire, with abaxial glands along entire length, midrib prominulous above, prominent beneath; primary veins inconspicuous above, slightly prominulous beneath. Inflorescences axillary, with 5–10 flowers, branches to 15 mm long, glabrous; bracts persistent, ovate, obtuse, glabrous, 0.5–0.7 mm long; pedicels 1–1.5 mm long, thickening toward apex. Sepals broad, rounded, 1.5 mm long, connate at base for lower half, glabrous except for minutely ciliate margin, sometimes with a rounded gland on outer surface. Petals white, thick, linear, 5–7 × 1–1.3 mm, glabrous, attenuate at apex, ending in a minute callose gland. Stamens 20, slightly shorter than petals, connate into a tube for lower half, filaments thick, papillose, of 3 lengths; anthers with a thick, lanceolate connective; thecae basal, ellipsoid, slightly hairy. Disc annular, 10-lobate with emarginate lobes, 1 mm high. Ovary pubescent at apex, 5-locular, the locules biovulate. Style erect, rigid, pubescent, 3 mm long, stigma capitate-lobate. Drupe ellipsoid.

Distribution and habitat. Thickets and low forest around savannas in seasonally flooded areas of Colombia, Venezuela, and northern Amazonian Brazil (Fig. 10.10).
Phenology. Flowering and fruiting throughout the year.
Local names and uses. None recorded.
Conservation status. Least Concern (LC).
Illustration. Cuatrecasas (1991) pl 10; Cuatrecasas and Huber (1999) fig. 635.

Selected specimens examined. COLOMBIA. GUAINÍA: Mun. Purto Inirida, Reserva Indigena Almidon-La Ceiba, 3°32′N, 67°51′W, 80 m, 21 Mar 1998, *Rudas* et al. *7171* (COAH, COL, MO), 20 Oct 1998, *Gordillo R.* et al. *313* (COL, MO). **VICHADA**: Mun. Cumaribo, Mataven, Caño Cajaro, 240 m, 4°31′56″N, 68°5′28″W, 16 Mar 2007 (fl), *Prieto* et al. *5337* (COAH). **VENEZUELA. AMAZONAS**: Dept. Cassiquiare, 20 km NW of Yavita, 3°01′N, 67°33′W, 120 m, 11 Feb 1981 (fl bud), *Huber & Medina 5958* (K, NY, US); 10 km W of Caño Pimichin, Río Guainía, 2°53′N, 67°44′W, 100 m, 61 Mar 1980 (fr), *Huber 4888* (MO, NY, US); Dept. Atabapo, 30 km W of Serranía El Tigre, Caño Yagua, 3°51′N, 66°27′W, 130 m, 22 Feb 1980 (fr), *Huber 4842* (INPA, K, NY, US); Río Atabapo, 30 km from confluence of Río Canare, 3°48′N, 67°52′W, 110 m, Nov 1989 (fr), *Yánez 256* (MO); Río Guasacán, 0.52 km S of Laja Suiza, 3°13′N, 67°03′W, 100 m, 22 Nov 1996 (fl), *Berry & Rosales 6499* (MO); N bank Caño Cumare, near mouth of Río Atabapo, 3°58′N, 67°41′W, 25 May 1993 (fl bud), *Berry* et al. *5489* (MO); Dept. Atures, 4°18′N, 67°28′W, May 1989 (fl), *Foldats & Velazco 9378* (MO, NY, US). **BRAZIL. AMAZONAS**: Campina do Patuá, PN Jaú, *Ferreira 07PNJ* (INPA, MO); Parque Estadual Matupiri, Canutama, −5.01, −61.27, 29 Oct 2010, *Prata 792* (INPA); Parque Nacional Jaú, Quadrante 5, 28 Aug 1998 (fr), *Vicentini 1298* (INPA, MO). **RORAIMA**: Manaus-Caracaraí, Km 422–424, near Novo Paraiso, 1°18′N, 60°35′W, 25 Aug 1987 (fl), *Cid Ferreira 9148* (F, INPA, MO, NY, US), Km 523,

São Luís, 29 Jun 1997 (fl), *Teixeira 1648* (INPA), Km 529–550, 15 Mar 1984, *Santos 716* (INPA, MO, NY, US).

10.5 Humiriastrum

Humiriastrum (Urb.) Cuatr., Contr. U.S. Natl. Herb. 35(2): 122. 1961; Cuatrecasas & Huber, Fl. Venez. Guayana (Steyermark et al.) 5: 631.1999. Type. *Humiriastrum cuspidatum* (Benth.) Cuatr.

Saccoglottis subgen. *Humiriastrum* Urb., Fl. Bras. (Martius) 12(2): 443. 1877.

Saccoglottis Sect *Humiriastrum* (Urb.) Reiche, Nat. Pflanzenfam. (Engler & Prantl) 3 (4): 37. 1890.

Sacoglottis Sect. *Humiriastrum* (Urb.) Winkl., Nat. Pflanzenfam. (Engler & Harms)19a: 128. 1931.

Humirium Benth., London J. Bot. (Hooker) 2: 373. 1843, pro parte.

Humirium Rich. ex Mart., Nov. Gen. Sp. Pl. (Martius) 2(2): 142. 1827.

Trees. Leaves alternate, petiolate, chartaceous to rigid-coriaceous, lamina margins entire or more frequently crenate-dentate. Inflorescences axillary or pseudoterminal, paniculate with trichotomous or dichotomous branching; bracts persistent or deciduous. Sepals 5, suborbicular, imbricate, united at the base. Petals 5, free, thick-membranaceous, aestivation quincuncial, contorted or cochlear. Stamens 20 in two alternating lengths, 10 longer ones antesepalous and antepetalous, alternating with 10 shorter, glabrous, filaments connate at base, subulate, complanate, smooth or spasely papillose; anthers ovate-lanceolate to oblong, attached near base, connective thick more or less lanceolate and acute at apex, thecae 2, unilocular, ellipsoid, subglobose, basal, or dorsifixed. Disc a dentate ring circling ovary or of more or less free scales. Ovary 5-locular, glabrous or pubescent, the locules uniovulate; carpels opposite the sepals; style short, stigma capitate, 5-lobed. Drupe ellipsoid or subglobose, smooth, exocarp fleshy, subcoriaceous when dry; endocarp woody, usually without resinous cavities, 5 foramina around apex and 5 oblong germinal opercula or valves on upper half, usually with 1–2 seed cavities developed; seeds oblong.

Distribution, Sixteen species mainly in lowland forests of South America with a single species, *Humiriastrum diguense*, entering into Central America as far north as Costa Rica.

Key to Species of Humiriastrum

1. Bracts persistent; petals hirsutulous, rarely glabrous; ovary glabrous.

    2. Leaf laminas obtuse or shortly obtusely apiculate, never acuminate or cuspidate; terminal branches hirsute or minutely hirtellous.

3. Leaf laminas obovate, 2.5–9 × 1.4–5.5 cm. apex rounded, truncate or obtuse to retuse, margin revolute and entire, lower surface hirsute; drupe 1.5–2 × 0.8–1 cm.. **10. H. obovatum**
3. Leaf laminas broadly elliptic or obovate-elliptic, 4–8.5 × 2.5–4.8 cm, apex abruptly contracted and shortly apiculate, margin entire or slightly crenulate, lower surface glabrous; drupe 3 × 2.5 cm..... **11. H. ottohuberi**

2. Leaf laminas acutely acuminate or cuspidate, young branches glabrous or pubescent.

4. Leaf laminas villous hirsute beneath, densely so on midrib; ovate-acuminate or cuspidate; young branches hirsute; sepals hirtellous .... **16. H. villosum**
4. Leaf laminas glabrous; young branches glabrous or rarely puberulous.

5. Leaf venation obsolete, inconspicuous, margin usually dentate; sepals glabrous with ciliate margins; drupe globose 1.7–2 cm diam .... **2. H. cuspidatum**
5. Leaf veins fine but conspicuous, margin crenate; sepals hispid or puberulous.

6. Young branches glabrous; sepals hispid-puberulous; drupe 2.4–3 × 1.8–2.5 cm... **12. H. piraparanense**
6. Young branches puberulous; sepals puberulous on exterior; drupe 3.5–4 × 2.2–3 cm.. **14. H. purusensis**

1. Bracts deciduous; petals pubescent on exterior except in *H. glaziovii*; ovary glabrous or pubescent.

7. Inflorescence terminal or subterminal, often as long or longer than leaves; leaves glabrous above, with sparsely minute, thin appressed inconspicuous hairs beneath.

8. Young branches winged, glabrous; leaves sessile or subsessile, ovary glabrous... **13. H. procerum**
8. Young branches pubescent-hirtellous, not winged; leaves with short petioles 1–3 mm long; ovary glabrous or pubescent.. **4. H. diguense**

7. Inflorescence mostly axillary, usually shorter than leaves except in *H. mapirense;* leaves glabrous or spreading-pilose.

9. Young branches hirsute or puberulous-hirtellous; peduncles and branchlets of inflorescence usually hirsute or hirtellous.

10. Leaf laminas pubescent or puberulous beneath; secondary veins and reticulation prominulous on both surfaces; branches very hirsute; sepals glabrous on exterior..3. **H. dentatum**
10. Leaf laminas glabrous or inconspicuously puberulous; venation inconspicuous above; sepals minutely pilose on exterior.
    11. Petioles 1 mm long; leaf lamina inconspicuously puberulous with small appressed hairs beneath; rigidly coriaceous; petals hirtellous...............**15. H. subcrenatum**
    11. Petioles 2–6 mm long; leaf laminas glabrous.
        12. Petioles 2–3 mm long; leaf laminas thin coriaceous, flexible; rounded to obtusely cuneate at base; petals puberulous hirtellous; drupe ellipsoid-ovoid 2–2.5 × 1.4–1.8 cm...................................**5. H. excelsum**
        12. Petioles 3–6 mm long; leaf laminas rigidly coriaceous, cuneate at base; petals glabrous; drupe globose, 1.5–1.6 cm diam....................**7. H. mapiriense**

9. Young branches glabrous, leaves glabrous.
    13. Leaf laminas rounded at apex or abruptly and obtusely acuminate; subsessile..........................**8. H. melanocarpum**
    13. Leaf laminas attenuate toward apex, acute, acuminate or cuspidate, petiolate.
        14. Ovary hispidulous; petioles 2–4 mm long...**1. H. colombianum**
        14. Ovary glabrous; petioles 4–11 mm long.
            15. Petals pubescent on exterior; drupe globose or slightly globose-elliptic, 3.6–4.2 × 3–3.5 cm...**9. H. mussunungense**
            15. Petals glabrous on exterior; drupe ellipsoid-ovoid, 1.9 × 2.6 cm.................................**6. H. glaziovii**

1. **Humiriastrum colombianum** (Cuatr.) Cuatr., Contr. U.S. Natl. Herb. 35(2): 134. 1961; Cuatrecasas & Huber, Fl. Venez. Guayana (Steyermark et al.) 5: 632. 1999. *Sacoglottis excelsa* var. *colombiana* Cuatr., Brittonia 8: 196. 1956. Type. Colombia: Santander, región de Carare, Cimitarra, Km 3 road to Ermitaño, 29 Jul 1954 (fl), *F. B. Lamb 141* (holotype, US 00101232; isotype, COL 000001808).

Medium to large trees, the young branches glabrous. Leaves with petioles 2–4 mm long, thickened at base; laminas elliptic to ovate-elliptic, 4–9 × 1.5–4 cm, thin-coriaceous, glabrous, papillose beneath, cuneate at base, acutely cuspidate at apex, margins crenulate, setae long; midrib prominulous above, prominent and glabrous beneath; primary veins 8–10 pairs, thin slightly prominulous or inconspicuous, venation inconspicuous. Inflorescences axillary, cymose-paniculate, shorter than leaves, lower branches trichotomous upper ones dichotomous; peduncle 1–1.5 cm long, rigid, striate, slightly puberulous: bracts ovate, 0.2–0.3 mm long, puberulous, deciduous, flowers almost sessile, pedicels 0.1–0.2 mm long, thickened, puberulous. Sepals 0.4–0.5 mm long, rounded, minutely papillose, puberulous, ciliolate on margins. Petals oblong, 2–2.1 × 1 mm, appressed-puberulous. Stamens 20, filaments papillose subulate, complanate, ten 1.1–1.2 mm long, antesepalous and antepetalous, alternating with ten shorter ones 0.7–0.8 mm long, anthers oblong-lanceolate, 0.7–0.8 mm long, basal, connective thick-lanceolate, thecae ellipsoid. Disc of several small weakly dentate free scales, 0.1–0.2 mm long, encircling ovary. Ovary globose, hispidulous, 5-locular, locules uniovulate; style 0.5 mm long, stigma capitate, 5-lobed. Drupe ellipsoid-ovoid, rounded at base, abruptly narrowed and subacute at apex, 2–2.25 × 1.2–1.5 cm; exocarp glabrous, smooth, thin; endocarp ovoid-ellipsoid, rounded at base, acuminate at apex, woody, hard, rugose, with 5 apical foramina and 5 subapical oblong descending opercula, ca 6 mm long.

Distribution and habitat. First discovered in the Magdalena valley of Colombia, but now found to be much more widespread to Venezuela and as far south as northern Peru in forest (Fig. 10.14).
Phenology. Flowering May to July and collected in fruit in April.
Local names and uses. Colombia: *Aceituno.*
Conservation status. Least Concern (LC).
Illustrations. Cuatrecasas (1961) figs. 26 k–l, 27 e–f.

Additional specimens examined. COLOMBIA. ANTIOQUIA: Anori, 7°18′N, 75°04′W, 400–700 m, 24 Feb 1977 (fr), *Alverson* et al *101* (COL, HUA, MO, NY). **SANTANDER:** Region de Carare, Cimitarra, Km 3 road to Ermitaño, 30 Jul 1954 (fl), *Lamb 145* (COL, US), 17 Jul 1954, *Lamb 170* (COL, US); Barranca Bermeja, 5 km from margin of Río Opón, 200 m, 28 Sep 1954, *Romero-Castañeda 4942* (COL, US); SE of Barranca Bermeja, 8 km from margin or Río Opón, 300 m, 31 July 1954, *Romero Castañeda 4785* (COL, US); Mun. Anorí, Corr. Providencia, 700 m, 5 Jun 1971 (fr), *Soejarto 2863* (COL); Mun. Amalfi, vereda Arenasblancas, 1250 m, 6°55′N, 74°55′W, 11 Apr 1994 (fr), *Fonnegra* et al. *47,611* (COL). **VALLE DEL CAUCA**: Buenaventura, 3°50′N, 76°35′W, 100 m, 14 Sep 1993, *Devia A.* et al. *4184* (MO). **VENEZUELA. AMAZONAS**: Rio Mawarinuma, 0°50′N, 66°05′W, 190 m, 24 Apr 1984 (fr), *Gentry & Stein 46,898* (K, MO, US); Dept. Río Negro, Cerro de Neblina, 150 m, 0°50′N, 66°07′W, 13 Feb 1984 (fl), *Liesner 15,876* (MO, NY, US). **PERU. LORETO**: Maynas, 4°08′S, 66°05′W, 9 May 1991, *Grández* et al. *2570* (MO). **HUÁNUCO**: 800 m, 29 Oct 1963, *Gutiérrez 134* (F, MO).

2. **Humiriastrum cuspidatum** (Benth.) Cuatr., Contr. U.S. Natl. Herb. 35(2): 130, 1961; Cuatrecasas & Huber, Fl. Venez. Guayana (Steyermark et al.) 5: 637. 1999. *Humirium cuspidatum* Benth. Hooker's J. Bot. Kew Gard. Misc. 5:101.1853; *Sacoglottis cuspidata* (Benth.) Urb., Fl. Bras. (Martius) 12 (2): 444. 1877; Ducke, Arch. Jard. Bot. Rio de Janeiro 3: 178. 1922. Type. Brazil. Amazonas: Barra de Rio Negro (Manaus), Jun 1851 (fl), *R. Spruce 1915*

(Lectotypfication by Cuatrecasas (1961). p. 132, K 000407332, isolectotypes, BM 000778557, FI-W, M, NY 00388389), Jun 1851 (fl), *R. Spruce 1715*, (possible isotypes, E 00326636, FH 00042823, F V0060768, F 012593, FI 006164, G 00368462, GH 00043823, OXF, P 01903265, P 01903266, TCD 0003836; B lost, photo F). Fig. 10.11.

Humiriastrum cuspidatum var. *glabriflorum* (Ducke) Cuatr., Contr. U.S. Natl. Herb. 35(2): 132. 1961. *Sacoglottis excelsa* var. *glabriflora* Ducke, Arq. Inst. Biol. Veg. 4: 25. 1938. Type: Brazil. Amazonas: Manaus, Cachoeira do Mindú, 8 Jul 1929, *A. Ducke RB 23436* (holotype, RB 00539055, isotypes, IAN, INPA, P 01903267, S-R10007, U, US 00101231).

Humiriastrum cuspidatum var. *subhirtellum* Cuatr., Contr. U.S. Natl. Herb. 35(2): 133. 1961. Type: Brazil. Amazonas, Rio Urubú, São Francisco, 4 Oct 1949 (fl), *R. L. Fróes 25,480* (holotype, IAN 051770; isotype, IAN 051770a, photos NY, US) (Fig. 10.12).

Medium-sized trees, the young branches glabrous or rarely hirtellous. Leaves with short petioles 2–6 mm long, thick, flattened; laminas ovate-elliptic, elliptic or obovate-elliptic, 4–11 × 2–6 cm, rigid-coriaceous, glabrous, cuneate at base, abruptly narrowed at apex, caudate or acuminate, the acumen 5–20 mm long, margins serrulate-dentate, setae long, thin, usually caducous, with small sparse glands near margin of both surfaces; midrib plane above, plane or prominulous and glabrous beneath, secondary veins and venation inconspicuous obsolete. Inflorescences axillary and subterminal, cymose-paniculate, shorter than or equal to leaves, lower branches trichotomous, upper ones dichotomous, sparsely hirtellous; peduncle, 1–2.5 cm long, robust, striolate, glabrous or rarely minutely puberulous; bracts ovate obtuse or subacute, 1 mm long, glabrous, persistent; bracteoles 0.4–0.8 mm long, puberulous, persistent; pedicels 0.2–0.4 mm, thick. Sepals orbicular, connate at base, 0.6 mm long, glabrous except for ciliate margin. Petals elliptic-oblong, 2.5–3 × 1.2–1.5 mm, puberulous or rarely glabrous. Stamens 20, filaments minutely papillose, lower third connate, ten 2–2.2 mm long, alternating with ten shorter ones ca 1.7 mm long, anthers 0.7–0.8 mm long, glabrous, connective thick, ovate acuminate, thecae short-elliptic, basal. Disc 0.5 mm high, formed of deeply bidentate united scales. Ovary globose, glabrous, 5-locular, locules uniovulate; style 0.6 mm long, stigma capitate, 5-lobed. Drupe globose to ellipsoid, 1.7–2.0 cm diameter; exocarp thin, smooth, glabrous; endocarp globose, rugose, with 5 foramina at apex and 5 elliptic-oblong descending opercula 7 mm long.

Distribution and habitat. Common in Central Amazonian Brazil on sandy riverbanks and periodically flooded ground and open campos on sandy soil, and more rarely on *terra firme* also in Amazonas and Bolívar, Venezuela southern Colombia and Loreto Peru (Fig. 10.13).

Phenology. Flowering mainly from June to August, fruiting from September to April.

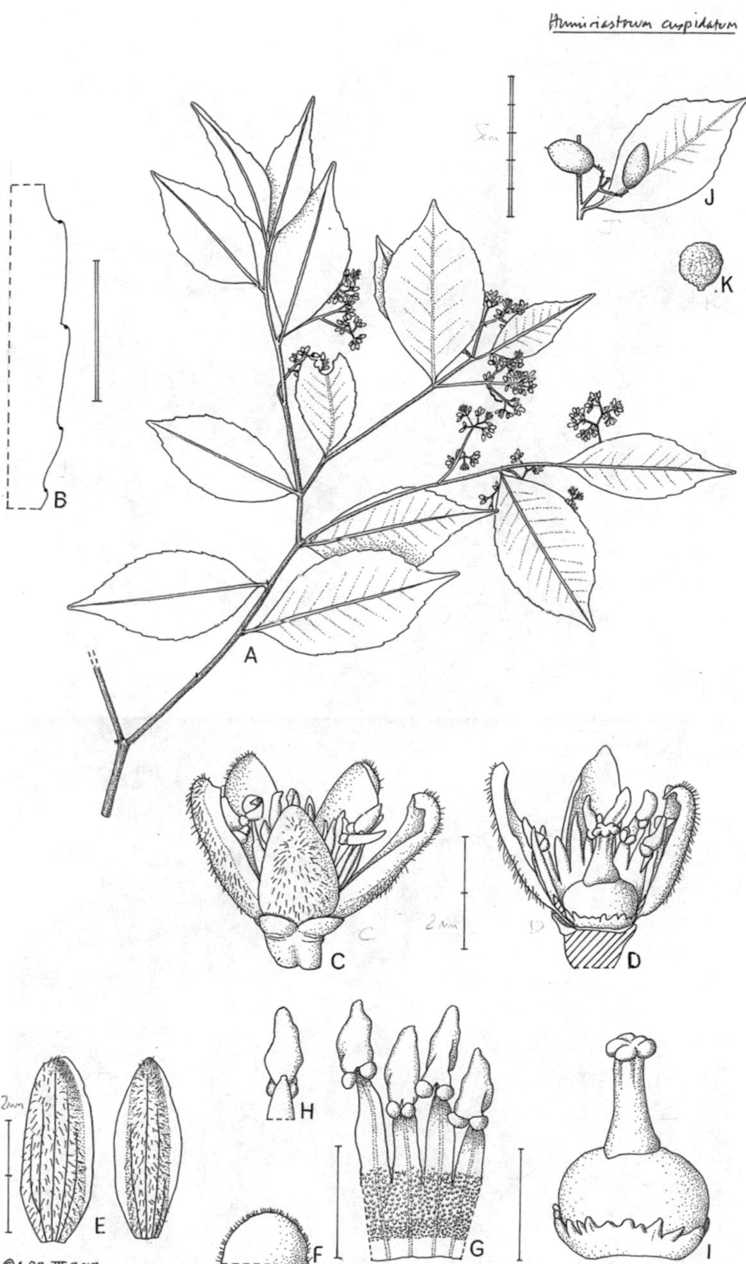

Humiriastrum cuspidatum del. Andrew Brown September 2017

Fig. 10.11 *Humiriastrum cuspidatum*. (**a**) Habit, flowering branch. (**b**) Abaxial leaf margin. (**c**) Flower. (**d**) Flower with 2 petals and sepals removed. (**e**) Outer surface of two types of petals. (**f**) Calyx lobe. (**g**) Part of stamina tube, inner surface. (**h**) Anther and filament tip, outer surface. (**i**) Pistil and disc. (**j**) Young fruits. (**k**) Mature fruits. Single bar = 1 mm, graduated single bar = 2 mm, double bar = 1 cm, graduated double bar = 5 cm. (A, B, K from *Ducke RB 23436*; J, from *Prance* et al. *3719;* C–I from *Zarucchi* et al. *3170*)

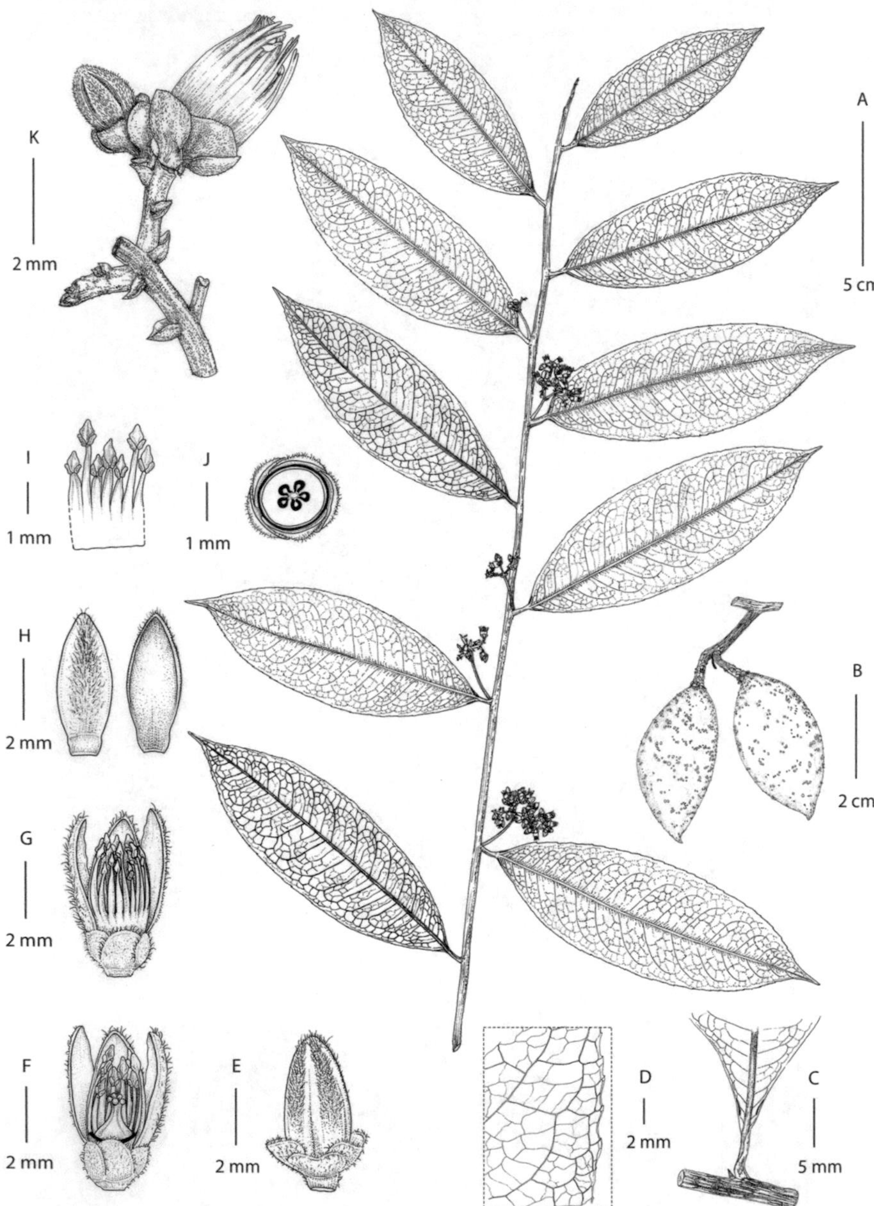

Fig. 10.12 *Humiriastrum purusensis*. (**a**) Habit, flowering branch. (**b**) Young fruit. (**c**) Leaf base and petiole. (**d**) Leaf venation and margin. (**e**) Unopened flower. (**f**) Flower with 2 petals and stamens removed. (**g**) Flower with 2 petals removed, showing stamens. (**h**) Petals. (**i**) Part of stems. (**j**) Section of ovary. K. Flower detail. (From the holotype)

Fig. 10.13 Distribution of six species of *Humiriastrum*

Local names and uses. Brazil: *Achuá*, Venezuela: *Guaco*, the fruit said to be edible, but astringent.

Conservation status. Least concern (LC).

Illustrations. Cuatrecasas (1961), fig. 27 a–d.

Selected specimens studied. **COLOMBIA. CAQUETÁ**: Río Caguán, La Argentina, 3 Mar 1977, *Roa 636* (COL, INPA, MG); Mun. Solano, Est. Biol. Puerto Abeja, Parque Nacional Natural Chiribiquete, 0°4′13″N, 72°27′27″W, 24 Nov 2000

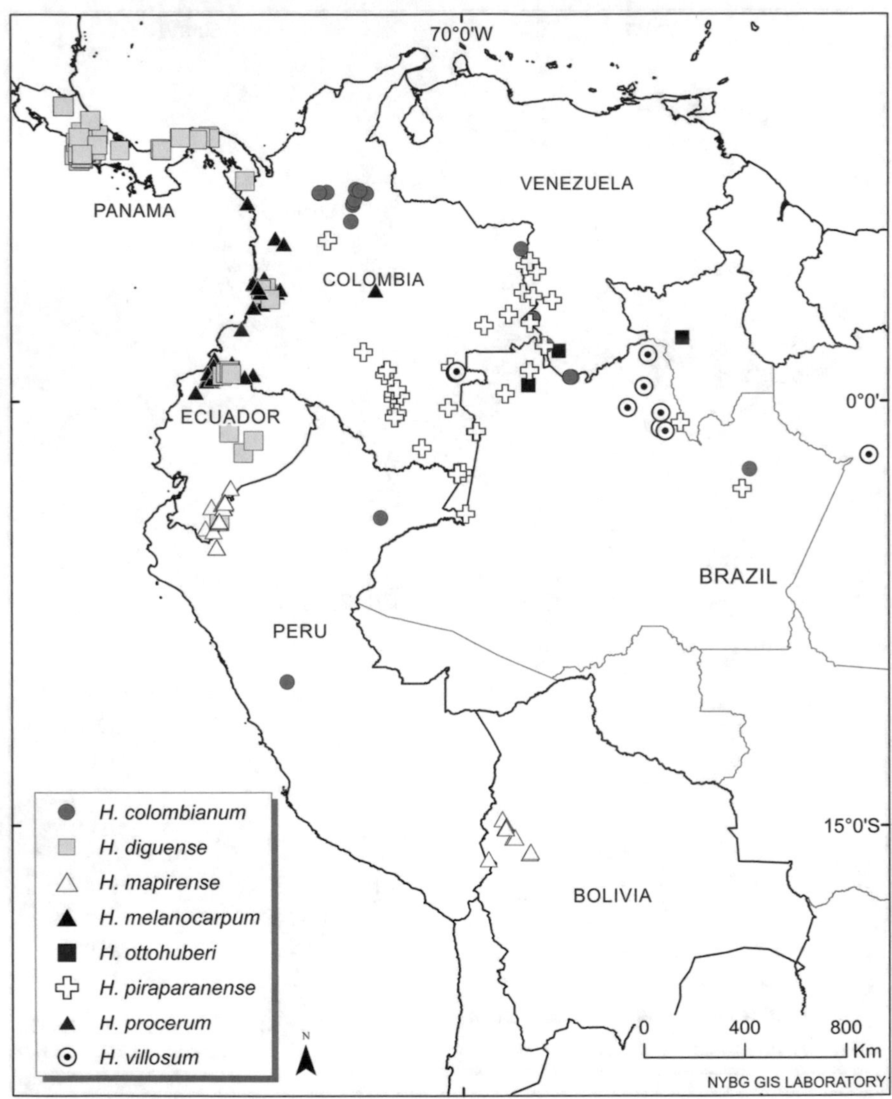

Fig. 10.14 Distribution of eight species of *Humiriastrum*

(fr), *Dávilla 505* (COAH). **GUAINÍA**: Panapaná, Campoalegre near Caño Guaviarito, 1°52′51,1356 N, 69°00′33,8940 W, 149 m, Feb–Mar 2016 (fr), *González* et al. *670* (COAH). **VAUPÉS**: Mun. Mitu, Laguna Marucuarí, 1°17′36″N, 70°11′37″W, 7 Jul 2012 (fl), *Castaño* et al. *3446* (COAH, NY). **VENEZUELA. AMAZONAS**: 1°55′N, 67°03′W, 1 Dec 1977, *Liesner 4102* (MO). **BOLÍVAR**: Piedra Marimare, E bank Río Orinoco, Opposite Isla El Gallo, 100 m, 20 Dec 1955 (fl, imm fr), *Wurdack & Monachino 40,881* (NY, US). **PERU. LORETO:** Jenaro Herrera Arboretum, 4°55′S, 73°45′W, 13 Aug 2005, *Meinich 25* (MO).

BRAZIL. AMAZONAS: São Paulo de Olivença, 23 Apr 1945 (fr), *Fróes 20,803* (IAN, K, NY); Itacoatiara-Mirim, São Gabriel da Cachoeira, Jul 2007, *Stropp & Assunção 475a* (EAFM); Mun. São Gabriel da Cachoeira, baia de Aparecida, 4 km NW of Ilha Açaí, 25 Jul 1991 (fl), *Martinelli 14,499* (RB); Panuré, Rio Uaupés, Oct 1852-Jan 1853 (fl), *Spruce 2424* (F, K, NY, P, S), *2443* (BM, F, GH, K, OXF, P); Rio Urubu, Manaus-Itacoatiara road Km 202, 19 Dec 1966 (fr), *Prance* et al. *3719* (F, INPA, K, MG, MO, NY, US); Mun. Maués, Rio Apoquitaua, above São Sebastian, 3°50′S, 57°55′W, 27 Jul 1983 (fl), *Cid Ferreira 4264* (F, INPA, K, MG, NY, RB, US); Rio Curicuriari, 4 Oct 1935 (fl), *Ducke RB 30126* (K, RB, S, U, US); Tunuí, Rio Içana, Igarapé Jacitari, 1°23.40 S, 68°09.264 W (fl bud), *Acevedo-Rodrigues 14,632* (INPA, K, NY, US); Mun. Presidente Figueiredo, 1°30′S, 59°30′W, 18 Sep 1986 (fr), *Cid Ferreira 8195* (F, INPA, MO, NY); Mun. Manaus, Igarapé Lajes, Manaus-Caracaraí, Km 130, 31 Aug 1974 (fl), *Prance 22,602* (F, INPA, MO, NY); Rio Urubú, Igarapé Cachoeira, 3 Dec 1956 (fr), *Rodrigues 296* (HEPH, INPA); Rio Cuieiras, Jacandá, 60 km from mouth, −2.696, −60.352, 9 Oct 1988 (fr), *Mori 19,278* (INPA, NY); Manaus Distrito Agropecuária, Reserva 1501, 2°24′S, 59°43′W, 11 Nov 1989 (imm fr), *Kukle & Boom 26* (INPA, K, MO, NY); Mun. Maués, lower Rio Pacoval, Igarapé Tijuca, −3.85, −58.116, 26 Jul 1983, *Zarucchi 3187* (FSL, INPA, MO, MG, NY, RB); lower Rio Pacoval, 3°48′S, 58°04′W, 26 Jul 1983 (fl), *Zarucchi 3170* (F, INPA, K, MG, MO, NY, RB, US); Rio Urupadi, Maués, 15 Nov 1987 (fr), *Kubitzki 87–37* (INPA, MG). **RORAIMA**: Km 350–355 Manaus-Caracaraí, near Equator, São Luís do Anauá, 21 Aug 1987 (fr), *Cid Ferreira 9086* (HAMAB, INPA, MIRR). **PARÁ**: Mun. Óbidos, 91 km from Oriximinã, Campo de Ariramba, 1°10′S, 55°35′W, 7 Dec 1987(fr), *Cid Ferreira 9805* (F, IEPA, INPA, K, MO, MG, NY, RB); Serras do Dedale da Igaçaba, 4 Sep 1907, *Ducke 8628* (BM, MG); Aramanai, Rio Tapajós Mun. Santarém, 5 Dec 1978 (fr), *Maciel & Cordeiro 173* (INPA, MG, NY). **ACRE**: Mun. Cruzeiro do Sul, BR307, 60 km from city, 7°53′S, 72°45′W, 10 Nov 1991(fl, imm fr), *Cid Ferreira 10,672* (INPA, MO, NY, UFAC); Road Cruzeiro do Sul-Benjamin Constant, Km 32, Mun. Cruzeiro do Sul, 23 Oct 1984 (fr), *Cid Ferreira 5205* (F, INPA, MO, NY, RB). **RONDÔNIA:** Porto Velho, Repressa Samuel, 9°05′S, 63°13′W, 14 June 1986 (fr), *Thomas 5088* (F, HFSL, INPA, MO, NY); Repressa Samuel, road parallel to dam, Km 46, 2 Aug 1987 (imm fr), *Mattos 28* (INPA, NY). **MATO GROSSO**: 4 km S of border with Rondônia State, BR364, Vila Bela da Santissima Trinidade, −12.9, −60.03, 3 Nov 1985 (fr), *Thomas 4793* (F, HFSL, INPA, MO, NY, US).

Cuatrecasas (1961) recognized three varieties in this species based on small differences of the pubescence of the petals and young branches. There is no geographic or environmental separation of these taxa and so I have not recognized them here. It is uncertain whether Spruce collections 1915 and 1715 are the same collection with erroneous numbering or separate collections. Cuatrecasas (1961) equated them but selected the K sheet of 1915 as the holotype and the M sheet as an isotype, and he indicated the US, GH, P, and NY collections of 1715 as isotypes. The mix up numbering of the labels is highly likely, but I hesitate to call the 1715 sheets as isotypes, and I consider Cuatrecasas' choice as a leptotypification of *Spruce 1915*. A further complication is that Spruce 1915 is also the type number of *Haploclathra paniculata* (Mart.) Benth. in the Guttiferae which confirms the mix up of labels.

3. **Humiriastrum dentatum** (Casar.) Cuatr., Cuatr. Contr. U.S. Natl. Herb. 35, 2: 136. 1961. *Humirium dentatum* Casar., Nov. stirp. bras. Decas IV: 39. 1842; *Sacoglottis dentata* (Casar.) Urb., Fl. Bras. (Martius) 12 (2): 444, pro parte. Type. Brazil. Rio de Janeiro; near Marica, 1840 (fl), *G. Casaretto s.n.* (lectotype of Delprete et al. 2019, TO).

Trees, the young branches pubescent-hirsute. Leaves with petioles 4–10 mm long, narrowly winged, pubescent-hirtellous; laminas ovate to elliptic-lanceolate, 4–13.5 × 1.5–5.5 cm, coriaceous, sparsely pubescent beneath, glabrous above except for densely pubescent midrib, cuneate at base, acuminate at apex, margins serrate, slightly revolute, setae present, small sparse abaxial glands near margin; midrib prominulous above, prominent, sparse hirsute beneath; primary veins ca. 10 pairs, prominulous beneath arcuate-anastomosing, venation reticulate and prominulous. Inflorescences axillary, cymose-paniculate, shorter than leaves, dichotomous, branchlets densely hirtellous; peduncle densely hirtellous; bracts ovate-oblong, 0.5–1 mm long, puberulous, deciduous; pedicels 0.5–0.8 mm long, thick, glabrous, articulate; glabrous or glabrescent, with peduncles 0.5–2 mm. Sepals orbicular, connate at base, glabrous except for ciliate margin. Petals oblong to subobtuse, 2.5 × 1 mm, thick, glabrous. Stamens 20, filaments 2–2.5 mm long, connate for lower third, glabrous; anthers oblong-lanceolate, glabrous, connective thick, thecae basal, oblong. Disc membraneous, 0.6 mm high, short dentate, surrounding ovary. Ovary ovoid, glabrous 1 mm high, 5-locular with uniovulate locules; style robust, 0.5–0.6 mm long, stigma capitate. Drupe globose, 2.5 × 2.4 cm, not tapered at ends or tapered at basal end; exocarp smooth, glabrous.

Distribution and habitat. Forests of eastern and central Brazil and south to Santa Catarina (Fig. 10.15).

Phenology. Flowering mainly from March to November and fruiting from January to June, but flower or fruit are found almost any time of year.

Local names and uses. Brazil: *Fruta de pedra, carne da vaca, umiri.*

Conservation status. Least concern (LC).

Illustrations. Cuatrecasas (1961), fig. 27 l–m.

Additional specimens examined. BRAZIL. DISTRITO FEDERAL: Fazenda Água Limpa, 8 Oct 1981 (fl), *Machado 54* (RB); Bacia do Rio Bartolomeu, 1080 m, 15°30′, 45°45′, *Pereira & Alvarenga 3335* (K). **MINAS GERAIS:** Parque Estadual do Rio Doce, Timóteo, −19.59, −42.56, 5 May 2004 (fr), *França & Raggi 545* (BM, CESJ, FUEL, K, L, OXF, P, U, US); Mina de Brucutu, Barão de Cocais, −19.94, −43.48, 24 Aug 2000, *Filho s.n.* (HUEFS85495); Condominio Aconchego da Serra, Itabirito, −20.253, −43.801, 13 May 1999, *Lombardi 2834* (MBM); Mun. de Marliéria, Parque Estadual do Rio Doce, Lake Helvécio, −19.766, −42.6, 18 Sep 1975 (fl), *Heringer & Eiten 15,003* (MO); Parque Estadual Rio Doce, Marliéria, −19.775, −42.603, 25 Oct 2006 (fl), *Giordano* et al. *2808* (BHCB, CESJ, RB, UB, UEC); Marliéria, trilha de Garapa Torta, 20 Jan 1999 (fr), *Bertoluzzi 441* (RB);

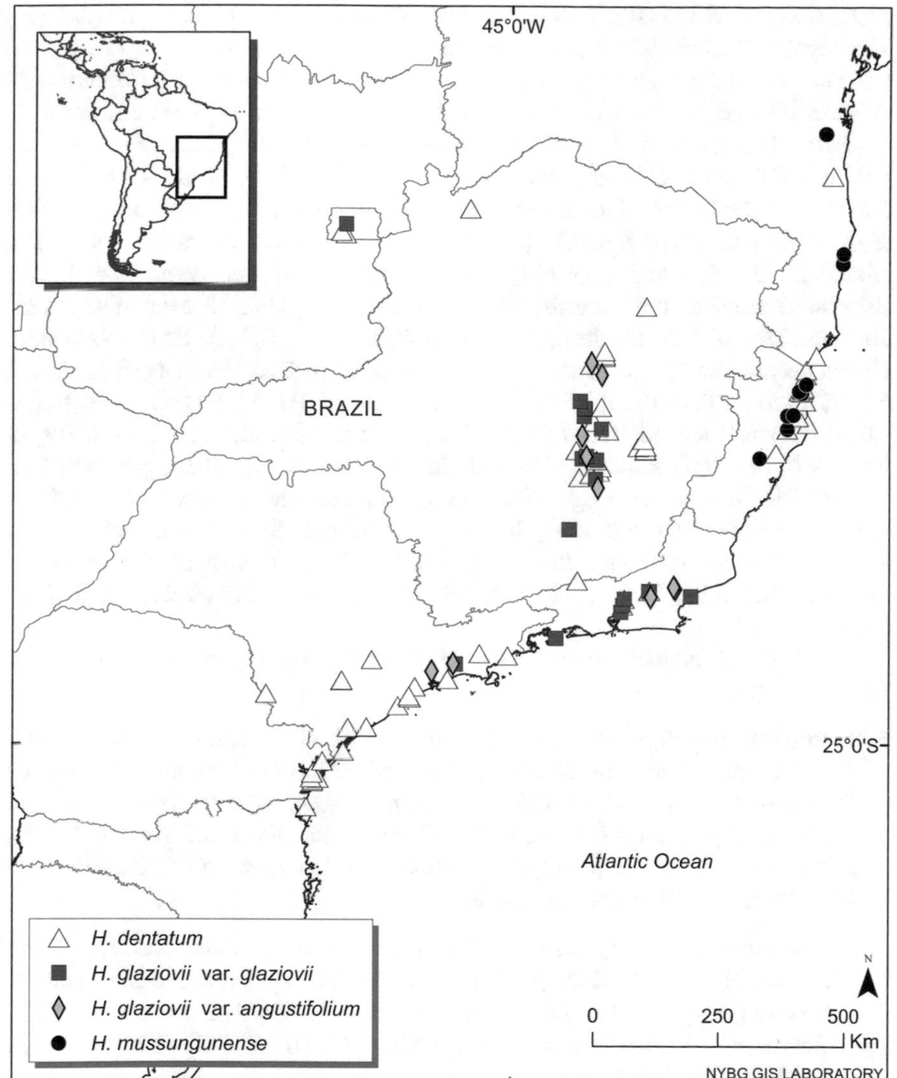

Fig. 10.15 Distribution of three species of *Humiriastrum*

Mun. Santana de Riacho, Serra do Cipó, 19°15′50" S, 43°32′48″W, 1278 m, 6 Mar 2002 (fl), *Pirani 5038* (F, K, NY, RB). **BAHIA: O**livença, Praia do Vila Rica, Ilhéus-14.545, −39.146, 6 May 2000 (fl bud), *Silva 368* (CEPEC, FUEL, HUEFS, JPB, RB); Mucuri-Nova Viçosa Km 8, −18.086, −39.550, 17 Feb 1995 (fr), *Pereira* et al. *5394* (VIES). **ESPÍRITO SANTO:** Conceição de Barra, −18.66, −39.802, 13 Jun 2012 (fr), *Menezes 2006* (VIES); Bairro Liberdade, São Mateus-Guriri, −18.889, −39.750, *Giaretta & Martins 201* (VIES); Itaúnas, Conceição da Barra −18.423, −39.73, 5 Feb 2010 (fr), *Giaretta* et al. *754* (VIES). **RIO DE JANEIRO**:

1891, *Glaziou 18,178* (BR, F, NY, P; B lost, photo F); *Herb. Richard s.n.* 1855 (P); Terezopolis, 27 Nov 1939 (fr), *Teixeira RB43676* (RB); Nova Friburgo, 1 Aug 1946 (fl), *Paes 112* (RB); Cidade das Meninas, Cambuaba, 6 Apr 1942 (fl), *Carcerelli 72* (RB). **SÃO PAULO**: Santos, Sorocaba, Jan 1875, *Mosén 3475* (P, S); Praia do Itaguaré, Bertoga, −23.85, −46.138, 26 Mar 1999 (fl bud), *Sampaio* et al. *105* (HUSC); Estação Ecológica Chauás, Iguape, −24.7, −47.55, 10 Jan 1999, *Kozera 826* (ESA, MBM, SPSF, UEC); Rio Preto, Itanhaém, −24.18, −46.78, 11 Mar 2006 (fl), *Garcia* et al. *SPSF07070* (SPSF); P Estadual Serra do Mar, Morro do Cuscuzeiro, Ubatuba, −23.43, −45.07, 19 May 2007, *Bertoncello & Pansonato 424* (UEC): Estação Ecológica Juréia-Itatins, Peruibe, −24.32, −46.99, 13 Mar 1992, *Rossi 1043* (SPSF); Ilha do Cardoso, 21 Sep 2008, *Assis 18* (ESA). **PARANÁ**: Mun. Guaraqueçaba road to Paruquara, −25.306, −48.32, 6 Dec 1991 (fl), *Hatschbach 56,145* (ASE, CESJ, FLOR, FUEL, HEPH, HUCS, HUEM, HUSC, ICN, INPA, MBM, MO, NY, RB, SJRP, SPSF); 5–6 km SW Pontal do Sul, −25.52, −48.509, 29 Nov 1993 (fl), *Hatschbach 59,751* (ALCB, E, HUEFS, K, MBM, NY, RB, UB, UPCB); Rio Guaraguaçú, road to Sambaqui, Pontal do Paraná, May 1977 (fr), *Silva & Campos 8451* (HCF, RB, UB); Ilha do Mel, Morro do Meio, Paranaguá, −25.52, −48.509, 19 Oct 1987 (fl), *Britez 1800* (UEC, UPCB). **SANTA CATARINA**: Reserva Volta Velha, Itapoá, −26,116, −48.616, 25 May 2013, *Grings* et al. *1760* (ICN, UFRGS).

The Richard herbarium collection at Paris is possibly a duplicate of the Casaretto type.

4. **Humiriastrum diguense** (Cuatr.) Cuatr., Contr. U.S. Natl. Herb. 35(2): 141. 1961; Gentry, Ann. Missouri Bot. Gard. 62: 37. 1975; Zamora, Manual de Plantas de Costa Rica 6: 11. 2007; *Sacoglottis diguense* Cuatr., Trop. Woods 96: 38.1950. Type. Colombia. Valle del Cauca: Río Dígua, 20 Aug 1943 (fl), *J. Cuatrecasas 14,956* (holotype, F V0060785; isotypes, COL 000001809, G 00368463, US 00101234, VALLE 000281).

   Humiriastrum diguense subsp. *diguense* var. *anchicayanum* Cuatrec., Contr. U.S. Natl. Herb. 35(2): 142. 1961. Type. Colombia. Valle del Cauca: Cordillera Occidental, Hoya del Río Anchicayá, 16 Apr 1943 (fl), *J. Cuatrecasas 14,418* (holotype, F V0060786; isotypes, COL 000001810, G 00368464, US 00101233, VALLE).

   Humiriastrum diguense subsp. *costaricense* Cuatr., Contr. U.S. Natl. Herb. 35(2): 142. 1961. Type: Costa Rica. Puntarenas: between Esquinas and Palmar Sur de Osa, 30 Jan 1951 (imm fr), *P. H. Allen 5812* (holotype, US 00101194, isotypes, BM 000617766, EAP, F 00043826, F V0060782, F V0060783, GH 00043826).

Large trees, the young branches puberulous or hirtellous. Leaves subsessile with short petioles 1–2 mm long, thickened at base; laminas obovate-elliptic 3.5–9.5 × 1.5–5.5 cm, coriaceous, slightly papillose above, glabrous or with sparse, appressed, scattered trichomes beneath, abruptly cuneate at base, rotundate and abruptly and obtusely acuminate to cuspidate at apex, the acumen 5–15 mm long, margins slightly to conspicuously crenate, abaxial glands along primary vein loops near margin, 1–3 pairs adaxial glands at margin and numerous small dotted glands throughout adaxial surface; midrib plane or prominulous above, prominent and with sparse appressed hairs beneath; primary veins plane, inconspicuous, with minute, fine strigose hairs beneath. Inflorescences terminal or subterminal, dichotomous or trichotomous-paniculate, corymbose, longer than leaves, 8–15 cm long, branches robust, pubescent; bracts small, ovate, ciliate, deciduous; flowers subsessile, pedicels very short. Sepals suborbicular, 1–1.2 mm long, glabrous or appressed-pubescent on exterior. Petals elliptic-oblong, 2.2 mm long, glabrous or sparsely appressed puberulous. Stamens 20, filaments connate at base, glabrous, unequal, ten longer 1.2 mm long, alternating with ten smaller, 0.9 mm long; anthers 0.7–0.8 mm long, connective thick, ovoid-lanceolate, obtuse, thecae 2, elliptic, basal. Disc of scales 0.2 mm long. Ovary subpyriform, 1 mm high, pilose; style short, stigma 5-lobed. Drupe elliptic-obovoid, obtuse at base and pointed at apex, 2.1–3.5 × 1.1–1.9 cm; exocarp coriaceous when dry, glabrous, 1 mm thick; endocarp 2–2.5 × 1.2–1.4 cm woody, with small resiniferous cavities, 5 deep holes at apex, alternating with 5 oblong, descending opercula 6–8 mm long, usually with 2 seeds developed.

Distribution and habitat. From Costa Rica to Ecuador and on slopes of Pacific coastal rainforests of Colombia, common on the Osa Peninsula of Costa Rica, up to 1200 m (Fig. 10.14).

Phenology. Mainly flowering from April to November and fruiting from January to March.

Local names and uses. Costa Rica: *Laurelito, níspero, chiricano*; *chiricano alegre.* Ecuador: *Chanul.* The wood has been much used for construction and so this species has become threatened in some areas.

Conservation status. Least concern (LC).

Illustrations. Cuatrecasas (1961), figs. 27. e–f, p. 29; Gentry (1975), fig. 1; Zamora (2007) p. 11.

Selected specimens examined. COSTA RICA. HEREDIA: Sarapiqui, 220 m, 10° 23′24″N, 84°06′00′W, 12 Nov 1971 (fl), *Lent 2220* (BM, MO, NY, RB). **LIMÓN**: Talamanca, 9°22′48″N, 82°56′24″W, 400 m, 14 Jul 1989, *Solis* et al. *28* (CR, MO). **PUNTARENAS**: Osa, Golfo Dulce Cantón, Los Mogos, 8°40′20″N, 83°22′40″W, 200 m, 10 Jan 1995 (fl), *Aguilar 4134* (INB, K, MO); Golfito Cantón, Osa, 8°34′N, 83°31′W, 200 m, 14 Mar 1995 (fr), *Aguilar 3794* (INB, K, MO, NY); 30–40 Km S of San Isidro de El general, 600 m, 16 Jan 1964 (fl), *Lems 5043* (NY). San Isidro de El General, El Pilar de Cajoón, 620 m, 18 Nov 1988 (fl) *Zamora* et al. *1528* (CR, K, MO, NY). **SAN JOSÉ**: Pérez Zeledon, El Pilar del Cajón, 600 m, 9°17′N, 83°37′W, 14 Dec 1990 (fr), *Hammel 18,014* (COL, K, MO). **PANAMA. CANAL ZONE:** Pipeline road, 24 Sep 1971 (fl), *Gentry 1938* (MO, SCZ). **COLÓN:** Donoso, 8° 53′04″N, 80°42′27″W, 150 m, 12 Apr 2009 (fl), *McPherson 20,914* (K, MO); Donoso, Botiju, 8°50′N, 80°39′W, 11 Sep 2012 (fl), *Van der Werff* et al. *24,445* (K, MO, NY). **DARIÉN**: Río Setigandi7°45′N, 77°42′W,

800–1100 m, 19 Apr 1980, *Gentry* et al. *28,611* (MO). **PANAMÁ:** Chepo, 9°18′40″N, 78°56′40″W, 200–500 m, 26 Mar 1973, *Liesner 1224* (MO). **SAN BLAS**: 9°21′N, 78°59′W, 250–350 m, 4 Nov 1984, *Nevers* et al. *4162* (MO). **COLOMBIA. VALLE DEL CAUCA**: Buenaventura, 3°56′N, 77°08′W, 500 m, 27 Mar 1986, *Gentry* et al. *53,677* (MO). Valle: Río Dígua, 20 Aug 1943 (fl), *Cuatrecasas 14,956* (COL, F, G, US, VALLE). **ECUADOR. CARCHI**: San Marcos de los Coaiqueres, trail Chical-Tobar Donoso 1°06′N, 78°16′W, 800 m, 8 Feb 1885 (fr), *Øllgaard* et al. *57,591* (AAU, K, MO, NY, QCA); Tulcán, 1°02′N, 78°15′W, 1150 m, 19–28 Feb 1993, *Grijalva* et al. *598* (MO, QCNE); Maldonaldo, Parroquia Tobar Donoso, Reserva Awá, 900 m, 0°55′N, 78°32′W, 22 Nov 1992 (fl), *Aulestia & Guanga 827* (F, NY); Tulcan, Reserva Awá, Parroquia Chical, 900 m, 1°02′N, 78°16′W, 18 Feb 1993 (fl), *Aulestia & Grijalva 1209* (NY); Tulcan, Parroquia Tobar Donoso, Reserva Awá, 650–1000 m, 19–28 Jun 1992 (fr), *Tipaz* et al. *1443* (NY). **PASTAZA**: Pastaza, 1°49′S, 77°47′W, 23 Sep 1993 (old fr), *Palacios 11,433* (MO, NY, QCNE). **ZAMORA-CHINCHIPE**: Nangaritza, 1450 m, –4.22666 -78.64111, 26 Feb 2003, *Quizhpe* et al. *580* (MO, QCNE).

Cuatrecasas (1961) recognized two subspecies and a variety of *H. diguense*. Gentry (1975) did not recognize the difference between subsp. *diguense* and subsp. *costaricense* because the Panamanian material was intermediate between the collections of Costa Rica and Colombia. There is considerable variation in the pubescence of the branchlets and of the petals, and I follow Gentry by recognizing this as a single widespread species of central America and western South America. This species is notable for its conspicuously crenate-dentate leaves. The fruit of this species is an important resource for birds, particularly *Thraupis episcopus* and *Ramphocelus costaricensis.*

5. **Humiriastrum excelsum** (Ducke) Cuatr., Contr. U.S. Natl. Herb. 35(2): 141. 1961. *Sacoglottis excelsa* Ducke, Arch. Jard. Bot. Rio de Janeiro 3: 178. 1922: *Humiriastrum cuspidatum* var. *excelsum* (Ducke) Cuatr., Contr. U.S. Natl. Herb. 35(2): 128. 1961. Type. Brazil. Pará: Belém, 20 Aug 1914 (fl), *A. Ducke MG 15459* (holotype, RB 00539058; isotypes, BM 000796054, G 00368453, MG, US 00101230).

Medium to large-sized trees, the young branches puberulous to hirtellous. Leaves with petioles 2–3 mm long, puberulous beneath or glabrous; laminas ovate to ovate-elliptic, 2.5–9 × 1.5–5 cm, thin-coriaceous, glabrous, broadly cuneate or roundish at base, acuminate to cuspidate at apex, the acumen 6–16 mm long, margins crenate, setae present, adaxial glands near primary vein loops toward margin; midrib plane above, prominent and glabrous beneath; primary veins 10–12 pairs, plane above, prominulous beneath, reticulate venation prominulous beneath. Inflorescences axillary, cymose-paniculate, trichotomous at base, dichotomous above, shorter than leaves, peduncle and branches minutely pubescent-hirtellous; bracts minute, early deciduous; pedicels 0.2–0.4 mm long, thick, pubescent, articulate with short, pubescent peduncles. Sepals rounded, 0.6–0.8 mm long, attenuate at apex, minutely pubescent, margins ciliolate. Petals linear-oblong, attenuate at apex, 2.5 × 1 mm, hirtellous-puberulous. Stamens 20; filaments connate at base, 1.5 mm long, slightly papillose, anthers 0.8 mm long, connective lanceolate, acute, thecae ellipsoid. Disc annular, dentate, 0.3–0.4 mm high. Ovary ovoid, glabrous, 5-locular, 5 ovulate; style 0.6 mm long, stigma capitate, 5-lobed. Drupe ellipsoid-ovoid, 2–3.5 × 1.4–1.8 cm, exocarp smooth, glabrous; endocarp woody with 5 foramina at apex and 5 oblong opercula ca 6 mm around apex; seeds usually 2.

Distribution and habitat. Forest on *terra firme* of Amazonian Colombia, Brazil, and Peru, also in French Guiana (Fig. 10.13).
Phenology. Flowering mainly from July to September.
Local names and uses. Brazil: *Achuá*. Peru: *Quinilla, hispi, uchu mullaca.*
Conservation status. Least concern (LC).
Illustrations. Cuatrecasas (1961) figs. 26 h-j, 27 g-h; Spichiger et al. (1990) fig. 3.

Selected specimens examined. **COLOMBIA. CAQUETÁ**: Mun. Solano, Reserva Aracuara, Chiribiquete, 0°17′56" S, 72°22′54″W, 15 Nov 1970 (fr), *Castro 10,694* (COAH, NY); Mun. Solano, región Araracuara, sectór Chiribiquete, Cerro Gamitana, 216 m, 0°10′33"S, 72°20′47″W, 3–10 Dec 2010 (fr), *Castro 10,837* (NY). **FRENCH GUIANA**. Saut-Bief, 3 km from Montagne Papillon, 29 Jan 1957, *Bena BAFOG1319* (U, US); Cayenne, 5°20′N, 53°00′W, 21 Sep 1996, *Sabatier & Prévost 4360* (CAY, MO); Saül, Monts La Fumée, 200–400 m, 3°37′N, 53°12′W, 15 Sep 1982 (fr), *Boom & Mori 1683* (NY, US). **ECUADOR. NAPO**: Lumbaqui, road to Bermejo, 1000 m, 28 May 1987 (fr), *Pennington & Chango 12,311* (K). **PERU. HUÁNUCO:** Tingo Maria, Road Huánuco-Pucallpa, 18 Jul 1946, *Burgos 85* (F, WIS); Alomía Roblos, Marona Alta-H de Delicias, 800 m, 17 Sep 1962 (fl), *Gutierrez R. 82* (K). **LORETO**: Maynas, 4°10′S, 73°30′W, 122 m, 9 Jun 1990, *Vásquez & Jaramillo 6845* (F, K, MO, NY); Requena, 125 m, Jenaro Herrera, 4°55′S, 73°45′W, 125 m, Mar 1984, *Spichiger* et al. *1745* (MO). **MADRE DE DIOS:** Tambopata, 12°50′S, 69°17′W, 17 Sep 2003, *Monteagudo 5877* (MO, USM). **BRAZIL. AMAZONAS**: Rio Cuieiras, 2 km below Rio Brancinho, 15 Sep 1973 (fr), *Prance* et al., *17,982* (INPA, NY, US). **PARÁ**: Belém, 16 Sep 1922 (fl), *Ducke RB 17780* (K, NY, P, RB, S, U, US; B lost, photo F); Santa Izabel, Belém-Bragança road, 18 Sep 1908, *Ducke MG 9672* (BM, MG, RB13675, US); Belém, Catu, 31 Aug 1944 (fl), *Ducke 1614* (IAN, MG, NY, US); Rio Mapuá between Emília and Boca do Mapuá, 18 Jul 1950, *Black & Ledoux 50–9811* (IAN, US); Vitória do Xingu, Belo Monte, −3.2203, −51.8851, 6 Feb 2013 (fl), *Silva 932* (RB); Rio Trombetas, Cachoeira Porteira, 5 Jun 1978 (fr), *Silva 4726* (MG, NY, PEUFR). **MARANHÃO**: Astonas, Turiaçú, −1.663, −45.371, 6 Dec 1976 (fl), *Rosa 2850* (NY, MG, RB).

6. **Humiriastrum glaziovii** (Urb.) Cuatr., Contr. U.S. Natl. Herb. 35(2): 137. 1961; *Sacoglottis glaziovii* Urb., Bot. Jahrb. Syst. 17: 503. 1893. Type: Brazil. Rio de Janeiro, Novo Friburgo, Alto Macahé, (fl), *A. F. M. Glaziou 18,964* (holotype, B lost, photo F; lectotype, RB00539059, **designated here**; isolectotypes, BR 05119967, C 10012850, F V0060788, K 000407331, P 01903262, P 01903263, P 01903264, R 000007787).

Small- to medium-sized trees, the young branches glabrous becoming rugose, lenticellate. Leaves with petioles 4–11 mm long; laminas elliptic-ovate to ovate-lanceolate, 3.5–10 × 1.5–6 cm, coriaceous, glabrous, with a single pair of adaxial glands at margin near base and few to many scattered small glands on adaxial surface, obtuse or short-cuneate at base, abruptly acute or acuminate at apex, the acumen 0–8 mm long, margins serrate, flat or slightly revolute, setae present; midrib plane or impressed above, prominent and glabrous beneath; primary veins 9–10 pairs, plane above, prominulous beneath, anastomosing near margin, venation reticulum prominent beneath. Inflorescences axillary, cymose-paniculate, shorter or longer than leaves, branches dichotomous (rarely trichotomous), glabrous or hirtellous-puberulous; bracts deciduous; flowers subsessile with short pedicels 0.2–0.3 mm long, glabrous, articulate, with 0.4–2 mm long glabrous peduncle. Sepals 1 mm long, rounded, glabrous except for ciliate margin. Petals oblong 3–3.5 × 1 mm, glabrous. Stamens 20, filaments 2.5–3 mm long, glabrous, connate at base, anthers glabrous, ovate-lanceolate, connective acute, thecae oblong. Disc annular, membraneous, circling ovary, dentate, 0.6–0.7 mm high. Ovary ovoid, glabrous, 8 mm high, 5 locular with uniovulate locules; style 0.7 mm long, stigma capitate. Drupe ellipsoid-globose, smooth, glabrous, ca 1.9 × 2.6 cm.

Key to Varieties of Humiriastrum Glaziovii

1. Leaves elliptic ovate to ovate-lanceolate, 3.5–10 × 1.5–6 cm, obtuse or short cuneate at base, thinly coriaceous, margin serrate.....................var. **glaziovii**
1. Leaves narrowly lanceolate, 2.7–6.5 × 1–2 cm, cuneate at base, rigidly coriaceous, margin crenate..var**. angustifolia**

6a. **Humiriastrum glaziovii** (Urban) Cuatr. var. **glaziovii.**

Small- to medium-sized trees. Leaves elliptic ovate to ovate-lanceolate, 3.5–10 × 1.5–6 cm, obtuse or short cuneate at base, thinly coriaceous, margins serrate.

Distribution and habitat. Atlantic rainforest of eastern Brazil at 1000 to 1700 m (Fig. 10.15).

Phenology. Flowering from November to April, fruiting from March to May.

Local names and uses. None recorded.

Conservation status. Least concern (LC).

Illustrations. Cuatrecasas (1961). fig. 27 i–j.

Selected specimens examined. **BRAZIL. DISTRITO FEDERAL:** Brasília Botanical Garden, Mata da Trilha Ecológica, −15.7798, −47.9297, 17 Dec 2009 (fr), *Jesus 337* (HEPH, RB), **MINAS GERAIS:** Mun. Santana do Riacho, road Belo Horizonte-Conceição do Mato Dentro, Km 117, Serra do Cipó, 19°15′50.8"S, 43°32′48″W, 1278 m, 6 Mar 2002 (fl), *Pirani 5038* (K, SPF); Serra do Cipó**,** Fazenda Inhame, 18°54′03" S, 43°47′07″W, 1084 m, 12 Mar 2014 (fr), *Zappi 2064* (K, RB, SPF); Barroso, −21.186, −43.975, 28 Jan 2004, *Assis 523* (CESJ, ESA, HUFU, MBM, RB, USP); Retiro do Barbado, Santana do Riacho, −19.169, −43.714, 22 Apr 1982, *Amaral s.n.* (MBM); Dist. Santana do Rio Preto, Parque Nacional Serra do Cipó, Itambé, −9.396, −43.402, 16 Mar 2008, *Santos & Serafim 328* (BHCB); Serra da Piedade, Km 1–5 road to top, Caeté, −19.92, −43.75, 2 Feb 1982 (fl), *Landrum 4262* (MBM, NY). **RIO DE JANEIRO**: Restinga de Maná, 30 Nov. 1896 (fl), *Glaziou 18,179* (K, NY, P, US); Porto da Estrella, 17 Nov 1825 (fl), *Ducke RB 19166* (K, P, RB, S, U, US); without locality, s.d., (fl), *Riedel s.n.* (K, P); Terezopolis, 14 Jan 1941 (fl), *Teixeira RB 60812* (RB). **SÃO PAULO**: 23°35′S, 46°00′W, 870 m, 4 Feb 1987, *Custodio Filho & Gentry 4682* (MO).

6b. **Humiriastrum glaziovii** (Urban) Cuatr., var. **angustifolium** Cuatr., Contr. U.S. Natl. Herb. 35(2): 138. 1961. Type. BRAZIL. Rio de Janeiro: Novo Friburgo, Alto Macahé, 6 Nov 1888 (fl), *A. F. M. Glaziou 16,724* (holotype, US 00101193; isotypes, A 00043827, BR 0000005119585, F 00043827, G 00368466, IAN 093893, K 000407330, MPU 020956, P 01903259, P 01903260, P 01903261, R 000007788, US 00921446, US 01101307).

Small trees. Leaves narrowly lanceolate, 2.7–6.5 × 1–2 cm, cuneate at base, rigidly coriaceous margins crenate.

Distribution and habitat. Montane areas of forest and campo rupestre in São Paulo and Minas Gerais, Brazil (Fig. 10.15).
Phenology. Flowering from January to April, fruiting in November.
Local Names and uses. None recorded.
Conservation status. Least concern (LC).

Selected specimens examined. **BRAZIL. MINAS GERAIS**: Mun. Ouro Preto, Parque Estadual Itacolomi, 1600–1700 m, 25 Feb 1980, *Peron 657* (K, RB); Serra do Espinhaço, Pico do Itambé, 1310 m, 13 Feb 1972 (fl), *Anderson* et al. *35,910* (MO, NY, UB, US); 5 km NE of Diamantina road to Mendoinha, 1300 m, 312 Jan 1969 (fl), *Irwin* et al. *22,926* (MO, NY, UB); Road Lagoa Santa-Conceição do Mato Dentro, Diamantina, Jaboticatubas, −19.166, −43.975, 9 Nov 1995 (fr), *Semir* et al. *4361* (UEC). **SÃO PAULO**: Estação Biologica, 5 Mar 1919, *Hoehne 3021* (F, NY, SP, UB); Estação Biológica, Biritiba Mirim, 29 May 1986, *Custodio Filho 2685* (RB, SPSF). **RIO DE JANEIRO**: Mun. Macaé, Glicério, acesso para Crubixás, 1100 m, 19 Nov 2002 (fr), *Giordano 2638* (RB); Macaé, Dist. De Frade, 22°12′26" S, 42°05′00″W, 1120 m. 14 Aug 2001 (st), *Giordano 2394* (RB); Mun. Macaé, Pico do Frale de Macaé, 1100 m, 14 Apr 1985 (fl), *Martinelli 10,675* (IPA, K, MO, RB).

This variety described by Cuatrecasas (1961) seems worth maintaining as its leaves are much smaller and narrower than in the typical variety, and it is found at slightly higher altitudes.

7. **Humiriastrum mapiriense** Cuatr., Contr. U.S. Natl. Herb. 35(2): 139. 1961; Type. Bolivia. La Paz: Sarampinni, near San Carlos, 600 m, 7 Mar 1927, *O. Buchtien 1518* (holotype, NY 00388401; isotypes, F V0060784, US 101102).

Large trees, the young branches puberulous. Leaves with thick petioles 3–6 mm long, slightly winged by confluent lamina, swollen at base; laminas obovate to oblanceolate, 3–5 × 1.5–4.1 cm, rigidly coriaceous, glabrous on both surfaces, with sparse abaxial glands, some large, toward margin, abruptly cuneate at base, acuminate at apex, the acumen 4–8 mm long, margins crenate; midrib plane or slightly immersed above, prominent and glabrous beneath; primary veins 8–9 pairs, plane and inconspicuous above, prominulous beneath, reticulate venation prominulous beneath. Inflorescences mostly axillary or terminal, cymose-paniculate, exceeding leaves in length trichotomous on lower portion, dichotomous above 1.5–2.5 cm long, branches hirtellous-pubescent; bracts ovate, 1 mm long, deciduous; pedicels thick, 0.3 mm long, hirtellous. Sepals ovate, 0.6 mm long, hirtellous on dorsal surface, margins ciliate. Petals elliptic-oblong, glabrous. Stamens 20, filaments united at base, alternately unequal, anthers oblong, 0.5 mm long, connective oblong-lanceolate. Disc of 10 free scales. Ovary ovoid, glabrous, 5-locular, locules uniovulate; style short, stigma capitate, 5-lobed. Drupe globose, 1.5–1.6 cm diam, exocarp smooth, endocarp globose, woody and hard, with resinous lacunae, 5 foramina at apex alternating with 5 oblong opercula, 6 mm long, surface rugose.

Distribution and habitat. In the upland pre-montane forests of Ecuador, Peru, and Bolivia to 1650 m (Fig. 10.14).
Phenology. Flowering from December to May.
Local names and uses. None recorded.
Conservation status. Least concern (LC).
Illustrations. Cuatrecasas (1961), pl. 13, figs. 26p-q, 28 g.

Selected specimens examined. **ECUADOR. ZAMORA-CHINCHIPE**: El Pangui, 1320 m, 3°34′44" S, 78°26′07″W, 5 Dec 2005 (fl), *Neill & Quizhpe 14,969* (ECUAMZ, F, K, MO, NY, QCNE); Nangaritza, 4°19′19" S, 78°38′43″W, 1200 m, 1 Mar 2003 (fl), *Quizhpe* et al. *612* (ECUAMZ, K, MO, NY, QCNE). El Pangui, 1312 m, 3°34′44" S, 78°26″W, 4 Apr 2006 (fr), *Croat 96,568* (AAU, GB, K, MO, NY, US); Yantzaza, 1050 m, 3°37′31" S, 78°31′50″W, 22 Mar 2006 (fl), *Quizhpe & Luisier 2060* (F, K, LOJA, MO, NY, QCNE); Nangaritza, 1620 m, 4°18′S, 78°40′W, 30 Jul 1993, *Gentry 80,778* (MO); Cordillera del Condor region, 1 km N of Río Machinaza, Las Peñas, 3°46′33" S, 78°29′46″W, 1640 m, 20 Mar 2008 (fl), *Neill & Quizhpe 16,271* (AAU, ECUAMZ, K, MO, NY, QCNE). **MORONA-SANTIAGO**: Limón Indanza, 1270 m, 18 Sep 2005 (fr), *Neill* et al. *14,653* (MO, NY, QCNE); Limón Indanza, Cordillera del Cóndor, 1150 m, 3°3′34" S, 78°14′45″W, 19 Dec 2005 (fl), *Morales* et al., *1607* (MO, NY, QCNE). **PERU. CAJAMARCA**: San Ignacio, 1609 m, 5°10′18" S, 78°44′54″W, 19 May 2007 (fr), *Perea & Flores 3144* (AAU, AMAZ, HUT, MO, MOL, NY, QCNE). **BOLIVIA. LA PAZ**: Larecaja, Copacabana, 850–950 m, 8 Oct 1939 (fl, fr), *Krukoff 11,270* (F, K, MO, NY, S, US), Larecaja, 1000 m, 15°25′S, 68°08″W, 8 Aug 2004 (fl), *Beck 29,502* (LPB, MO); Bautista Saavedra, 1373 m, 15°05′08" S, 68°29′17″W, 9 Apr 2010, *Loza* et al. *1435* (LPB, MO, USZ); Nor Yungas, 1500 m, 15°58′S, 67°37′W, 12 Feb 1983, *Solomon 9560* (LPB, MO).

Cuatrecasas (1961) had only two specimens of *Humiriastrum mapiriense* from the Mapiri region of Bolivia, but this has proved to be a much more widely distributed species, reaching as far north as Ecuador.

8. **Humiriastrum melanocarpum** (Cuatr.) Cuatr., Contr. U.S. Natl. Herb. 35(2): 145. 1961. *Sacoglottis melanocarpa* Cuatr. Trop. Woods 96: 37. 1950. Type. Colombia: Valle del Cauca, Buenaventura, 24 Feb 1946, *J. Cuatrecasas 19,989* (holotype, F V0060790; isotypes, COL 000001811, COL 000001812, G 00368465, VALLE, US 22701280, W).

Large trees, the young branches rugulose, glabrous. Leaves subsessile with very short petiole, laminas obovate, 4–7 × 2.5–4.5 cm, coriaceous, glabrous, cuneate at base, abruptly short, obtuse acuminate at apex, margins slightly crenate with small setae; midrib plane above, prominent and glabrous beneath; primary veins prominulous beneath, anastomosing. Inflorescences axillary or subterminal, cymose-paniculate, 2–4 cm wide, 2–6 cm long, as long as the leaves, branches dichotomous, articulate, glabrous; bracts ovate-triangular, acute, clasping, ciliate, sparsely puberulous, later deciduous; pedicels 0.1 mm, glabrous, articulate to short peduncle 0.1 mm. Sepals rounded, 0.7 mm long, glabrous except for ciliate margin. Petals oblong-elliptic, sparsely puberulous on exterior, 3–3.5 mm long. Stamens 20, filaments smooth, connate at base, ten long, 2.6 mm and ten shorter 2 mm long, anthers ca 0.7 mm, cordate-lanceolate, connective thick, subobtuse, thecae 2, suborbicular, divergent at base. Disc formed of several long dentate scales, 0.4 mm long. Ovary ovate-orbicular, 1.1 mm high, glabrous, 5-locular, locules uniovulate, stigma capitate, 5-lobed. Drupe ovoid to ovoid-oblong, 1.6–1.8 × 1–1.3 mm, apiculate, acute, black and lustrous; endocarp woody, acute at apex with 5-minute foramina alternating with 5 small oblong, descending opercula.

Distribution and habitat. Confined to the rainforest of Pacific coastal Colombia (Fig. 10.14).
Phenology. Collected in flower in February.
Local names and uses. Colombia: *Chanú*.
Conservation status. Listed as endangered (EN) in 2006 IUCN Red List, see Calderon (1998). Vulnerable (VU) B1ab(iii,v).
Illustrations. Cuatrecasas (1961), figs. 26r–t, 28 h–i.

Additional specimens examined. COLOMBIA. CHOCÓ: 100 m, 5°32′N, 76°20′W, 13 Jun 1982, *Gentry & Brand 36,787* (MO); Río Atrato, Loma de Belén, Mun. Guayabal, 24 Apr 1982 (fl), *Prance 28,055* (COL, NY). **VALLE DEL CAUCA**: Buenaventura, Quebrada de San Joaquín, 22 Feb 1946, *Cuatrecasas 19,909* (COL, F, G, K, VALLE, W, paratypes); Buenaventura, 50 m, 3°50′N, 77°10′W, 13 May 1987, *Faber-Langdoen & Renteria 530* (MO), *547* (MO), *876* (MO); Buenaventura, 4°0′N, 77°15′W, 17 Feb 1983 (fl), *Gentry* et al. *40,421* (COL, MO); Bajo Calima, 10 km N of Buenaventura, 50 m, 3°56′N, 77°08′W, 9 Dec 1981 (st), *Gentry 35,500* (COL).

9. **Humiriastrum mussunungense** Cuatr., Phytologia 75: 235.1993 [1994]. Type. Brazil. Espirito Santo: Reserva Forestal CVRD, Linares, −19.391, −40.072, 27 Aug 1991 (fl), *D.A. Folli 1393* (holotype, US 00611191, isotypes, CVRD, RB 00620811).

Trees 15 m tall, the young branches glabrous, lenticellate. Leaves with thick petioles, 4–10 mm long; laminas elliptic to ovate-elliptic, 7–11 × 3.5–5.5 cm, chartaceous, glabrous, sometimes with a single pair of adaxial glands near base, cuneate to subobtuse at base, acuminate at apex, margins crenate, setae present; midrib plane above, prominent and glabrous beneath; primary veins 8–9 pairs, prominent on both surfaces, veins reticulate and prominulous beneath. Inflorescences axillary, shorter than leaves, 3–4 cm long, cymose-paniculate, congested, branches minutely hirtellous; peduncle 2–3 cm long, articulate, minutely hirtellous; bracts ovate, 0.5–1 mm long, deciduous; pedicels to 1 mm long, minutely hirtellous. Sepals suborbicular, 1.2–1.8 × 1.2–2 mm, minutely hirtellous on abaxial surface, margins ciliate. Petals oblong-elliptic, acute, 3.5–4 × 1.3–1.8 mm, sparsely hirtellous above, glabrous below. Stamens 20, filaments connate at base, 1–1.5 mm, five longer, 3 mm long and 15 shorter, 2–2.5 mm long, anthers 1 mm long; connective triangular, acute, thecae globose-elliptic. Disc tubular, 1.2–1.5 mm high, denticulate, surrounding ovary. Ovary oblong-ovate, 1.4–1.4 mm, glabrous; style thick, 0.8–1.1 mm long, glabrous, stigma capitate, 5-lobed. Drupe globose or slightly elliptic, 3.6–4.2 × 3–3.5 cm, exocarp smooth, 5 mm thick, coriaceous; endocarp woody, with resinous cavities, opercula obscure.

Distribution and habitat. Atlantic rainforest of Bahia and Espírito Santo, Brazil (Fig. 10.15).

Phenology. Flowering in August, fruiting in March.

Local names and uses. *Casca dura.*

Conservation status. Near threatened (NT) B1 ab(iii,v). Restricted distribution, but mainly found in a reserve.

Selected specimens examined. **BRAZIL. BAHIA:** Mun. Porto Seguro road from BR367 to Arraial d'Ajuda, 30 km W of Porto Seguro, 20 Oct 1978 (fl), *Mori* et al. *10,874* (CEPEC, K, NY, RB); Mun. Santa Cruz de Cabrália, 2–4 km W. of Cruz de Cabrália, 21 Aug 1978 (fl, fr), *Mori* et al. *10,906* (CEPEC, K, NY, RB). **ESPÍRITO SANTO:** Linares, −19.3911, −40.072, 8 Oct 2013, *Correia 3* (CVRD), 13 Oct 1990 (fl), *Menandro 277* (CVRD, RB), 19 Apr 2006, *Siquiera 218* (CEPEC, CVRD), Reserva Natural Compania Vale do Rio Doce, Estrada Flamengo Km 45, 19°09′09" S, 40°02′16″W, 3 Dec 2003 (fr), *Giordano 2683* (CVRD, RB); Reserva Natural da Vale do Rio Doce, Av Flamengo, Mussununga, −19.138, −39.961, 23 Jun 2009 (fr), *Meireles 616* (NY, RB), Floresta de Mussununga, Reserva CVRD, 25 Mar 1991 (fr), *Folli 1299* (CVRD, RB, US611190, paratypes); Bairro Liberdade, road São Mateus-Guriri, −18.750, −39.81, 27 Sep 2008 (fl), *Monteiro & Oliveira 72* (VIES); Área 157 da Aracruz Cellulose, Conceição da Barra, −19.133, −38.35, 22 Sep 1992, *Pereira* et al. *3874* (RB, VIES); Mun. Santa Teresa, Nova Lombardia, Reserva Biológica. Augusto Ruschi, 13 May 2003 (fl), *Vervloet 2399* (RB) (Fig. 10.16).

This species is noted for the large size of the fruit, and bats have been observed eating the fruit (Lima et al. 2016).

10. **Humiriastrum obovatum** (Benth.) Cuatr., Contr. U.S. Natl. Herb. 35(2): 125. 1961; Cuatrecasas & Huber, Fl. Venez. Guayana (Steyermark et al.) 5: 632. 1999; *Humirium obovatum* Benth., London J. Bot. 2: 373. 1843; *Sacoglottis obovata* (Benth.) Urb., Fl. Bras. (Martius) 12 (2): 443. 1877. Type. Guyana, *Rob. Schomburgk 166* (holotype, K 000407334; isotypes, BM 000796047, OXF, P 01903258).

Small trees to 15 m tall, the young branches pubescent-hirsute, waxy. Leaves subsessile with short, winged petiole; laminas obovate to elliptic-obovate, 2.5–9 × 1.4–5.5 cm, rigid-coriaceous, glabrous above except on midrib and margin, hirsute beneath at least on midrib, with few scattered abaxial glands, cuneate at base, rounded or truncate-emarginate at apex or obtuse, margins entire, strongly revolute; midrib plane above, prominent beneath, thick, hirsute; primary veins 6–8 pairs, prominulous beneath, venation prominent and lax-reticulate. Inflorescences axillary in upper leaves, corymbose-paniculate, shorter than leaves, peduncle and branches pubescent-hirtellous; bracts persistent, clasping, triangular to ovate, acute, hirsute, 0.5–1 mm long; pedicels 0.5 mm long, hirsute, articulate with 0.4–1 mm peduncles. Sepals 0.7 mm long, hirtellous, connate at base, rounded at apex. Petals thick, oblong, subacute to subobtuse at apex, 2 × 0.8 mm. Stamens 20, filaments 0.8–1.2 mm long, connate toward base, anthers oblong, glabrous, ca 0.6 mm long, connective thick, sublanceolate, thecae subglobose. Disc annular, 0.4 mm high, 20-denticulate, surrounding ovary. Ovary subglobose, glabrous, 0.7–0.8 mm high, 5-locular, locules uniovulate, ovules oblong, 0.4 mm long; style thick, 0.6 mm long, stigma capitate. Drupe narrowly ovoid, 1.5–2 × 0.8–1 cm, pointed at apex; exocarp smooth, glabrous.

Distribution and habitat. Forests of Guyana and Venezuela up to 900 m altitude (Fig. 10.13).

Phenology. Flowering from May to October, fruiting from February to April.

Local names and use. Guyana: *hurihi, hooroheballi* (Arawak); Suriname: *bofroe-oedoe*. Fruit eaten by parrots.

Conservation status. Least concern (LC).

Illustration. Sabatier (1987), fig. 10.

Selected specimens studied. **VENEZUELA. BOLÍVAR**: Río Icaburú, 450–850 m, 7 Jan 1956 (fl), *Bernardi 2814* (NY); Dist Piar, base of Amaruay-tepuí, 500 m, 5°56′N, 62°17′W, 5 May 1986 (fl, imm fr), *Holst & Liesner 2791* (F, INPA, K, MO, NY, US); 6°32′N, 63°08′W, 13 May 1987 (fl), *Stergios 11,045* (MO, NY). **GUYANA**. S of Timeri, 17 Oct 1979 (fl), *Maas & Westra 3602* (COL, K, MO, NY, US); E Berbice-Corentyne, Canje River, 5°41′N, 57°15′W, 11 Apr 1987 (fr), *Pipoly 11,426* (F, K, MO, NY); Butukari, 20–21 Jul 1921 (fl), *Gleason 729* (GH, NY, US); Moraballi Creek, Essequibo River, 14 Nov 1938 (fl), *Davis 600* (FD2720) (K); Demerara-Berbice, Butahari Landing, 5°46′N, 58°35′W, 20 Jun 1995 (fl), *Chanderbali & Gopaul 98* (BRG, K, MO, NY, U, US). **SURINAME**. Kabalebo Dam, Nickerie District, 30–130 m, 31 Oct 1981 (fl), *Lindeman & de Roon 721* (K, MO, NY).

11. **Humiriastrum ottohuberi** Cuatr., Phytologia 68: 260. 1990; Cuatrecasas & Huber, Fl. Venez. Guayana (Steyermark et al.) 5: 632. 1999. Type. Venezuela. Amazonas: Río Casiquiare, Solano, 100 m, 8 Apr 1970 (fl), *J.A. Steyermark & G.S. Bunting 102,442* (holotype, US 00289004; isotypes, NY 03376851, US 00512762, VEN 251891).

Trees 4–10 m tall, the young branches densely, minutely hirsute, glabrescent. Leaves subsessile with petiole 1–3 mm long; laminas broadly elliptic to obovate-elliptic, 4–8.5 × 2.5–4.8 cm, rigid-coriaceous, glabrous on both surfaces except for midrib, lower surface minutely glandular-papillose and with adaxial glands near margin, cuneate at base merging into short petiole, abruptly obtuse-apiculate at apex, margins entire or inconspicuously crenate, setae present; midrib puberulous on young leaves, plane above. Inflorescences cymose-paniculate, 3–6 cm long, shorter than leaves; peduncle 1–2 cm long, branches and peduncle minutely and densely pilose-subvelutinous; bracts persistent, ovate to triangular, 0.5–1 mm long, sparsely pilose, margins ciliate; flowers sessile or subsessile. Sepals 0.8 mm long, connate at base, glabrous, resinous punctate on exterior, margins minutely ciliate. Petals oblong, 3–3.2 × 1.4 mm, thick, apex obtuse to subacute, glabrous on axial surface, abaxial surface hirsute, glandular. Stamens 20, ten longer, ca 2 mm alternating with ten shorter ones ca 0.5 mm, connate at base into short tube, anthers 0.8 mm long, connective thick, acuminate, thecae elliptic, basally attached. Disc annular, thick, membranaceous, dentate, 0.7 mm high, surrounding ovary. Ovary globose, glabrous, 5-locular, locules uniovulate, ovules oblong-elliptic, 0.5–0.6 mm high; style thick, 0.6 mm long, stigma capitate. Drupe ellipsoid to rounded, 3 × 2.5 cm, exocarp thick; endocarp woody, with resiniferous cavities.

Distribution and habitat. Only known from a small area of Amazonas State, Venezuela, and adjacent Brazil on seasonally flooded caatinga forest on white sand and sandy riverbanks (Fig. 10.14).

Phenology. Collected in flower in April and in fruit in April and young fruit in November.

Conservation status. Near Threatened, (NT) B1ab(iii).

Additional specimens examined. VENEZUELA. AMAZONAS: Dept. Río Negro, Río Pacimoni between mouth and Laguna Buridajao, 19 Apr 1985 (fr), *Stergios* et al. *8327* (MO, US); Río Pacimoni, (fr), *Stergios* et al. *15,534* (MO, US). **BRAZIL. AMAZONAS**: Rio Içana, above mouth of Rio Cubate, 0.500, −67.633, 3 Nov 1987 (fl, imm fr), *Maas 6905* (INPA, K, MG, MO, NY, RB).

12. **Humiriastrum piraparanense** Cuatr., Contr. U.S. Natl. Herb. 35(2): 127. 1961; Cuatrecasas & Huber, Fl. Venez. Guayana (Steyermark et al.) 5: 637. 1999. Type. Colombia, Vaupés: Río Apaporis, 9 Mar 1952 (fr), *R. E, Schultes & I. Cabrera 15,922* (holotype, US 00101191; isotypes, F 00043825, GH 00043828).

Small- to medium-sized trees to 20 m tall, the young branches glabrous. Leaves with petioles 2–8 mm long, thick, flattened above, narrowly winged on sides; laminas ovate-oblong to elliptic-oblong or ovate, 7–15 × 3–6 cm, rigid-coriaceous, glabrous, papillose beneath, with 1–2 pairs of adaxial glands near base and more scattered ones on lower portion; short abruptly cuneate at base, acuminate and cuspidate at apex, the acumen 7–10 mm long, margins crenate, flat; midrib plane above, prominent and glabrous beneath; primary veins 13–16 pairs, slender and inconspicuous on both surfaces. Inflorescences cymose-paniculate, axillary, shorter than leaves, peduncle puberulous branches trichotomous below and dichotomous above, hirsute; bracts persistent, clasping, ovate, obtuse to slightly acute, 0.5–1 mm long, margins ciliolate; pedicels thick, short, 0.2–0.3 mm long, minutely pubescent. Sepals 0.7–0.8 mm long, connate at base, rounded, hispid-puberulous, margin ciliate. Petals thick, elliptic-oblong, subobtuse, 3 × 1.2 mm, pubescent on outside. Stamens 20, filaments oblong, lower third connate, longer ones 1.7 mm, shorter ones 1.1–1.2 mm long, anthers 0.7 mm long, glabrous, minute, connective thick, angular, lanceolate, thecae globose or elliptic, basal. Disc annular, 0.5 mm high, 20-denticulate, surrounding ovary. Ovary 0.8 mm high, glabrous or sparsely pilose, 5-locular with locules uniovulate; style 0.5 mm, stigma capitate, 5-lobed. Drupe ellipsoid or ovate-ellipsoid, 2.4–3 × 1.8–2.5 cm, subacute at apex, 5 foramina below apex alternating with 5 oblong, descending opercula 7–11 mm long; seeds oblong, ca. 1.5 mm long.

Distribution and habitat. Seasonally flooded riverbanks in black water areas in Amazonian Colombia and Venezuela and northwestern Brazil (Fig. 10.14).

Phenology. Flowering from August to November, fruiting in March, May, August.

Local names and uses. Colombia: *Titida* (Curripaco). The bark is aromatic, and when young and green it is chewed to relieve toothache in Colombia (*Schultes & Cabrera 15,922*) and in Brazil by the Makuna Indians (Schultes and Raffauf 1990).

Conservation status. Least concern (LC).

Illustrations. Cuatrecasas (1961), figs. 26e–g.; Cuatrecasas and Huber (1999), fig. 539.

Additional specimens examined. COLOMBIA. CAQUETÁ. Araracuara, near mouth Río Caquetá, 0°36′S, 72°24 ‘W, 19 Nov 1993 (fl), *Cárdenas* et al., *4123* (COAH, COL, MG, MO); Araracuara, 10–22 Nov 1982 (fl), *Idrobo* et al. *11,468* (COAH, COL, NY); Serra Chiribiquete, 1°05′N, 72°40′W, 560 m, 25 Aug 1992, *Palacios* et al. *2609* (COAH, COL, MO), same loc., 17 Aug 1992 (fr), *Franco 3698* (COL, MO, NY). **AMAZONAS:** Región de Araracuara near airstrip, 250 m, 0°25′S, 72°20′W, 20 0ct 1990 (fl), *Restrepo* et al. *140* (COL). **VAUPÉS**: Río Piraparaná tributary of Apaporis, Loma Buc-chi, 250–600 m, 28–31 Aug 1952 (fl), *García Barriga 14,287* (COL, NY, US, paratypes); Río Piraparaná, 9 Mar 1952 (fr), *Schultes & Cabrera 15,922* (US); Río Kubiyo, 200 m, 25 Mar 1970 (fl fr), *Soejarto & Lockwood 2392* (COL, F, K); Solano, 0°14′S, 72°26′W, 16 Mar 1998, *Arbeláez* et al. *977* (COAH, HUA, MO). **GUAINÍA**: Mun, Inírida, Resguardo Atabapo, San Juan, 3°40′80″N, 67°27′21″W, 17 Dec 2005 (fr), *Cárdenas* et al. *18,017* (COAH). **PUTUMAYO**: Mun. Orito, Gran Jardin de la Serra, near Nuevo Mundo, 0°40′ 35.49″N, 77°5′53.18″W, 953 m, (−13 Jan 2014 (fl), *Montoya J.* et al. *2009* (COAH). **VENEZUELA. AMAZONAS**: Río Guainía between Comunidad & Santa Rita, 8 July 1959 (fl), *Wurdack & Adderley 43,355* (K, NY, RB, US); Río Cau between Isla

Picure and Raudal Ceguera, 4°54′N, 67°34′W, 9 May 1998 (imm fr), *Castillo 5533* (MO); Río Sipapo between laja Pendare & Tinaja, 4°44′29″N, 67°44′10″W, 27 Feb 2000 (fr), *Castillo 2000* (MO); San Carlos de Río Negro, 10 Apr 1981 (fr) *Clark & Christenson s.n.* (INPA, MO, US); Bajo Caíto Yagua near mouth with Río Orinoco, 3.53, −66.76, 15 Jan 1979 (fl), *Huber 3082* (INPA, US). **BRAZIL. AMAZONAS**: Rio Vaupés, Panure, 15 Nov 1947 *Pires 1030* (IAN); Rio Içana, Estirão Santana, 15 m, 22 Mar 1952 *Fróes 27,985* (IAN), 3 Apr 1952, *Fróes 28,407* (IAN); Rio Cuiuini, 0°46′07" S, 62°13′15″W, 100 m, 13 Aug 1996 (fr), *Acevedo-Rodíiguez 8257* (INPA, K, NY, US).

This species is very closely related to *Humiriastrum cuspidatum* and should perhaps be united with it. It differs in the oblong rather than globose fruit and in the larger elongate and less dentate leaves.

13. **Humiriastrum procerum** (Little) Cuatr., Contr. U.S. Natl. Herb. 35(2): 143. 1961; *Humiria procera* Little, J. Wash. Acad. Sci. 38: 93. 1948; *Sacoglottis procera* (Little) Cuatr., Trop. Woods 96: 40. 1950. Type. Ecuador, Esmeraldas: Playa de Oro,1 May 1943, *E. L. Little 6412* (holotype, US 00101196, isotypes, K 000407327, K 000407326, WIS 000000509MAD, NY 00388397).

Large trees to 40 m, trunk buttressed at base, the young branches glabrous, distinctly marked with londitudinal wings from decurrent leaves. Leaves sessile or subsessile; laminas elliptic or ovate-elliptic, 5.5–12 × 3–7.5 cm, rigid-coriaceous, glabrous, with abaxial glands along loops between primary veins and also scattered, rounded, or obtuse and clasping at base, acuminate at apex, the acumen 3–8 mm long, margins sinuate-crenate, flat or slightly revolute; midrib plane above, prominent and glabrous beneath; primary veins plane and visible above, venation inconspicuous above, reticulate and conspicuous beneath. Inflorescences terminal or subterminal, cymose-paniculate, corymbiform; peduncle robust, rigid, short pubescent, branches trichotomous below, dichotomous above, compressed pubescent; bracts later deciduous, ovate-triangular, acute, clasping, ciliate, sparsely puberulous; pedicels 0.1 mm, glabrous, articulated to short peduncle 0.1 mm. Sepals 0.7 mm long, rounded, glabrous except for ciliate margin. Petals elliptic-oblong, 3–3.5 mm, sparsely puberulous on exterior. Stamens 20, filaments smooth, connate at base, ten longer, 2.6 mm long alternating with ten shorter, 2 mm long, anthers 0.9 mm long, connective lanceolate, thick, angulate, cuspidate at apex, thecae basal, globose-ellipsoid. Disc 0.6–1 mm high. Ovary globose, 1 mm high, pubescent toward apex, glabrous below, 5-locular, the locules uniovulate; style thick, 0.6 mm long, stigma capitate, 5-lobed. Drupe oliviform, rounded at base, subacute at apex, 2.8–3.8 × 1.8–2.3 cm; exocarp coriaceous, resinous-granulose, 1.5 mm thick; endocarp woody, lacking resinous cavities, surface rugose and pitted, 5 deep holes at apex alternating with 5 oblong descending opercula, ca 8 mm long; 1 or 2 seeds developing.

Distribution and habitat. Pacific coastal forests of Colombia and Ecuador (Fig. 10.14).

Phenology. Flowering from September to February, fruiting from April to December.

Local names and uses. Colombia: *Chanú, chanúl, chanó, aceituno, batea*. Ecuador: *Chanúl*. Fruit edible; the popular wood is much used for general construction, parquet, furniture, house pillars and for canoes and boats.

Conservation status. Widespread in a threatened area, but several records are from reserves. This species is listed as critically endangered in the Colombian Red data book (CR A2acd) because it is endangered by overuse (Camacho & Montero, 2006). Vulnerable (VU) A4cd.

Illustrations. Cuatrecasas (1961), figs. 26 r–t, 28 h–i.

Selected specimens studied. COLOMBIA. CHOCÓ: Bajo San Juan, *Mahecha s.n.* (UDBG); Between Curiche & Alto Curiche, 10-100 m, 31 Jan 1967 (fr), *Duke 9634* (COL). **VALLE DEL CAUCA**: Río Calima, La Trojita, 1 Mar 1944 (fl), *Cuatrecasas 16,615* (COL, F, G, K, WIS, NY, VALLE); Río Cajambre, Barco, 5–80 m, 27 Apr 1944 (fr), *Cuatrecasas 17,186* (COL, F, G, K, WIS, VALLE); Buenaventura, 3°55′N, 77°0′W, 100 m, 20 Feb 1985 (fl, fr), *Monsalve 689* (COL, MO, NY). **NARIÑO**: Tumaco, Bajo Mira secor Carlos Ama, 80 m, 3 Mar 2005, *López 10,016* (COL, UDBC). **ECUADOR. CARCHI:** Tulcán, 1600 m, 0.8833, −78.3333, 10 Mar 1991, *Tipaz* et al. *256* (MO). **ESMERALDAS**: 2 km S of San Lorenzo, 10 m, 21 Apr 1943 (fl bud), *Little 6320* (F, K, WIS, NY, RB, US); 40 m, 25 Sep 1965 (fr) *Little & Dixon 21,148* (F, MO, NY); San Lorenzo, 1°08′N, 78°33′W, 21 Sep 1992, (fr), *Aulestia* et al. *548* (MO, NY); San Lorenzo, 200 m, 1°12′N, 78° 34′W (st), *Neill* et al. *11,829* (MO); Eloy Alfaro, 130 m, 0°43′N, 79°04′W, Oct 2008 (fr), *Palacios 16,486* (MO, QCNE), Eloy Alfaro, 200 m, 0°43′N, 78°53′W, 20–31 Sep 1993 (fr), *Tirado* et al. *471* (MO, QCNE), 8–14 Dec 1993 (fr), *Tirado* et al. *825* (MO, QCNE).

Humiriastrum procerum is distinguished by its large, coriaceous, sessile leaves.

14. **Humiriastrum purusensis** Prance, sp. nov. Type. Brazil. Amazonas: Beruri, Reserva de Desenvolvimento Sustentável Piagaçú-Purus, 4°27′28" S, 61°9′95″W, 9 Jul 2017 (fl fr), *B. G. Luize 521* (holotype, INPA). Fig. 10.12.

Ab *Humiriastrum pirparanense* ramis juvenilibus puberulis, sepalis puberulis, drupis maioribus 3.5–4 cm long (haud 2.4–3 cm) differt.

Trees, the young branches appressed puberulous. Leaves with petioles 3–5 mm long, slightly winged by confluent lamina, sparsely pubescent; lamina chartaceous, narrowly oblong, 8.5–13 × 3.2–4.5 cm, glabrous beneath except on midrib, narrowly cuneate at base, finely acuminate at apex; the acumen 10–15 mm long, the margins undulate and slightly crenate; midrib prominulous above, prominent beneath, pubescent on both surfaces; primary veins 14–18 pairs, thin, plane but conspicuous above, prominulous beneath. Inflorescences of axillary cymes, 2–2.5 cm long, peduncle 0.8–1.3 cm long, bracts oval, triangular, persistent; flowers sessile with short pedicel below articulation. Sepals little rounded, fused near base, pubescent on exterior; petals narrowly oblong, ca 2.5 mm long, pubescent on exterior; staminal tube united for lower third of length. Stamens 18–20, of two lengths, short and long, usually 5 long the rest shorter; thecae 2, oblong, basal. Disc surrounding ovary, ciliate on margin. Ovary globose-conical, glabrous, 5-locular; style short, glabrous, stigma capitate, 5-lobed. Drupe oblong, tapered at both ends when young and pointed and curved at apex; exocarp 0.5 mm thick, dotted with conspicuous lenticels, 3.5–4 × 2.2–3 cm; endocarp woody, with conspicuous cavities and single locule in center developed.

Distribution and habitat. Riverine forest in the Purus river basin (Fig. 10.13).
Phenology. Flowering in July and August, fruiting from October to December.
Local names and uses. None recorded.
Conservation status. Least Concern (LC).

Additional material: BRAZIL. AMAZONAS: Beruri, Reserva de Desenvolvimento Sustentável Piagaçú-Purus, Planice de Paraná do Tapagem e Paraná do Chibuí, setor Itapurú, 4°27′34" S, 61°91′96″W, 9 Jul 2017 (fl), *Luize 514* (INPA), 2°42′30" S, 61°45′44″W, 15 Dec 2010 (imm fr), *Luize 297* (INPA); same loc., Setor Cauá, 4°24′82" S, 61°76′96″W, 29 Nov.2009 (fr), *Luize 231* (INPA); Piagaçú-Purus, setor Itapurú, 4°21′91"S, 62°02′22″W, 5 Aug 2009 (fl bud), *Luize 117* (INPA); entorno Reserva de Desenvolvimento Sustentável Piagaçú-Purus, right margin of Lago Matíasmirim, 4°26′36" S, 61°74′48″W, 8 Aug 2009 (fl), *Luize 130* (INPA); Rio Ituxi, vicinity of Bôca do Curuquetê, 9 Jul 1971 (fl), *Prance* et al. *14,032* (INPA, NY); Mun. Lábrea, Reserva Extractivista Ituxi, margin of Rio Ituxi, 7°57′53" S, 65°13′17″W, 75 m, 25 Oct 2014 (fr), *Almeida 3894* (INPA).

This species is closely related to *Humiriastrum piraparanense,* another riverside species, but differs in the larger fruit, the pubescent sepals and young branches and in the thinner chartaceous leaves.

15. **Humiriastrum subcrenatum** (Benth.) Cuatr., Contr. U.S. Natl. Herb. 35(2): 138. 1961; *Humirium subcrenatum* Benth., London J. Bot. 2: 374. 1843; *Humiria subcrenata* (*Benth.*) Urb., Fl. Bras. (Martius) 12 (2): 442. 1877; *Sacoglottis subcrenata* (Benth.) Urb., Sitzungaber. Ges. Naturf. Freunde, Berlin 5: 1878. Type. French Guiana, Cayenne, s.d. (fl), *J. Martin s.n.* (holotype, K 000407329; isotype, US 00101189).

Trees, the young branches hirtellous. Leaves subsessile, petioles 1 mm long, hirtellous; laminas subelliptic, 2.6–4.5 × 1.7–2.8 cm, rigid-coriaceous, glabrous above except for minute hairs on midrib, sparsely puberulous beneath, with 1 pair of adaxial glands at margin near base, cuneate at base, acuminate at apex, the acumen 1–4 mm long, margins subentire or slightly crenate near apex; midrib prominent on both surfaces, puberulous; primary veins 8–10 pairs, thin and inconspicuous. Inflorescences axillary, shorter than leaves, cymose-paniculate, dichotomous, above branchlets alternate, angular; peduncle 9–12 mm long, hirtellous; pedicels 0.2–0.4 mm long, hirtellous; bracts deciduous. Sepals free, 0.5 mm long, truncate-rounded, minutely ciliate on margins, minutely puberulous on exterior. Petals thick, linear, acute, 2.1–2.23 × 6 mm, hirtellous on exterior. Stamens 20, glabrous, filaments connate near base, alternating length of 1.1 and 1.4 mm, anthers thick, acute, 0.6 mm long, thecae minute, basal, 0.15 mm wide. Disc cupular, 0.7 mm high, 20-denticulate. Ovary ellipsoid, minutely hirtellous, 5-locular, locules uniovulate, ovules deltoid, acute at apex; style 0.7 mm long, glabrous, stigma capitate. Drupe not seen.

Distribution and habitat. Only found in lowland rainforests of French Guiana (Fig. 10.13).
Phenology. Collected in flower in November and February.
Local names and uses. *Afiwa* (Palikur).
Conservation status. Vulnerable (VU), B1ab(iii.v).
Illustrations. Sabatier (1987), fig. 11.

Additional specimens examined. FRENCH GUIANA. Cayenne, 6 Nov 1998 (fl bud, fr), *Grenand 3086* (CAY, MO, NY); Camp Mbaípouri, Piste d'Apatou, 24 Feb 1984, *Sabatier 1020* (CAY, MO); Piste Saint-Elie, 5°20′N, 53°0′W, 15 Nov 1988 (fl), *Sabatier & Prévost 2269* (NY); Riviére Sinnamary, above Petit Saut, 5°0′N, 53°1′W, 4 Sep 1993 (st), *Mori* et al. *23,572* (NY); Crique Plomb, 20 m, 13 Jul 1992 (fl), *Denys 1789* (CAY, NY).

This species has the smallest leaves in the genus.

16. **Humiriastrum villosum** (Fróes) Cuatr., Contr. U.S. Natl. Herb. 35(2): 126. 1961; *Sacoglottis villosa* Fróes, Bol. Técn. Inst. Agron. N. 20: 53. 1950. Type. Brazil. Amazonas: Rio Negro, Padauiry, Rio Pitima, 21 Oct 1947 (fl), *R. L. Fróes 22,644* (holotype, IAN 32443; isotypes, A 00043831, F 00043831, IAN 32443A, INPA 14556, NY 00388412, NY 00388413, P 01903256, P 01903257, RB 00539061, SP 000789, UB 1869).

Medium-sized trees, the young branches hirsute. Leaves subsessile or short-petiolate; laminas ovate-acuminate or ovate-lanceolate, 3–6 × 2–3.8 cm, coriaceous, glabrous above, except on midrib, villous-hirtellous beneath, with 1 pair of adaxial glands at margin near base, shortly cuneate at base, narrowed and acuminate at apex, the acumen 10–15 mm long, margins slightly crenate or almost entire; midrib prominulous above, prominent beneath, pubescent on both surfaces; primary veins immersed and obsolete. Inflorescences axillary and subterminal, half length of leaves, cymose-paniculate, lower branches trichotomous, upper ones dichotomous, branches densely hirtellous; peduncle 1–1.5 cm long, densely hirsute; pedicels thick, 0.4–0.5 mm long, hirtellous; bracts ovate-oblong, persistent, hirtellous, 0.5–1.5 mm long. Sepals ovate-orbicular, 0.6 mm long, short-connate at base, hirtellous. Petals elliptic-oblong, subobtuse, ca. 3 × 1–1.3 mm, hispidulous. Stamens 20, filaments of 2 alternating lengths 1.2 or 1.6 mm long, connective fleshy, ovate-acuminate, thecae ellipsoid. Disc of oblong, bi- or tri-dentate scarcely adherent scales. Ovary globose, glabrous, 5-locular, locule uniovulate; style 0.5 mm long, stigma capitate, 5-lobed. Drupe (immature only seen) oblong, narrowed to base; exocarp smooth, glabrous; endocarp hard, woody, rugose, usually with one or two cavities developing.

Distribution and habitat. Riverine forest of northwestern Amazonia of Colombia and Brazil (Fig. 10.14).
Phenology. Flowering from October to November, fruiting in July.

Local names and uses. The bark is used as a powerful purgative. The Kubeo Indians near Mitú indicate that it is used in a tea only when quick action is needed following food poisoning (Schultes, 1979).
Conservation status. Least Concern (LC).
Illustration. Cuatrecasas (1961), pl. 11; Schultes (1979 fig.1a.

Additional specimens examined. COLOMBIA. VAUPÉS: Río Cubiyú, tributary of Río Vaupés, 350 m, 9 Oct 1952, *Humbert & Schultes 27,363* (P, US); Río Cubiyú, Cerro Kañendá, 15 km above mouth 1°0′N, 7°15′W, 10 Nov 1953 (fl), *Schultes & Cabrera 18,372* (BM). **BRAZIL. AMAZONAS**: Rio Aracá, 29 Oct 1952, *Fróes & Addison 29,144* (IAN); Mun. Barcelos, Rio Jauari, near mouth with Rio Aracá, 0° 30′N, 63°30′W (imm fr), 3 Jul 1985, *Sette Silva 207* (INPA, K); Rio Negro, 20 km below Barcelos, 24 Nov 1978 (fl), *Madison 6139* (NY). **PARÁ**: Óbidos, campina da Serra do Valho-me Deus, 20 Jul 1912, *Ducke MG 12030* (BM, MG, RB13673, US).

Distinguished from related species such as *Humiriastrum cuspidatum* by the hirsute young branches, petioles and midrib.

10.6 Sacoglottis

Sacoglottis Mart., Nov. Gen. et Sp. Pl. 2: 146. 1827; Bentham & Hooker, Gen. pl. 1: 247. 1862; Cuatrecasas Contr. U.S. Natl. Herb. 35(2): 161–180. 1961; Zamora, Manual de Plantas de Costa Rica 6: 112–13. 2007; *Sacoglottis* Sect. *Eusacoglottis* Urb., Fl. Bras. (Martius) 12 (2): 442, 448. 1877; *Sacoglottis* Sect. *Eusacoglottis* (Urb.) Reiche, Nat. Planzenfam. (Engl. & Prantl) 3 (4): 37. 1890; *Sacoglottis* Sect. *Eusacoglottis* (Urb.) Winkl., Nat. Pflanzenfam. (Engler & Prantl) 19a: 128. 1931. Type: *Sacoglottis amazonica* Martius.
Saccoglottis Endl., Gen. Pl. 1040. 1840; Baill., Adansonia 1: 208.1866., orth. var.
Aubrya Baill., Adansonia 2: 265. 1862.
Houmiri Sect. *Aubrya* (Baill.) Baill., Adansonia 10: 370. 1873.
Houmiri Sect. *Sacoglottis* (Mart.) Baill., Adansonia 10: 370. 1873.

Large or small trees. Leaves alternate, petiolate, coriaceous, lamina margins usually slightly crenate. Inflorescences axillary or subterminal with dichotomous or trichotomous branching; bracts persistent or deciduous. Sepals 5, connate at base, imbricate. Petals 5, free, thick-membraneous, aestivation cochlear or quincuncial. Stamens 10, glabrous, 5 long and antesepalous alternating with 5 ahorter antepetalous ones, filaments thick, complanate, united at base; anthers ovoid or ovoid-oblong, attached dorsally near to base, connective ovate-acuminate, acute with a thick apical protrusion, thecae 2, unilocular, ellipsoid, affixed at lower side, dehiscing by detachment. Disc cupular, dentate, surrounding ovary. Ovary 5-locular, usually glabrous less frequemtly pubescent or villous, the locules uniovulate; carpels opposite the sepals; style as long as or exceeding stamens, stigma capitate, 5-lobed. Drupe medium- to large-sized, smooth, exocarp fleshy, subcoriaceous when dry; endocarp

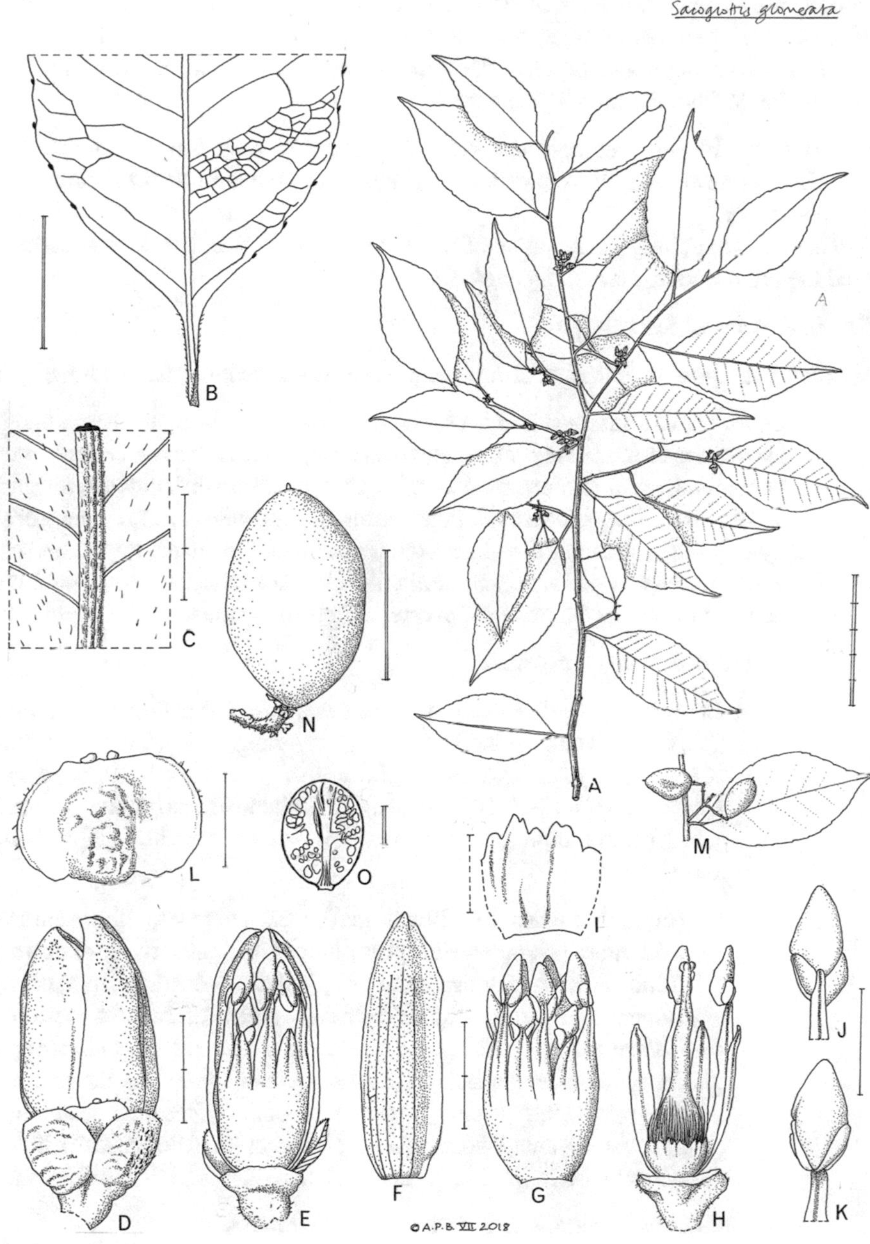

Sacoglottis glomerata del. Andrew Brown July 2018

Fig. 10.16 *Sacoglottis perryi. (***a**) Habit, flowering branch and inflorescence. (**b**) Adaxial leaf surface. (**c**) Adaxial leaf surface showing sparse hairs. (**d**) Side view of unopened flower. (**e**) Flower in D after removal of 2 sepals and 2 petals. (**f**) Outer surface of petal. (**g**) Androecium. (**h**) Section of androecium. (**i**) Part of disc around ovary. (**j**) Outer surface of anther. (**k**) Inner surface of anther. (**l**) Detail of calyx lobe. (**m**) Young fruit. (**n**) Fruit. (**o**) Fruit cross section. Dashed bar = 500 μm, single bar = 1 mm, graduated single bar = 2 mm, double bar = 1 cm, graduated double bar = 5 cm. (A–L from *McDowell 2993* M, N from *Prance* et al.*4736*; O from *McDowell* et al. *2920*)

woody, more or less bullate, with 10 narrow, usually slightly apparent furrows filled with resinous vacuous cavities; usually with 1 or 2 oblong seeds developing, on germination developing embryos push off broad, thick, oblong-elliptic valves which stretch almost from the apex to the base.

Distribution. Eleven species widespread throughout tropical South America and extending northward to Costa Rica, with a single species in lowland forests of West Africa.

The correct spelling of the name of this genus is *Sacglottis*, but several authors used the orthographic variant *Saccoglottis*.

Key to Species of Sacoglottis

1. Bracts deciduous or minute (0.2 mm) and leaves not exceeding 3.5 cm long

    2. Leaves small, laminas 2–3.5 cm long, obovate, rigid-coriaceous, smooth above, apex rounded to obtuse; secondary veins scarcely more conspicuous than minute, prominulous reticulum beneath; petals hispidulous on exterior; sepals hispidulous-pubescent, bracts minute, persistent........7. **S. maguirei**
    2. Leaves larger, laminas 4.5–22 cm long, apex acute to cuspidate; secondary veins usually prominent and tertiary veins lax-reticulate, prominulous beneath or inconspicuous in *S. perryi*.; bracts deciduous.

        3. Inflorescence glomerulate.

            4. Leaves 4.5–8 × 2.5–3.2 cm, apex cuspidate, the acumen 1–1.8 cm; primary veins inconspicuous; young branches hirtellous..10. **S. perryi**
            4. Leaves 7–22 × 3.5–10 cm, apex acuminate or cuspidate primary veins prominulous or inconspicuous; young branches glabrous or hirtellous.

                5. Young branches hirtellous; leaf apex cuspidate, the acumen 8–21 mm; primary veins prominulous beneath, obscure above; drupe narrowly oblong, acute or apiculate, 4–5 × 1.3–2 cm; calyx caducous in fruit, cupular; leaves rigid-coriaceous; petioles 5–13 mm ....2. **S. ceratocarpa**
                5. Young branches glabrous; leaf apex acute or acuminate, the acumen 5–10 mm; primary veins prominent on both surfaces; drupe oblong-ellipsoid or subglobose; calyx persistent at fruit base; petioles 5–9 mm ....6. **S. holdridgei**

        3. Inflorescence cymose-paniculate, not glomerate.

6. Leaves rigid-coriaceous, thick, sharply nerved beneath; drupe ovoid, 5–5.5 × 2.5–4.5 cm, dry exocarp 6–7 mm thick.........9. **S. ovicarpa**
6. Leaves thin-coriaceous, flexible; drupe ellipsoid or subglobose, 2.7–6 × 2–3.5 cm, dry exocarp 1.2–2 mm thick.
    7. Sepals glandular at margin; petals 4–4.5 × 1.5 mm, subobtuse; disc dentate; endocarp 10-sulcate...................1. **S. amazonica**
    7. Sepals eglandular at margin; petals 5–7 × 1.2–3 mm, acute; disc dentate or fimbriate; endocarp 5 or 10-sulcate.
        8. Petals pubescent on exterior, 6–7 × 1.5–2 mm; disc fimbriate; endocarp 10-sulcate..............................4. **S. gabonensis**
        8. Petals glabrous, 5–6 × 1.2–1.6 mm; disc dentate; endocarp 5-sulcate; ovary villous pubescent...............11. **S. trichogyna**

1. Bracts persistent; leaves 5–17 cm long.
    9. Leaves thin-coriaceous, flexible, prominently lax-reticulate on both surfaces; endocarp smooth or tuberculate.
        10. Petals glabrous or sparse puberulous; sepals 0.6 mm, glabrous; bracts 0.6–1 mm. not clasping; endocarp smooth.........8. **S. mattogrossensis**
        10. Petals sericeous; sepals 1.5 mm, pubescent toward base; bracts ca 2 mm, clasping; endocarp bullate.............................4. **S. gabonensis**
    9. Leaves rigid-coriaceous; minute inconspicuous reticulum beneath; endocarp smooth or tuberculate.
        11. Endocarp smooth; drupe oblong, narrowed at base; leaf veins slightly immersed to plane above; leaf margins not revolute..5. **S. guianensis**
        11. Endocarp tuberculate; drupe globose, not narrowed at base; leaf veins plane or prominulous above; leaf margins revolute...3. **S. cydonioides**

1. **Sacoglottis amazonica** Mart., Nov. Gen. Sp. Pl. 2: 146. 1827; Cuatr., Contr. U.S. Natl. Herb. 35(2): 169. 1961; Cuatrecasas & Huber, Fl. Venez. Guayana (Steyermark et al.) 5: 634. 1999. Type. Brazil. Pará: Tefé, Tagipurú, −3.3547, −64.7114, s.d., *C. E. P. von Martius s.n.* (holotype, M; isotypes, FI-W; K 000407315).

Trees, the young branches glabrous. Leaves with petioles 5–12 mm long, subterete, slightly winged, sulcate above, thickened at base; laminas oblong-elliptic, sublanceolate, 6–15 × 2.5–5.7 cm, coriaceous, glabrous, with small glands near margin, obtusely cuneate or obtuse at base, acuminate-cuspidate at apex, the acumen 0.8–2 cm long, margins slightly crenate; midrib plane above, prominent and striolate beneath; primary veins 9–12 pairs, prominulous above, prominent beneath, reticulate venation prominulous. Inflorescences axillary, cymose-paniculate, much shorter than the leaves, branches alternate, striolate, glabrous or sparsely pilose; bracts deciduous, ovate-triangular, acute, 1 mm long; pedicel short, thick, glabrous articulate to short peduncles 1 mm long. Sepals broadly orbicular, 1–1.5 mm long, thickened, glabrous on exterior, the margins ciliate and with small glands. Petals linear-oblong, subobtuse, 4–4.5 × 1.5 mm, greenish-white, glabrous. Stamens 10, filaments complanate, glabrous, connate for lower third, 5 shorter 3 mm long, 5 lower ones ca. 4 mm long, anthers ovate-oblong, 1 mm long, connective thickened, angulate, sublanceolate, thecae 0.5 mm. Disc surrounding ovary, dentate, 0.6–0.7 mm high. Ovary ovoid, 1.5 mm high, glabrous; style 2.5–3 mm long, stigma capitate, 5-lobed. Drupe oblong-ellipsoid, 5–6 × 3–3.5 cm; exocarp smooth, glabrous, coriaceous when dry, 1.2–2 mm thick; endocarp woody, slightly and irregularly 10-sulcate with 5 thin septa, bullate, filled with numerous resinous cavities, seed oblong, usually single.

Distribution and habitat. Forest in seasonally flooded areas from Trinidad to the Guianas, Venezuela, Ecuador, and Peru to northern and central Amazonian Brazil (Fig. 10.19).

Phenology. In Venezuela and the Guianas, flowering from March to May and fruiting from May to October; in Amazonia flowering August to October and fruiting October to April.

Local names and uses. Brazil: *Uchiran*a. Guyana: *Fuuyu*. Suriname: *Makararan* (Tirio); Trinidad: *Cojón de burro;* Venezuela: *Nabaru* (Guarauno). International: Bubblenut, *grenade pod, blister pod* on account of the blister-like cavities of the exocarp which is often found in drift. The wood is used locally for construction. The fruits are edible. Smoke from the burning bark was inhaled to relieve asthma by the Makunas (Schúltes 1979, Schultes and Raffauf 1990) and the fruit is used by the Guarauno to treat diarrhea (*Wurdack 293*) and for coughs (*Wibert* et al. *213*).

Conservation status. Least concern (LC).

Illustrations. Cuatrecasas (1961), pl. 20, figs. 1d–f, 34 g, 35 a–c, 36a–g; Cuatrecasas and Huber (1999), fig. 54.

Selected specimens examined. COLOMBIA. **AMAZONAS**: Tarapacá basin, Caño Agua Blanquilla, trib. Río Allegría, 2°36′42"S, 70°5′18″W, 50 m, 12 Apr 2000 (st), *López-C* et al. *6348* (COAH). **CAQUETÁ**: Mun. Solano, PNN Serrania de Chiribiquete, Puerto Abeja, mouth of Río Mesay, 0°5′13″N, 72°30′11″W, 127 m, Sep 2016 (imm fr), *Restrepo* et al. *955* (COAH). **VICHADA**: Territorio Faunísitico Tuparro, 5 km W of administrative center, 100 m, 27 May 1979 (fr), *Vincelli 1160* (COL). **VENEZUELA. BOLÍVAR:** Río Caura, Raudal Sejiato, 5°35′N, 64°17′W, 120 m, 9–26 May 1988 (fr), *Stergios & Delgado 12,961* (MO). **DELTA AMACURO**: Dep. Antonio Díaz, Caño Joburo, 7 Mar 1955 (fl), *Wurdack 293* (NY, US, VEN); Caño Joba-Suburu, 50 m, 21 Oct 1977 (fr), *Steyermark* et al. *115,190* (MO); Dep. Tucupita, Río Winikina, 0–5 m, 9°00′N, 61°00′W, 18 Aug 1983 (fr), *Wilbert 88,365* (MO). **ST VINCENT.** *Guilding s.n.* (K). **TRINIDAD**. Cedros forest, 18 Aug 1896,

Lunt s.n. (BM, OXF, US); Irois, Jun 1896 (fl), *Lunt 5984* (K, NY, OXF); Irois forest, 18 Aug 1896 (fl), *Hart s.n.* (K, NY). **GUYANA.** Takutu U, Essequibo Region, Acarai Mts, 530–610 m, 1°20′N, 58°50′W, 3 Nov 1996 (fl), *Clarke 2913* (K). **FRENCH GUIANA**. Oyapock, Islet Yacarescin, 10 Dec 1965 (fr), *Oldeman 1725* (K, US). **ECUADOR**. **SUCUMBIOS**: Lago Agrio, Reserva Cuyabeno, Laguna Grande, 0°00′, 76°11′W, 230 m, 15 Nov 1991 *Palacios* et al. *9371* (ECUAMZ, MO, NY, QCNE). **PERU. LORETO**: Maynas, Yanomomo Explorama Camp, −3.4666, −72.8333, 9 Jul 1983 (st), *Gentry 42,837* (MO); Maynas, Allpahuayo, −4.1666, −73.5000, 24 Mar 1992 (st.), *Vásquez 18,114* (MO). **BRAZIL. AMAZONAS**: Esperança, mouth of Rio Javari, 30 Oct 1943 (fl), *Ducke 1055* (IAN, K, MG, NY, RB, US); Manaus-Porto Velho road, Bom Futuro, 85 km N of Humaitá, 7° S, 63° W, 7 Apr 1985 (fr), *Cid Ferreira 5386* (INPA, K, MO, NY, RB); Manaus-Caracaraí Km 60, Reserva Campina, 16 Sep 1974 (fl), *Pennington 9950* (UEC). **PARÁ**: Belém, 24 Aug 1922 (fl), *Ducke RB 17781* (K, RB, S), 13 Dec 1942, *Archer 7964* (IAN, K, US); Belém, Utinga 13 Jul 1945 (fl), *Ducke 1723* (A, F, IAN, K, MG, NY, US); Rio Guama, between São Miguel & Acary, 31 Oct 1948 (fr), *Black & Foster 48–3393* (IAN); Monte Dourado, Rio Jarí, Mun. Almeirim, −1.523, −52.581, 25 Apr. 1986, *Pires & Silva 897* (INPA, MG); road Porto Trombetas-Terra Santa, −2.104, −56.486, 3 Sep 2009 (fl), *Ribeiro 1279* (INPA, RB).

Cuatrecasas (1961) provisionally included two sterile collections from Costa Rica in this species (*Dayton and Barbour 3004* and Merker et al. *3041*). They, in fact, are *Sacoglottis trichogyna* which was described later by Cuatrecasas in 1972. This species is difficult to distinguish from *S. guianensis* unless one has a complete collection. *S. amazonica* is a species of the flooded várzea and igapó forests, whereas the much more widespread *S. guianensis* is a *terra firme* species, and so I have kept them tentatively apart. The persistent bracts of *S. guianensis* is different from *S. amazonica*. Cuatrecasas (1961) pointed out that the fruits of *S. amazonica* are often found stranded on beaches and that this was first mentioned by Clusius (1605) without an identification and then discussed by Morris (1889, 1895) in articles in *Nature* as an unidentified "drift fruit."

2. **Sacoglottis ceratocarpa** Ducke, Bol. Tecn. Inst. Agron. N. 4: 13. 1945; Cuatr., Contr. U.S. Natl. Herb. 35(2): 165. 1961; Cuatrecasas & Huber, Fl. Venez. Guayana (Steyermark et al.) 5: 634. 1999. Type. Brazil, Amazonas: Manaus, 23 Jan 1943 (fr), *A. Ducke 1174* (lectotype, MG 018263, **here designated**; isolectotypes, IAN 10133, K, MG 018262, MO 1835898, NY 00388404, R 000054711, RB 00539063, RB 00542497, US 00101237).

Small- to medium-sized trees, 6–10 m tall, the young branches puberulous, glabrescent. Leaves with petioles 5–13 mm long, robust, subterete, slightly winged, flat above, thickened at base, glandular; laminas ovate-oblong or ovate-elliptic, rigid-coriaceous, glabrous, 8–12(−22) × 3.5–10 cm; subcuneate, rounded or obtuse at base, acuminate or cuspidate at apex, the acumen 5–21 mm long, margins slightly crenate or entire, with apair of glands at base of lower surface; midrib plane above, prominent beneath ending in minute apiculate glandular tooth; primary veins 10–12 pairs, obscure above, prominent beneath; reticulate venation prominulous; stipules minute, narrowly triangular, early deciduous. Inflorescences axillary, cymose-paniculate, dichotomous, sessile, glomerulate, peduncle almost absent; branchlets angulate, short-hirtellous pubescent; bracts deciduous, ovate-triangular, acute, sub-glabrous, 0.5–1 mm long; pedicels short-articulate. Sepals, thick, orbicular, 1.5 mm long, glabrous except for ciliate margin. Petals greenish, thick, linear, subacute, glabrous, 5 mm long, 1 mm broad. Stamens 10, connate for lower half, glabrous, filaments flattened, 2.5 and 3 mm long, alternating, anthers ovoid-oblong, 1.2 mm long, connective thick, ovoid-sublanceolate, subacute, thecae elliptic, 0.5 mm long. Disc membranaceous, denticulate, surrounding ovary, 0.7–0.8 mm high. Ovary ovoid, glabrous, 5-locular, 1 mm high, the locules uniovulate; style 2–3 mm long, erect, glabrous, stigma capitate, 5-lobed. Drupe elliptic-oblong, 4–5 × 1.3–2 cm, elongate, narrowed at both ends, almost fusiform, apex acute and often apiculate and long pointed when young, calyx caducous; exocarp smooth, subcoriaceous when dry, 1 mm thick; endocarp woody, narrowly oblong, acute at apex, resinous-lacunose, surface slightly bullate.

Distribution and habitat. Widely distributed through the Amazon basin in Colombia, Peru, Venezuela, and Brazil east of Pará, mainly in floodplain seasonally flooded várzea forest (Fig. 10.19).

Phenology. Flowering from August to November, fruiting from January to June.

Local names and uses. Brazil: *Uchirana*. Colombia: *Nee-saw-kaw'kě-too* (Maku), *cákuna* (Huitoto). The bark is burned and the smoke inhaled by locals suffering from recurrent coughing (Schultes 1979).

Conservation status. Least concern (LC).

Illustrations. Schultes (1979) fig. 2; Cuatrecasas (1961), figs. 34d–e, 35:i–k.

Selected specimens examined. COLOMBIA. CAQUETÁ: Río Coemaní, 22 Feb 1977 (st), *Roa 605* (COL). **AMAZONAS:** Mun. Leticia, Corregiamento Tarapacá, Parque Nacional Amacayacu, 100 m, 3°11′S, 70°20′W, 1 Jun 1991 (fr), *Rudas* et al. *2677* (FMB, MO, US); Río Igará-Paraná. Pumayo, 45 km from Chorrera, 12 Aug 1977 (st), *Rodríguez 12* (INPA). **AMAZONAS-VAUPÉS:** Caño Oogö-dja, Jinogojé, 26 Aug 1952 (fl), *Schultes & Cabrera 17,045* (NY, US). **VICHADA**: Mun. Cumaribo, Parque Nacional Natural El Tuparro, Río Tomo, 5°20′42″N, 68°19′46″W, 20 Jul 2009 (fr), *Clavijo & Trujillo-Y 1383* (COAH). **VAUPÉS**: between Mitú and Javaraté, Igarapé Murutinga, 14–24 Sep 1952, *Schultes & Cabrera 19290a* (US); Río Pacá, Río Papurí tributary, Uacaricuari, 1–3 Jun 1953, *Schultes & Cabrera 17,253* (COL, NY, US); Mun. Taraira, Estación Biológica Mosiro Itajura, 1°4′21.8"S 69°31′2.9″W, 2 Apr 2004 (fr), *Calvijo-R & Tanimuka 703* (COAH, COL). **GUAINÍA**: Mun. Puerto Inirida, Reserva Indigena Almidón-La Ceiba, 100 m, 23 Mar 1998, *Murillo 1038* (COL, MO). **VICHADA**: Mun. Santa Rosalía, Reserva Tomon Grande, 200 m, 4°50.228 N, 70°16.493 W, 16 Jun 2009 (fl), *Clavijo R. & Tanimuka 703* (COL): Gaviotas, Omaipia, Río Vichada. 10 Mar 1973 (fr), *Cabrera 2724* (COL). **VENEZUELA. AMAZONAS**: Upper Río Baria, 0°55′N, 66°16′W, 9 May 1984 (fr), *Gentry & Stein 47,295* (K, MO, NY, US): Dep. Río Negro, Río Marawinuma, 160 m, 0°60′N, 66°02′W, 26 Apr 1984 (fr), *Thomas 3209* (MO, NY, US). **PERU. LORETO**: Prov. Requena, Mun. Sapuena, Jenaro

Herrera, 18 Nov 1988 (fr), *Daly* et al. *5703* (NY). **BRAZIL. AMAZONAS**: Manaus, Pensador, 20 Jul 1935 (fl), *Ducke 12* (syntypes A 00043829, F V0360262, FH, K 000370367, MO 1835898, MO 1836724, NY 00388403, NY 00388404, R 000075281, S R-9880, US 101236); Rio Tibahá, Santa Isabel do Rio Negro, 18 Aug 2000 (fl), *Lima 169* (INPA); Mun. Presidente Figueiredo, BR 147, km 113, Igarapé das Lajes, −2.03, −60.025, Sep 2004 (fl), *Souza* et al. *30,024* (ESA, HPL); Manaus, Reserva Ducke, 2°53′S, 59°58′W (fl), *Ribeiro* et al. *1 134* (INPA, K, MG, MO, NY, RB); Santa Isabel de Rio Negro, Rio Maraiuá, 22 Sep 2000 (fl), *Silva* et al. *936* (INPA, MO); Rio Papury, Vaupés, Rio Negro, 17 Oct 1945, *Fróes 21,192* (F, IAC, K, NY, US); Rio Cauaburi, between Rios Maturacá & Yá, 2 Feb 1966 (fr), *Silva & Brazão 60,947* (NY, RB); Manaus-Itacoatiara km 90, Rio Preto, 21 Jul 1961 (fl), *Rodrigues & Chagas 3024* (INPA, NY); Cachoeira baixa do Tarumã, Manaus, 12 Sep 1966 (fl), *Prance* et al. *2276* (F, INPA, K, NY, RB); Manaus, Rio Tarumã, 7 Aug 1949, *Fróes 24,934* (IAC, IAN); Road BR319, Manaus-Porto Velho, km 319, 8 Jul 1972 (fl), *Silva 273* (INPA); Manaus-Porto Velho road, Km 190, 11 Oct 1974 (fr), *Prance* et al. *22,808* (INPA, NY). **PARÁ**: Mineração Rio Norte, near Igarapé Sacazinho, Mun. Oriximinã, 10 Nov 1987 (fr), *Cid-Ferreira 9521* (INPA, NY, RB, US); Mun. Almeirim, Monte Dourado, Gleba Itapeura, −0.58, −52.65, 17 Dec 1987 (st), *Pires-O'Brien & Silva 1888* (K, NY). **ACRE**: Rio Antimani, Floresta Estadual Antimani, Bujari, −9.411, −68.123, 9 Mar 1997 (fl), *Daly 9405* (NY, UFACPZ).

3. **Sacoglottis cydonioides** Cuatr., Contr. U.S. Natl. Herb. 35(2): 183. 1961; Cuatrecasas & Huber, Fl. Venez. Guayana (Steyermark et al.) 5: 636. 1999. Type. Suriname. Brownsberg Reserve, 28 Jun 1924 (fr), *B.W. Boschwezen 6495* (holotype, U 0002481; isotype, US 101235).

Large trees, the young branches glabrous, lenticellate. Leaves with petioles 3–6 mm long, thickened at base; laminas ovate-elliptic, ovate-oblong, obovate-lanceolate, 6–13 × 2.5–6 cm, rigidly coriaceous, glabrous, rounded or obtuse at base, acuminate or cuspidate at apex, the apex 6–20 mm long, the margins slightly serrate, revolute, with a pair of glands near base of lower surface; midrib prominulous above, prominent beneath; primary veins 10–12 pairs, impressed above, prominulous beneath, reticulate venation obsolete above, prominulous or impressed and conspicuous beneath. Inflorescences axillary and terminal, cymose-paniculate, dichotomous, shorter than the leaves; peduncle 0.5–1.5 cm long, stout, striate, puberulous, branches minutely hispid-puberulous; bracts persistent, 0.5–1.3 mm long, clasping, ovate, acute, sparsely ciliate; pedicels thick, glabrous, 0.4–0.5 mm long. Sepals ovate, 0.6–0.7 mm long, obtuse, thick, glabrous except for ciliate margin. Petals oblong, attenuate toward apex, 2.8–3 mm long, minutely hispidulous on upper half. Stamens 10, glabrous, filaments flattened, connate at base, longer ones 1.9–2 mm alternating with shorter ones 1.4 mm long, sparsely papillose, connective thick, ovate, compressed, cuspidate at apex, thecae narrowly oblong on lower sides, 0.4 mm long. Disc 0.5 mm high, scales united, thick, denticulate. Ovary ovoid, glabrous, 5-locular, the locule uniovulate; style 0.5–0.7 mm long, stigma capitate, 5-lobed. Drupe globose, 1.5–2 cm diam., exocarp smooth or somewhat granular, glabrous, 35 mm thick, 2-layered, outer layer coriaceous when dry, resinous-granular, inner layer densely fibrous; endocarp woody, resinous-lacunose, tuberculate, 8–12 mm diam., 1–3 seeded.

Distribution and habitat. *Terra firme* and riverine forest in the Venezuela, the Guianas and northeasterm Amazonian Brazil (Fig. 10.20).

Phenology. Flowering from July to November, fruiting from November to June.

Local names and uses. Brazil: *Uxi de morcego*. Venezuela: *Trompillo, ponsique montanero.* Guyana: *Duhuria.* Suriname: *Awarraballi, kwatta sirie, sort buffel-hout*. French Guiana: *Bofo-oudou, boliquin* (Paramaka), *mahot cochon.*

Conservation status. Least concern (LC).

Illustrations. Cuatrecasas (1961), fig. 38 k, pl. 24.

Additional specimens examined. VENEZUELA. AMAZONAS: Dept. Casiquiare, Caño San Miguel, 2°37′N, 67°7′W, 18 Apr 1991 (fr), *Aymard 8930* (MO, NY); Dep. Atabapo, San José del Orinoco, 3°5′N, 65°35′W, 27 Feb 1990 (fr), *Aymard & Delgado 8320* (MO, NY). **BOLÍVAR**: 80–90 km S of El Dorado, 30 Mar 1956 (fr), *Bernardi 3033* (K, NY, VEN); border with Delta Amacuro, near Río Grande o Toro, E of Upata, 300 m, 8°14′N, 61°44′W, 6 Apr 1967 (fr), *de Bruijn 1645* (F, K, MO, NY, WAG); same locality, 10 Mar 1966 (fr), *Breteler 4956* (NY, U, VEN); Cerro Cotorre, S side Río Paragua, between La Paragua & San Pedro de Las Bocas, 5 Aug 1960 (fr), *Steyermark 86,872* (F, NY, US). **DELTA AMACURO**: Near Río Grande o Toro, Upata, 300 m, 8°14′N, 61°44′W, 10 Apr 1984 (fr), *De Bruijn 1684* (MO, NY). **GUYANA**. NW District, Matthew's Ridge, 26 Aug 1976 (fr), *Mori & Bolten 8196* (K, MO, NY); Cuyuni River, Akarabice Creek, 28 Jul 1933 (old fl), *Tutin 421* (BM, K, U, US695442); Kabrora Backdam, Moruca River, 2 Oct 1997 (st), *Van Andel* et al. *2347* (U); Takutu-U-Essequibo Region, Kuyuwini River trail to Kassikaitu River, 1°58′N, 59°09′W, 11 May 1997 (fr), *Clarke 4497* (K, MO, NY, US), Crabwood Creek, at confluence with Rupununi River, 3°10′N, 59°24′W, 30 May 1996 (fr), *Clarke 1829* (K, NY); Demerara-Berbice Region, Kuruduni Creek to Shirakura Landing, 4°56′10″N, 58°13′29″W, 29 Apr 1995 (fr), *Mutchnick 1208* (MO, NY, U, US); E Berbice-Corentyne, Upper Canje River, 5.6, −57.5833, 50 m, 13 Apr 1987 (fl), *Pipoly 11,497* (US). **SURINAME**. Watramiri, 21 Jun 1920 (fl), *Boschwezen 4720* (IAN, K, U, US, paratypes), 8 Oct 1918, *Boschwezen 4038* (U, US); Nassau Mts, (fr), *Lanjouw & Lindeman 2194* (K, NY, U); Sectie O, Nov 1944 (fr), *Stahel 263* (A, K, MO, NY, U, US); Brownsberg National Park, 26 Oct 1977 (fl), *Teunissen-Werkhoven LBB16254* (NY). **FRENCH GUIANA**. Without locality, 1863, *Melinon s.n.* (P); Cayenne, *Martin s.n.* (K, P); Serpent River 800 m from mouth, 20 Jul 1953 (fr), *BAFOG 35 M* (P, U); Chantier Fosima, 1 km S of Margot, 7 Dec 1953 (fr), *BAFOG 124 M* (P, U); Route St. Laurent to Cayenne, Km 18, 10 Dec 1953 (fr), *BAFOG 131 M* (P); Route de Cayenne, Km 14.1, 29 Jan 1957 (fr), *BAFOG 7656* (NY, U); Piste Mataroni, 4°05′N, 52°06′W, 1 Dec 2010 (fr), *Sabatier & Smock 5782* (MO); Oyapock, Saut Armontabo, 5 Jul 1969 (fl), *Oldeman T304* (P, U, US): Route Regina-St Georges, Approuague Basin, 4°2′N, 52°1′W, 2 Dec 1995 (st), *Granville & Cremers 13,200* (CAY, NY, P, U); Crique Plomb, Sinnamary Basin, 5°0′N, 52°54′W, 7 Mar 1994 (fr), *Bordenave 780* (CAY, P, U). **BRAZIL. AMAZONAS:** Distrito Agropecuária, Fazenda Porto Alegre, 2°22′S,

59°57′W, 10 Apr 1992 (fr), *Dick 91* (INPA, K, NY). **PARÁ**: Trombetas, Rio Aminá, Lago Salgado, 22 Apr 1917, *Ducke RB 16809* (MG, RB). **AMAPÁ**: Rio Oiapoque, 1 Feb 1950 (fr), *Fróes 25,783* (IAC, IAN), 15 Oct 1950 (fl), *Fróes 26,636* (IAN). **ACRE**: Mun. Brasiléia, Reserva Extrativista Chico Mendes, Seringal Porongaba, 10 Jul 1991(fl, fr), *Cid Ferreira 10143A* (INPA, MO, NY).

The epicarp of the fruit has a strong smell of *Cydonia* (quince) fruit, hence the specific epithet. This species can be recognized by the distinctly revolute leaf margins and the rigidly coriaceous leaves. Henry et al. (2000) reported that the fruit of this species is eaten by lowland tapirs *T. terrestris* (L.) in French Guiana.

4. **Sacoglottis gabonensis** (Baill.) Urb., Fl. Bras. (Martius) 12 (2): 449. 1877; Reiche Nat. Pflanzenfam. (Engler & Prantl) 3 (4): 37. 1890; Cuatr., Contr. U.S. Natl. Herb. 35, 2: 172. 1961; *Aubrya gabonensis* Baill. Adansonia 2: 266. 1862; *Houmiri gabonensis* Baill., Hist. Pl. (Baillon) 5: 52. 1874. Type. Gabon, Jul 1853, *C. E. Aubry-Lecomte s.n.* (lectotype, **here designated**, P 00389405; isolectotype, P 00389406).

Large trees to 23 m tall; trunk with small buttresses, the young branches glabrous. Leaves with petioles 6–10 mm long, narrowly winged toward lamina, thickened at base; lamina ovate-elliptic or oblong-elliptic, 7–14 × 3–7 cm, coriaceous, glabrous, suddenly and obtusely narrowed at base and slightly decurrent onto petiole, acuminate at apex, the acumen 8–15 mm long, margin slightly crenate, setae present, with a pair of glands at basal margin and scattered on lower surface; midrib plane above, prominent and glabrous beneath; primary veins ca 12 pairs, prominulous above, prominent beneath, reticulate venation prominulous beneath. Inflorescences axillary, cymose-paniculate, dichotomous, branches minutely pubescent; pedicels 0.5–1.5 mm long thick, angulate, puberulous or hirtellous-pubescent, articulate to short peduncle or sessile; bracts persistent, triangular, 2 mm long, clasping, pubescent and ciliate. Sepals broadly orbicular, 1.5 mm long, puberulous at base, cilate at margin. Petals linear, thick, 6–7 mm long, 1.6–2 mm broad, sericeous. Stamens 10, filaments thickened, flattened, connate at base, 5 opposite petals oblong, acute, 3.5 × 0.5–0.6 mm, alternating with 5 longer linear 4–5 × 0.3–0.4 mm, anthers 0.6 mm long; connective thick, lanceolate, compressed on upper part, those of shorter filaments 2 mm long, those of longer filaments 1.5 mm long, thecae ellipsoid. Disc membranaceous, 0.8–1.5 mm high, laciniate. Ovary ovoid, glabrous, 5-locular, locules opposite petals and uniovulate, 2 mm high; style thick, erect, 3 mm long, stigma capitate. Drupe ellipsoid or almost globose, 2.7–4 × 2.5–3 cm; exocarp fibrous-fleshy, hard when dry, 2–2.5 mm thick; endocarp woody, slightly 10-sulcate, bullate, with resinous cavities. 2n = 12.

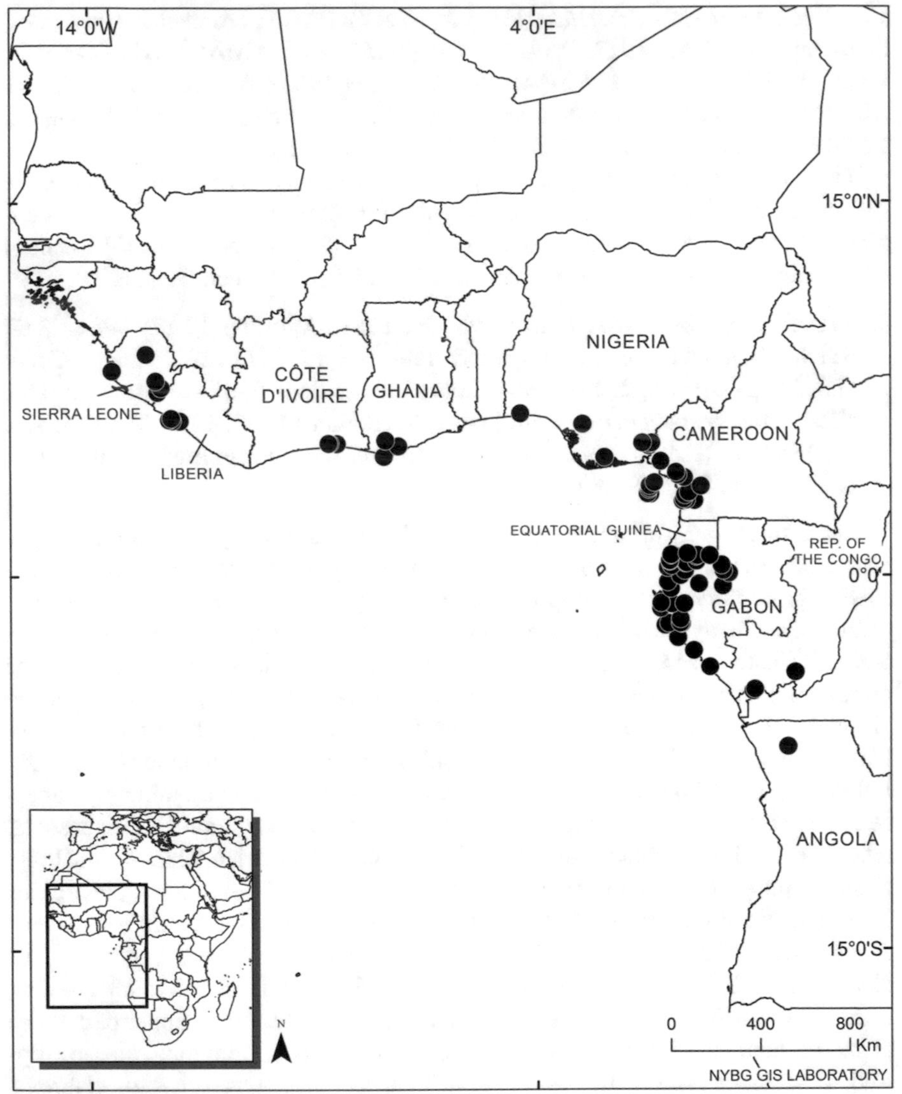

Fig. 10.17 Distribution of *Sacoglottis gabonensis*

Distribution and habitat. Widely distributed in the rainforests of West Africa from Sierra Leone to Angola in both flooded and non-flooded areas, mainly in wetter areas and riversides (Fig. 10.17).

Phenology. Flowering mainly from January to June, fruting mainly from September to November, but found around the year and often bearing flowers and fruit together.

Local names and uses. English: *Bitter bark tree*. Liberia: *Daush, Liberian cherry;* Ivory Coast: *Aguapo*; Sierra Leone: *Kpou-wuli*; Nigeria: *Atala* (Yoruba).

Cameroon: *Bidou* (Ewondo); Gabon: *Ozouga;* Equatorial Guinea: *Bidu* (Fang), *esua* (Combe). The wood has many uses outdoors for bridges, poles, piers and indoors for furniture, boards, flooring etc. Infusions of the bark are commonly used to treat fever, diarrhoea and abdominal pain.

Conservation status. Least concern (LC).

Illustration. Hutchinson and Dalziel (1928), fig. 114; Cuatrecasas (1961), figs. 34a–c, 35f–h.

Selected specimens examined. SIERRA LEONE. Kambui Forest Reserve, 29 Oct 1937 (fr), *Edwardson 181* (BM, FHO); N Region, Tonkolili Distr, 220 m, 8°56′N, 11°43′W, 5 Mar 2010 (fr), *Kanu 86* (BR, K); Gola Forest, 26 May 1965 (fl), *Fox 99* (K). **LIBERIA.** Vicinity of Dukwai River, 10 Apr 1926 (fr), *Cooper 68* (A, BM, FHO, GH, K, WIS, NY, US, USw); 1929 (imm fr), *Cooper 274* (A, BM, FHO, GH, K, WIS, NY, US, USw); 20 miles N of Sinoe, 16 Jan 1969 (fl), *Jansen 1088* (K, BR, NY); Duport, 15 Nov 1926 (fl), *Bequart 1469* (A 00242160). **IVORY COAST**. Abidjan, 7 May 1929 (fl), *Aubréville 92* (A, BR, K, P); road Abidjan-Dabou, 40 km W of Abidjan, 14 Aug 1963 (fr), *De Wilde 661* (BR, K). **GHANA**. W Province, Ankasa, Fuale, Jan 1942 (fr), *Adjimang 4849* (A, FHO); Bimpong Forest Reserve, 5.71, 1.466, 16 Feb 1972 (fl), *Enti 72* (K). **BENIN.** Atlantique, 6.372058, 2.434697. 22 May 2011, *Quiroz-Villarreal 9* (WAG). **NIGERIA**. Oban, 5.5 N, 8.58E *Talbot 1744* (BM); Calabar, Uwet, 30 Jul 1959 (fr), *Binuyo FHI 41418* (BR, FHO, K); Benin Prov, Sapoba, Jamieson River, 3 Nov 1950 (fl), *Keay FHI 28077* (BR, K); Cross River State 25 km N of Oban road to Ekang, 20 Jun 1981 (fl), *Gentry & Pilz 32,860* (BR, K, MO). **CAMEROON**. Bipinde, Urwaldgebiet, 3°14′, 10°02′, *Zenker 1249* (A, BM, K, M, S); 1898, *1624* (A, BM, BR, K, M, NY, S); near Masore, 8 km NW of Ekondo Titi, 4.650, 8.983, 6 Jun 1976 (fl), *Letouzey 15,099* (K); 9 km N of Kribi, 3°N, 9°56′E, 4 Jun 1969 (fl), *Bos 6137* (BR, K, LD). **EQUATORIAL GUINEA**. Malabo-Luba, 15 Jun 1987 (fl), *Fidalgo de Carvalho 2944* (K, MA, NY, WAG); Bioco, Malabo-Luba, Rio Tiburones, 8 Feb 1989 (fl fr), *Fernández-Casas 11,322* (BR, K); Bioco, *Mann 1417* (K). **GABON**. 180 km SE of Port Gentili, Lake Anengue, between Kouroue and Edjiwe Rivers, 1°09′S′, 9°30′ E, 21 Jun 1930, *Krukoff 121* (A, NY, USw); Nyanga on Maambi River 3°01′S, 10°22′ E, 7 Aug 1992 (fr), *Wieringa 1362* (K); Estuaire: Remboué River, S. of Estuaire, 0°00′, 9°50′E, 1 May 1992 (fr), *McPherson 15,716* (MO); Moyen-Ogooue: Lope National Park, 0°12′S, 11°34′ E, 20 Apr 2006 (fr), *Leal* et al. *1155* (MO, WAG); near Lake Azingo, 0°29′S, 10°03′W, 20 Jun 2014 (fr), *Stévart 4805* (BR, BRLU, L, LBV, MO, P); Ngoounie: SE of Lambaréné, Mabounié, 0°48′S, 11°01′ E, 14 May 2001 (fl bud), *Stévart* et al. *34,759* (BR, L, LBV, MO, P); Nyanga: Mabamba Plain, 4 km from Kwassa Camp, 3°40′S, 11°01′ E, 3 May 1998 (fr), *Stone* et al. *3283* (BR, MO); Ogooue-Maritime: Region of Lac Alombié, 0°49′S, 9°27′ E, 15 Oct 2014 (fl,fr), *Lachenaud* et al. *1934* (BR, BRLU, G, LBV, MO, P, WAG). **REPUBLIC OF THE CONGO**. Bongou Forest, 18 Dec 1969 (fl), *Attims 265* (BR, K); Brazaville, Mayombe-Holle Forest, 26 Feb 1954, *Koechlin 2632* (U). **ANGOLA**. Sumba, near

Zaire River, 1923, –6.15, 12.61, *Gossweiler 8751* (BM, US); Belize on Luali River, –4.65, 12.76, 12 Feb 1918 (fr), *Gossweiler 8182* (BM).

Cuatrecasas (1961) mentioned an Aubry-Lecomte collection as the type but did not designate one of the two sheets of this at Paris as the type, and so a leptotypification was needed here.

White (1994) showed that the fruits of this species are important in the diet of elephants in Gabon and elephant migrations are linked to the fruiting time. Elephants move over large distances to feed on *Sacoglottis gabonensis*. The buoyant fruit of this common tree of African riversides is a good indication of its long-distance dispersal and evolutionary divergence from other species on the South American continent. It is closely related to *Sacoglottis amazonica* the fruits of which are often found in drifts even as far away as Europe. The bark of this species is rich in two chemicals bergenin and gallic acid, with antimicrobial properties, Tchouya et al. (2016). The stem bark is commonly used as an additive to palm wine in Nigeria (Maduka and Okoye 2005).

5. **Sacoglottis guianensis** Benth., Hooker's J. Bot. Kew Gard. Misc. 5: 103. 1853; Urb., Fl. Bras. (Martius) 12 (2): 4481877, pro parte; Cuatr., Contr. U.S. Natl. Herb. 35(2): 174. 1961; Cuatrecasas & Huber, Fl. Venez. Guayana (Steyermark et al.) 5: 636. 1999. Type. Guyana, Roraima. *Rob. Schomburgk 574* (= *Rich Schomburgk 842*) (holotype, K 000407314; isotypes, BM, F V0060789, G 00368591, NY 00388405, OXF, P 01903250; US 282417, B lost, photo F OBN012599). Fig. 10.18.

Sacoglottis amazonica Benth., Hooker's J. Bot. Kew Gard. Misc. 5: 104. 1853, non Mart.

Sacoglottis guianensis forma *dolichocarpa* Ducke, Arch. Jard. Bot. Rio de Janeiro 3: 179. 1922. Type. Brazil. Pará: Rio Trombetas, Lago Salgado. Type unknown.

Sacoglottis guianensis Benth, var. *maior* Ducke, Arq. Inst. Biol. Veget. 4: 27. 1938. Type. Brazil. Amazonas: Manaus, 2 Oct 1932 (fl), *A. Ducke RB 23818* (holotype, RB 00539066; isotypes, INPA 230792, K 000407313, NY 01131932, P 01903251, S-R9882, US 00101228, U 0002487).

Sacoglottis guianensis Benth. var. *hispidula* Cuatr., Contr. U.S. Natl. Herb. 35(2): 180. 1961. Type. Brazil. Amazonas: Mun. Humaitá, Livramento, Nov 1934 (fl), *B. A. Krukoff 6653* (holotype, US 00101229; isotypes, A 00043830, BM, BR 05119219, F 00043830, IAN 38935, K 000407311, K 000407312, MO 2796457, NY 00388406, S R-5531, U, US 01269236, WIS 00000511MAD).

Sacoglottis guianensis forma *glabra* Cuatr., Contr. U.S. Natl. Herb. 35(2): 174. 1961. Type. Brazil. Pará: Ilha Collares, 29 Dec 1953, *R. L. Fróes 30,670* (holotype, IAN 82301).

Large- or medium-sized trees, the young branches glabrous or minutely hirtellous puberulous. Leaves with petioles 4–13 mm long, glabrous or minutely puberulous, sulcate above; laminas elliptic, ovate-elliptic, subovate or oblong-ovate, 5–17 × 3–6.5 cm, rigid-coriaceous, glabrous or with scattered minute appressed

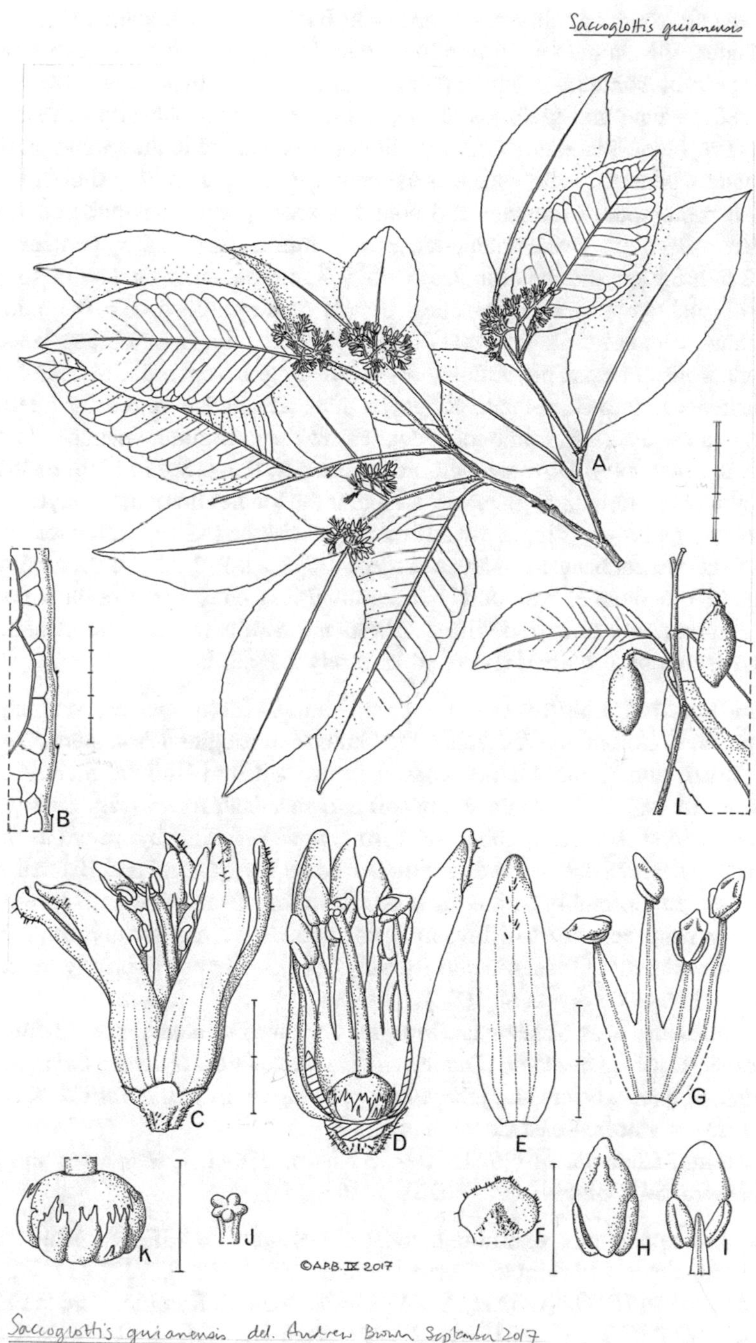

Fig. 10.18 *Sacoglottis guianensis*. (**a**) Habit, flowering branch. (**b**) Abaxial leaf margin. (**c**) Flower. (**d**) Flower with 2 petals and sepals, part of stamina tube removed. (**e**) Petal, outer surface. (**f**) Sepal outer surface. (**g**) Part of stamina tube, inner surface. (**h**) Anther, inner face. (**i**) Anther, outer face with filament. (**j**) Stigma. (**k**) Disc. (**l**) Fruit. Single bar = 1 mm, graduated single bar = 2 mm and 5 mm, graduated double bar = 5 cm. (A–K from *Ratter* et al. *9815*; L from *Vicentini 361*)

hairs beneath, rounded, obtuse, or cuneate at base, narrowed, acuminate, or cuspidate at apex, the acumen 4–16 mm long, margins slightly crenate, persistent setae usually present, sometimes with a pair of glands at basal margin of lower leaf surface; midrib plane and glabrous above, thick, prominent beneath; primary veins 10–12 pairs, inconspicuous above, prominulous beneath; reticulate venation slightly prominulous beneath. Inflorescences axillary, cymose-paniculate, shorter than the leaves, dichotomous, peduncles and branches shortly hirtellous-pubescent; bracts persistent, clasping, ovate-triangular, 0.5–1 mm long, ciliate; pedicels short, 0.1–0.2 mm long, glabrous articulate with 1–3 mm peduncles below or sessile on pilose or glabrate terminal branches. Sepals ovate-rounded, 0.6–0.7 mm long, puberulous, minutely ciliate. Petals greenish, thick, linear or oblong-lanceolate, acute, glabrous or hispid-puberulous, 3–4.5 mm long, 1.5 cm broad at base. Stamens 10, filaments flattened, connate for lower half, 2.5 and 3 mm long alternating, anthers ovate-lanceolate, 1 mm long, connective thick, trigonous-lanceolate, thecae elliptic, 0.4 mm long. Disc annular, membraneous, 0.5–0.6 mm high, denticulate. Ovary globose, 1 mm high, glabrous, 5-locular the locules uniovulate; style 2–3 mm long, erect, glabrous, stigma weakly capitate, 5-lobed. Drupe ellipsoid-oblong, attenuate at base, subacute or obtuse at apex, 1.5–3 × 0.9–1.32 cm; exocarp smooth or slightly tuberculate, resinous, 1–1.5 mm thick; endocarp woody, smooth or slightly bullate and furrowed with 5 valves and 5 thin septa, acute at both ends, resinous-lacunose, usually one-seeded, or rarely 2–3 seeds.

Distribution and habitat. A widespread and variable species, ranging from Amazonian Colombia, Venezuela, the Guianas, throughout Amazonian and central Brazil and in the Atlantic coastal forest and into Bolivia, mainly in non-flooded forest, campinas on sandy soil and savannah forest (Fig. 10.20). It has been divided into a number of forms and varieties by previous authors (Cuatrecasas 1961). The seeds are dispersed by bats (Gastal and Bizerril 1999).

Phenology. In Colombia to the Guianas, flowering from June to December and fruiting from January to May; in Amazonia, flowering mainly from June to November in the dry season and fruiting from October to February in the rainy season (Pires-O'Brien et al. 1994).

Local names and uses. Venezuela: *Peru-yek*. Guyana: *Dukuria* (Arawak). Suriname: *Apesiya* (Carib), *buffelhout* (Dutch), *doekoelia, dukulia, dukuria, hubu-diamaro, wirimiridja uara* (Arawak*), japopalli, gannasagon*. Brazil: *Achuá, uchi torrado*.

Conservation status. Least Concern (LC).

Illustrations. Cuatrecasas (1961), figs. 37 a–m, 38a–d; Cuatrecasas and Huber (1999) fig. 542; Bhikhi et al. (2016), p. 167.

Selected specimens examined. **COLOMBIA. CAQUETÁ**: Mun. Solano, Araracuara, Setor Chiribiqute, Cerro Gamitana, 0°10′33"S, 72°29′47″W, 3 Dec 2010 (fr), *Castro 10,837* (COAH). **GUAVIARE**: Mun. El Retorno, Vereda El Treno, 230 m, 2°22′33″N, 72°38′31″W, 12 Nov 2002 (fl), *López C. 7915* (COAH). **VAUPÉS**: Río Apaporis, Jirijirimo, 250 m, 25 Nov 1951 (fl), *García-Barriga 13,681* (COL, US); Estación Biológica Caparú, 1°05′S, 69°32′W, 15 Jan 1996 (fr), *Palacios & Rodriguez 103* (COAH, K). **GUAINÍA**: Campamento Cerro Nabuquén,

2°47′N, 68°46′W, 4 Mar 1995 (fr), *Córdoba* et al. *913* (MO); Río Guaviare, 31 May 1976, *Roa 436* (INPA); Río Inirida, 5 May 1976, *Roa 394* (INPA). **VICHADA**: Parque Nacional Natural El Tuparo, Caño Peinille, 18 km W of Centro Administrativo, 5°18′N, 69°01′W, 90 m, 6 Mar 1985 (fr), *Zarucchi & Barbosa 3596* (MO, NY), lower area of granitic outcrop, 90–110 m, 5°20′N, 68°02′W, 7 Mar 1985 (fr), *Zarucchi & Barbosa 3616* (COAH, K, MO, NY). **VENEZUELA. AMAZONAS**: Cerro Moriche, Río Ventuari, 200 m, 16 Jan 1957 (imm fr), *Maguire* et al. *30,967* (K, NY, US), 300–1000 m, 13 Jan 1957 (fr), *Maguire* et al. *30,844* (K, NY, US); Rio Orinoco, N base of Cerro Duida, 200 m, 13 Oct 1950 (fl), *Maguire* et al. *29,416* (NY, US, VEN); Río Atabapo, Caño Temi, 125 m, 20 Oct 1950 (fl), *Maguire* et al. *29,337* (NY, US); Dep. Átures, Río Coromoto, 35 km S. of Puerto Ayacucho, 5°22′N, 67°33′W, 6 Apr 1984 (fr), *Plowman & Guanchez 13,522* (F, K, MO, US); Dep. Atabapo, Cerro Mahedi, 600 m, 2°58′N, 64°45′S, 4 Mar 1980 (fr), *Huber 4930* (COL, INPA, K, NY, US), San José del Orinoco, 4°44′N, 61°44′W (imm fr), *Castillo 7209* (MO); Dep. Río Negro, Río Varía, 1°25′N, 66°24′W, 90 m, Apr 1991 (fr), *Velazco 1766* (MO), Cerro Vinilla, 30 km SSW of Ocano, 440–600 m, 1 Mar 1984, *Steyermark* et al. *130,398* (MO, US); Dep. Cassiquiare, 20 km SE of San Fernado de Atabapo, 110 m, 3°50′N, 67°47′W, 10–16 Jan 1998, *Aymard* et al. *6508* (MO, US); Dep. Atures, Río Sipapo, between Ríos Cuao & Autana, 90 m, 4°45′N, 67°42′W, May 1989 (fr), *Foldats & Velazco 9005* (MO, NY); Caño Yutaje, S base of Serranía de Yutaje, 110–120 m, 5°38′N, 66°06′W, 16 Feb 1987 (fl), *Liesner & Holst 21,181* (MO, US) Río Cassiquiare, 3 km S. of El Porvenir, 132 m, 2°02′N, 66.28′W, 2 Feb 1992, *Aymard* et al. *9986* (MO); Río Matacuni, 220–450 m, 3°15′N, 64°56′W, 29 Jan 1990 (fr), *Stergios & Velazco 14,666* (MO, NY); Pie de Cerro Duida, Orinoco, 3°08′N, 65°53′W, Jan-Feb 1969 (fr), *Fariñas* et al. *478* (F, MO, NY). **BOLÍVAR**: above Salto Humito, 7 Jan 1956 (fl), *Wurdack & Monachino 41,149* (F, K, NY, RB, US). Río Karuai, base of Sororopán Tepuí, W of La Laja, 1220 m, 29 Nov 1944, *Steyermark 60,756* (F, NY, VEN); Río Apacará, Mun. Urimán, 500 m, 28 Aug 1954, *Bernardi 1751* (NY); Río Icaburú, Río Hacha, 450–850 m, 7 Jan 1956, *Bernardi 2813* (NY, VEN); Dist. Heres, Guaiquinima, 5°51′N, 63°33′W, 900 m, 6 Dec 1987 (fl), *Huber 12,386* (COL, K, MO, NY); Piar, Río Acanán, Ibana-merú rapids, 5°54′N, 62°16′W, 480 m, 5 May 1986 (fr), *Liesner & Holst 20,628* (MO, US); 10 km SW Karaurin Tepuí, confluence Ríos Karaurín & Asadon, 900–1000 m, 5°19′N, 61°03′W, 29 Apr 1988 (fr), *Liesner 24,017* (MO); Mun. Raul Leoni, Río Túriba & Caño La Miel, 800 m, 6°34′N, 66°23′W, Jun 1989 (fl fr), *Fernández 5587* (MO, NY); Mun. Raul Leoni, 64 km SE Pijiguaos, 550 m, 6°09′N, 66°23′W, Jul 1989 (fr), *Delgado 279* (MO). **GUYANA**. Roraima, Oct 1842, *Schomburgk 571* (K, P), (prob = 574 of NY, P), *Rich. Schomburgk 842* (M, US); Gunn's Landing, Essequibo River, 1°39′N, 58°38′W, 240–260 m, 30 Sep 1989 (fl), *Jansen-Jacobs* et al. *1898* (F, INPA, K, NY, US); NW Distr., Kaboru Backdam, Moraca River, 9 Oct 1997, *van Andel* et al. *2393* (K); Courantyne area, Espira River, 4°58′N, 57°30′W, 100 m, 6 May 1991 (fl), *Polak 310* (NY); Takutu-U, Essequibo, Rewa River, 0–5 km S of Kwitaro River 3°°17′22″N, 58.45′7″W, 25 Feb 1997 (fr), *Clarke 3936* (MO, NY); 8 miles E of Onoro Creek mouth, 1000 feet, 30 Sep 1952 (st),

Guppy 308 (NY). **SURINAME.** Coppenam River, 23 Sep 1944 (fl), *Maguire 24,836* (BM, COL, F, GH, K, MO, NY, RB, U, US, VEN); Tafelberg, 24 Sep 1944 (fl), *Maguire 24,844* (COL, MO, NY); Nassau Mts, 22 Mar 1949, *Lanjouw & Lindeman 2869* (U); Zanderij I, 27 Jun 1919, *Boschwezen 2974* (IAN, K, MG, MO, U, US), Sep 1942, *Stahel 18* (A, IAN, K, NY, U, US); Sipaliwini Savannas, 11 Oct 1968 (fl), *Oldenburger* et al. *261* (F, K); Sipaliwini, 2°19′N, 56°47′W, 6 Apr 2005 (fr), *Hoffmann 6145* (K, US). **FRENCH GUIANA**. 1862, *Melinon 584* (NY, P), 1863, *Melinon s.n.* (NY, P, US); Lac des Americaines, Isle de Cayenne, 4°51′N, 52°21′W, 11 Jan 1987 (imm fr), *Granville 9121* (BM, CAY, MO, NY, P, U, US); Sinnamary, Petit Saut, 5°3′N, 53°3′W, 22 Feb 19,898 (st), *Sabatier 2373* (CAY, MO, NY); Saül, 3°37′N, 53°12′W, 314 m, 4 Sep 1994 (fl), *Mori* et al. *23,788* (NY); Station des Nouragues, 4.50, −52.7, 25 Sep 1990, *Larpin 962* (US). **PERU. LORETO:** Jenaro Herrera, *Spichiger 4562* (MBM), Maynas, −3.3333, −72.91666, 24 Feb 1991, *Pipoly 13,625* (MO). **BRAZIL. AMAZONAS**: Mun. São Gabriel da Cachoeira, Igarapé Pamaali, −0.130, −67.089, Jul 2007, *Stropp & Assunção P75* (EAFM); Mun. Cucui, Rio Xé, 0°57′N, 67°10′W, 24 Oct 1987 (fl), *Daly 5459* (INPA, MO, NY, RB); Mun. Novo Japurá. Vila Bittencort, Rio Apaporis, 1°14′S, 69°25′W, 21 Nov 1982 (fr), *Cid Ferreira* et al. *3762* (INPA, K, MO, NY, RB); Mun. Humaitá, between Rios Livramento and Ipixuna, 7–18 Nov 1934 (imm fr), *Krukoff 7082* (A, BM, F, IAN, K, MO, NY, RB, S, U, US); Rio Solimões, Mun. São Paulo de Olivença, creek Belém, Oct-Nov 1936 (imm fr), *Krukoff 8757* (BM, F, K, MO, NY, P, S, U, US). Manaus, Cachoeira do Mindú, 25 Sep 1929 (fl), *Ducke RB 23433* (K, RB, S, US); Manaus, Flores, 14 Sep 1945 (fl), *Ducke 1756* (A, IAN, K, MG, NY, US). Rio Urubú, Cachoeira Iracema, 22 Sep 1949 (fl), *Fróes 25,369* (IAN, US); Rio Negro, Rio Demití, 0°10′N, 66°53′W, 1 Nov 1987 (fl), *Maas* et al. *6894* (INPA, K, MO, NY, RB); Manaus, Distrito Agropecuária, Reserve 1502, 50–125 m, 2°24′S, 59°43′W, 18 Nov 1988 (imm fr), *Boom* et al. *8525* (INPA, MO, NY); Reserva Ducke, Manaus, −2.883, −59.966, 29 Sep 1994 (fl), *Sothers 192* (INPA, K, MG, NY, RB); Manaus-Caracaraí, Km 97 road to Balbina, 4 Sep 1979 (fl), *Cid Ferreira 1000* (INPA, MG, MO, NY, RB); Rio Canumã, Nova Olinda do Norte, Nov 1976 (fl), *Monteiro 76–1299* (INPA). **RORAIMA:** road Caracarai-Boa Vista, 3 Mar 1948 (fl), *Fróes 22,940* (IAC, IAN); 80 km S. of Boa Vista, 31 Jan 1969 (fr), *Prance* et al. *9540* (F, INPA, K, MG, NY); Ipiranga, Rio Mucajaí, between Pratinha & Rio Apiaú, 26 Jan 1967 (imm fr), *Prance* et al. *4114* (F, INPA, K, MG, NY, US); Igarapé Capivara, below Rio Preto, Caracaraí, 1.816, −61.128, 17 Sep 2005 (fl), *Barbosa 174* (INPA); Campinha, Rio Branco, Sep 1913, *Kuhlmann MG 3510* (K, MG); Mucajaí-Caracaraí road Km 17, 2.8197, −60.6733, 8 Nov 1977 (fl), *Coradin 1009* (INPA, NY). **PARÁ**: Breves (fl, fr), *Ducke RB 17784* (IAN, U); Cajatuba, 18 Jan 1932, *Monteiro da Costa 281* (F, IAN, P); Peixeboi, 16 Jul 1907 *Siqueira MG 8281* (BM, MG, US); Óbidos, Serra da Boa Vista, 24 Dec 1913, *Ducke MG15234* (MG); Rio Tapajós, Mangabal cataracts, 31 Aug 1916, *Ducke MG 16419* (MG, P, RB13666, US); Alter do Chão, Santarém, −2.443, −54.708, Oct 1999, *Magnusson 4665* (INPA); Mun. Monte Alegre, Rio Maicurú, trail Cáaussu Balança, 16 Sep 1953, *Fróes 32,084* (IAN, US); Santarém, Aug 1850, *Spruce s.n.* (BM, K, MG, NY, S), *Spruce 1009* (K, P); Alto Ariramba, 21 Dec 1906, *Ducke MG 8042* (BM, MG, RB); Cachimbo,

Cuiabá-Santarém Km 831, 16 Feb 1977 (fr), *Kirkbride & Lleras 2836* (F, INPA, K, MO, NY, RB); Cachoeira, BR22, Km 98, 24 Aug 1964 (fl), *Prance & Silva 58,838* (F, K, MO, NY, RB): Reserva Gorotire, Rio Fresco, 220–500 m, 7°47′S, 51°07′W, 21 Jan 1983 (fr), *Gottsberger & Posey 24–22,183* (MO); Belém-Brasília km 105, 7 Aug 1963 (fl), *Maguire* et al. *56,047* (F, MO, NY, RB); Rio Pacajá, 15 Oct 1965 (imm fr), *Prance* et al. *1648* (COL, F, IAN, K, NY); Rio Xingu, Largo do Passari, Altamira, −4.3220, −52.7416, 31 Jan 1987, *Souza 921* (NY). **AMAPÁ**: Camaipi, 0°10′N, 51°37′W, 21 Sep 1983 (fl), *Mori* et al. *16,489* (HAMAB, IEPA, K, MO, NY); Ramal do Camaipi, Mazagão, −0.115, −51.289, 2 Dec 1985 (imm fr), *Rabelo 3296* (HAMAB, IEPA, INPA); road to Matapi, behind Porto Grande, Matapi, 13 Sep 1995 (fr), *Austin* et al. *7070* (INPA, NY, RB, UEC). **MARANHÃO**: Turiaçu, 6 Jun, 1978 (fl), *Rosa & Vilar 2870* (F, MO, NY); P. I. Guajá, Rio Turiaçu, Monção, −3.12, −46.08, 29 May 1987, *Balée 3127* (NY); Garajahú, Rio Mearim, *Lisboa MG2330* (BM, MG); Parque Estadual do Mirador, −6.370, −51.616, 21 Sep 1983 (fl), *Orlandi 869* (ALCB, HUEFS, HST, INPA, JPB, MBM, RB, UESC, UFBA, UFPB, UFRPE); Acampamento SUDENE, Alto Turi, 300 km from BR22, km 2, −2.89669, −45.635248, 12 Nov 1962 (fr), *Teixeira 113* (HST, IPA). **RONDÔNIA**: Rio Preto, Rio Machado, 15 Oct 1997, (RON); Transecto 04 módulo Caiĝçara, Porto Velho, 8 Feb 2012, *Pereira-Silva 16,029* (CEN, HUEFS, INPA, NY, RB, RON, UNIR); Chapada dos Parecis, 29 km from Vilhena, 28 Oct 1979 (fr), *Vieira 736* (INPA, MG, RB); Road RO-399, 13 km from Vilhena, 12°45′S, 60°10′W, 3 Nov 1979 (fr), *Vieira 889* (INPA, RB). **TOCANTINS**: 10 km S of Gurupí, Belém-Brasília road, 30 Aug 1964 (fl), *Prance & Silva 58,954* (F, K, NY); meeting of rios Curicaca & Ribeira, Darcinópolis, −6.702, −47.708, 200 m, no date, *Pereira-Silva 14,742* (CEN); UHE Novo Acordo, Rio Lajeado, 9°59′43" S, 47°37′09″W, 28 Dec 2008 (fr), *Haider & Santos 1138* (RB); Córrego Mosquito, Novo Acordo, −10,097, −47.415, 261 m, 27 Dec 2008 (fr), *Santos 1789* (HUTO); Parque Estadual do Jatapão, São Felix do Tocantins 19 km from Mateiros, −10.418, −46.49, *Cavalcanti 2884* (CEN); Estrada da fazenda Maria Bonita, Fildelfia, *Amaral-Santos 3069* (CEN). **MATO GROSSO**: Xavantina-Cachimbo road, 18 Nov 1967 (imm fr), *Philcox 3131* (K, MO, NY, RB); 1 km N of base camp, 12°49′S, 51°40′W, 16 Dec 1967 (fr), *Philcox & Ferreira 3545* (K, NY, RB). **PIAUÍ**: Piracuruca, −4.09, −41.688, 23 Jul 2007, *Matos* et al. *62* (UB). **PERNAMBUCO**: Tamandaré, −8.75972, −35.1047, 16 May 1999 (fr), *Lucena740* (PEUFR, UFRPE). **SERGIPE**: Mun. Santa Luzia do Itanhi Crasto, 5 Oct 1993 (fl), *Carvalho* et al. *4346* (ALCB, ASE, CEPEC, HUEFS, K, MBML, MO, NY); Fazenda Trapsa, Itaporanga d'Ajuda, −10.998, −37.310, 7 Jul 2008 (fr), *Matos 45* (ASE); Km 2, Crasto-Santa Luzia do Itanhi, −11.350, −37.448, 5 Oct 1993, *Amorim 4346* (ASE, CEPEC, K, MBM, NY, UFBA, UFS). **DISTRITO FEDERAL**: Estação Ecológica do Jardim Botánico, −15.8666, −47.85, 3 Aug 1995 (st), *Silva* et al. *40* (HEPH, JBB); in front of dam, −15.86667, −47.85, (fr), *Caetano 40* (CEN); Fazenda Água Limpa, −15.95444, −47.941389, 10 Feb 2013 (fr), *Oliveira & Arcela 87* (CEN, UB). **BAHIA**: Formoso do Rio Preto, Estação Ecológica Serra Geral, −11.073889, −45.30722, 23 Oct 2011 (fr), *Barbosa-Silva 48* (HEPH, UFMA). **BOLIVIA. SANTA CRUZ**: Velasco, Parque Nacional Noel Kempff Mercado, Laguna Bella Vista, 13°36′S, 61°33′W,

200 m, 9 Aug 1995 (fl), *Guillén & Soliz 3845* (F, K, MO, NY, USZ); Campamento, Las Gamas, 850 m, 14°48′S, 60°23′W, 19 Feb 1997 (imm fr), *Uslar* et al. *601* (MO, USZ); 850 m, −13.8947, −60.8127, 28 Oct 1995 (imm fr), *Rodriguez & Surubí 543* (MO, NY, USZ).

This is an extremely variable polymorphic species and has been divided into number of varieties and forms by Ducke (1922–1933) and by Cuatrecasas (1961), which were based on pubescence of various organs and the shape of the fruit. It is often impossible to decide to which taxon a collection belongs as they are often missing the crucial part. It seems preferable at this stage to recognize this as a single taxon as was done by Cuatrecasas and Huber (1999), but it would make a good molecular study for the future.

The fruits are eaten by *Callithrix penicillata* monkeys in central Brazil. Gunn and Dennis (1972) reported and illustrated stranded fruit of this species washed up on the Carolina coast of the United States.

6. **Sacoglottis holdridgei** Cuatr., Cíencia (México) 23: 138. 1964; Zamora, Manual de Plantas de Costa Rica 6: 12. 2007. Type. Costa Rica, Isla del Coco, Jan 1963 (fl), *L. Holdridge 5164* (holotype, US 00101223; isotypes, US 00101224, US 00130679).

Medium-sized trees to 25 m tall, the young branches terete, glabrous. Leaves with petioles 5–9 mm long, flattened above, glabrous; laminas ovate-elliptic or oblong-ovate, rigid-coriaceous, glabrous, 7–22 × 4.2–9 cm, rounded or subcordate at base, decurrent onto petiole, finely acuminate or acute at apex, the acumen 5–10 mm long, margins entire or slightly crenate and conspicuously undulate with minute glands when young, and 1–2 pairs of basal glands on lower surface; midrib prominulous and striolate above, prominent and glabrous beneath; primary veins 8–10 pairs, prominent on both surfaces, reticulate venation prominulous and conspicuous on both surfaces. Inflorescences axillary, cymose-paniculate, sessile or subsessile, glomerulate, branches thick, angular, hirsute; bracts deciduous, ovate, clasping, ca 2 mm long; pedicels short, 0.5 mm long. Sepals 1.7–2 mm long, glabrous on exterior except for minutely ciliate margins. Petals green or pale yellow, fleshy, linear-oblong, 5.5 mm long, 2 mm broad, glabrous. Stamens 10, filaments glabrous, connate for lower third, 5 opposite sepals 4 mm long, 5 opposite petals 3 mm long, anthers ovate, 1 mm long, connective thick, obovate, with rounded apex, thecae elliptic, attached below middle. Disc annular, membranaceous, 0.8 mm high, margin denticulate. Ovary ovoid, 1.5 mm high, glabrous, 5-sulcate; style 2.5 mm long, stigma stellate-capitate, 5-lobed. Drupe oblong-ellipsoid, 3.5–5.5 × 2.5–3 cm, the apex mucronate-apiculate, base with persistent calyx remaining attached; exocarp 1 mm thick, smooth, glabrous, coriaceous when dry; endocarp woody, irregularly 5 septate, bullate, the cavities resinous; seeds 1–3 cm, oblong, ca 1.2 cm long.

Distribution and habitat. Endemic to Isla del Coco, Costa Rica, where it is the dominant tree on the island (Fig. 10.20). The population status of this species was studied in detail by Acosta-Vargas (2015) who showed it to be the most important tree species on the island but that it has a low rate of regeneration.
Phenology. Flowering from March to August, fruiting from June to October.
Local names and uses. Costa Rica: *Palo de hierro.*
Conservation status. Least concern (LC) even though distributed in a small area.
Illustrations. None found.

Selected specimens examined. COSTA RICA. PUNTARENAS: Isla del Cocos, between Punta del Cerro Pelón and Punta de Cerro Iglesias, 5°31′56″N, 87°04′ 41″W, 20 Jun 1997 (fl, imm fr), *Rojas 3631*(INB, K, MO, NY); Isla del Cocos, *Pittier 16,260* (US); behind Wafer Bay, 15 m, 5°32′31″N, 87°03′37″W, 15 m, 14 Aug 1973 (fl), *Dressler 4467* (MO, NY); Bahía de Wafer, 5°32′08″N, 87°03′39″W, 150 m, 1 Aug 1981, *Gómez-Laurito 6934* (MO); Parque Nacional Isla del Coco, 5°32′24″N, 87°03′00″W, 634 m, 13 Jun 1994 (fl), *Lépiz 328* (CR, MO); SE of Chatham Bay, 5°33′00″N, 87°01′48″W, 9–11 Apr 1979 (fl), *Foster 4126* (K, MO, NY); SW of Chatham Bay, 100–300 m, 5°33′03″N, 87°03′03″W, *Wiggins 18,988* (MO); Camino de Bahía Chatham, 1–100 m, 5°32′24″N, 87°03′00″W, 14 Aug 1996 (fl), *González 1149* (INB, K, MO, NY).

7. **Sacoglottis maguirei** Cuatr., Contr. U.S. Natl. Herb. 35(2): 163. 1961; Cuatrecasas. & Huber, Fl. Venez. Guayana (Steyermark et al.) 5: 636. 1999. Type. Venezuela. Amazonas: Cerro Yapacana, 1200 m, 3 Jan 1951 (fl), *B. Maguire* et al. *30,693* (holotype, US 00101221; isotypes, NY 388410, VEN 44513).

Small trees, the young branches minutely hirtellous, lenticellate. Leaves with petioles 3–6 mm long, sulcate above, thickened at base; laminas obovate, 2–3.5 × 1.5–3 cm, coriaceous, glabrous, abruptly cuneate at base; rounded or obtuse at apex, margins obtusely crenate; midrib prominulous above, prominent beneath; primary veins inconspicuous on both surfaces, reticulate venation prominulous above. Inforescences axillary, shorter than leaves; peduncle 0.8–1.5 cm long, stout, striate, minutely hirtellous; branches dichotomous, short, minutely hirtellous; bracts minute but persistent. Sepals rounded, 0.5 mm long, hispid-pubescent. Petals oblong, acute, hispidulous, 1.3 mm long. Stamens 10, filaments glabrous, complanate, papillose, connate for lower half, 5 longer, 1.6 mm alternating with 5 shorter 1.2 mm long. Anthers oblong, 0.5 mm long, connective thick, subobtuse, thecae oblong. Ovary ovoid, glabrous; style 0.4 mm long, stigma capitate, 5-lobed. Young fruit globose, glabrous, with numerous resinous cavities.

Distribution and habitat. Endemic to Parque Nacional Cerro Yapacana, Venezuela, at above 1000 m. Known only from the type of collection (Fig. 10.20).
Phenology. Collected in flower in January.

Local names and uses. None recorded.
Conservation status. Data deficient (DD).
Illustrations. Cuatrecasas and Huber (1999) fig. 540.

8. **Sacoglottis mattogrossensis** Malme, Ark. Bot. 22A (7): 9. 1928; Cuatr., Contr. U.S. Natl. Herb. 35, 2: 181. 1961; Cuatrecasas & Huber, Fl. Venez. Guayana (Steyermark et al.) 5: 636. 1999. Type. Brazil. Mato Grosso: Santa Ana da Chapada, 10 Aug 1902 (fl), *G. O. A. Malme 2237* (holotype, S-R7853; isotypes, LD 1771937, S-12-22,748, US 00101220).

Sacoglottis guianensis forma *sphaerocarpa* Ducke, Arch. Jard. Bot. Rio de Janeiro 3: 178. 1922. Type. Brazil. Pará: Monte Alegre, 26 Mar 1928, *J. G. Kuhlmann RB2128* (holotype, RB 00539067; isotypes, S R-9881, U, US 00101227).

Sacoglottis mattogrossensis var. *mattogrossensis* forma *glabra* Cuatr. Contr. U.S. Natl. Herb. 35(2): 182. 1961. Type. Brazil. Pará, Faro, 11 May 1911 (fl), *A. Ducke MG 11653* (holotype, US 00101219; isotypes, BM, IAN 40651, MG 011653, P 01903247, RB 00539070).

Sacoglottis mattogrossensis var. *subintegra* (Ducke) Cuatr., Contr. U.S. Natl. Herb. 35(2): 183. 1961; *Sacoglottis guianensis* var. *subintegra* Ducke, Arq. Inst. Biol. Veg. 4: 27.1938. Type. Brazil. Amazonas, Manaus, 2 Aug 1937, *A. Ducke RB 23820* (holotype, RB 00539068, isotypes, INPA, K 000407310, K 000543567, MG, P 01903248, S-R9883, U 0002489, US 00101226. US 00921447), probable isotypes, 2 Aug 1937 *Ducke s.n.* (F V0092286, SP 000788).

Sacoglottis mattogrossensis var. *subintegra* forma *puberula* Cuatr., Contr. U.S. Natl. Herb. 35(2): 183. 1961. Type. Brazil. Pará: Canutá, *A. Ducke MG 16286* (holotype, MG016286_1; isotypes, MG 016286_2, BM 000559606).

Trees up to 20 m tall, the young branches glabrous or hirtellous, lenticellate. Leaves with petioles 5–8 mm long, terete, puberulous or glabrescent, thickened at base; laminas oblong-elliptic or elliptic-lanceolate, 5–15 × 2–6.5 mm coriaceous, flexible, glabrous, sometimes with two glands at base, rounded, obtuse or cuneate at base, acuminate or cuspidate at apex, the acumen 5–15 mm long often curved, margins obtusely serrate or almost entire; midrib prominulous above, prominent and glabrous or with sparse hairs beneath; primary veins 8–10 pairs, thin, prominent on both surfaces, reticulate venation prominulous and conspicuous on both surfaces. Inflorescences axillary, cymose-paniculate, dichotomous, shorter than the leaves; peduncle 2–5 cm long, stout, striate, hirtellous, branches short, hipidulous; bracts persistent, clasping, ovate-triangular, 0.6–1 mm long, acute, minutely puberulous, ciliate; pedicels thick, 0.7–1 mm long, glabrous. Sepals ovate, rounded at apex, thick, ca. 0.6 mm long, glabrous except for ciliate margin. Petals linear, narrowed toward apex, 4 × 1–1.2 mm, cochlear, glabrous or puberulous. Stamens 10, glabrous, filaments flattened, long ones 3.2 mm long alternating with shorter ones 2.5 mm long, connate at base into a tube, anthers ovate, 0.7–0.9 mm long, connective thick, acute or subacute, thecae oblong. Disc annular, 0.4 mm high, dentate.

Fig. 10.19 Distribution of three species of *Sacoglottis*

Ovary ovoid, glabrous, 0.8 mm high; style 2–3 mm long, stigma short, capitate, 5-lobed. Drupe globose 1.7–2.8 cm diam., exocarp resinous, 1–2 mm thick, coriaceous and granular when dry; endocarp woody, smooth, slightly bullate and sulcate, resinous-lacunose within, usually 1-seeded or 2–3.

Distribution and habitat. A widespread species of forest on *terra firme* from the Guianas, Amazonian Brazil to central Brazil and in the Atlantic rainforest and in sandy restingas and widespread in Bolivia (Fig. 10.19).

Phenology. Mostly flowering from July to November and fruiting mainly from November to June.

Local names and uses. Bolivia: *Chaquillo;* Brazil: *Achuá.* The fruit is edible.

Conservation status. Least concern (LC).

Illustrations. Cuatrecasas and Huber (1999), fig. 540. 1999; Cuatrecasas (1961), pls. 22, 23, fig. 37n–p, 38i–k.

Selected specimens examined. COLOMBIA. GUAVIARE: Mun. San José del Guaviare, Verda Caracol, 2°25.765 N, 72°47.749 W, 240 m, 21 Mar 2000 (fr), *Ramírez* et al. *7318* (COAH). **GUAINÍA**: Serrania de Naquen, Mun. Maimachi, 2°13′N, 68°14′W, 700 m, 3 Aug 1992 (fr), *Córdoba* et al. *226* (COAH). **VICHADA**: Gaviotas, Cumaribo, 17 Mar 1973 (fr), *Cabrera 2778* (COL). **VENEZUELA. AMAZONAS:** Isla del Raton, Río Orinoco, 90 m, 5°2′N, 67°46′W, 23 Nov 1965 (fl), *Breteler 4798* (COL, F, WAG). **GUYANA**. Takutu-U. Essequibo Region, Rupununi River 2.5 km S. of Karanambo Ranch, 3°44′6″N, 59°18′24″W, 90 m, 19 Feb 1992 (fr), *Hoffman 1002* (K, MO, NY, US). **FRENCH GUIANA.** Riviére Oyapock, islets Yacarescin, 3.161666, −52, 10 Dec 1965 (fr), *Oldeman 1725* (K, P). **PERU. MADRE DE DIOS**: Manu Prov, Puerto Maldonaldo, Los Amigos Biological Station, 7 km from mouth of Río Los Amigos, −12.5682, −70.1, 1187 m, 1 Apr 2004 (fr), *Macedo 1293* (K). **BRAZIL. AMAZONAS:** Manaus, Reserva Ducke, 21 Nov 1966, *Prance* et al. *3614* (INPA, K, MO, NY, US)**;** Mun. Humaitá, 115 km from Humaitá, Reserva Indigena Tenharim, 6°58′S, 62°08′W, 13 Apr 1985 (fr), *Cid Ferreira 5446* (F, INPA, K, MBM, MO, NY, RB); Tres Casas, Mun. Humaitá, 7 Oct 1934 (fl), *Krukoff 6506* (GH, K, NY, U); Distr. Agropecuário Reserva 1501, −2.315, −59.685, 26 Oct 1991, *Oliveira 182* (INPA, NY); Reserva Ducke, Manaus, −2.883, −59.966, 3 Mar 1964 (fr), *Rodrigues & Osmarino 5729* (HEPH, INPA, RB); Reserva Ducke, Manaus-Itacoatiara Km 26, *Ribeiro* et al. *1881* (BM, INPA, K, MBM, UB, UEC); Rio Japurá, Matipiri, 21 Apr 1986 (fr), *Cid-Ferreira 7084* (F, HST, INPA, K, MO, NY); Manaus-Porto Velho road, km 220, 24 Mar 1974 (fr), *Campbell P20892* (INPA, NY). **RORAIMA:** Serra dos Surucucus, Alto Alegre, Feb 1995 (fr), *Prance* et al. *9966* (INPA, NY, US); Parque Nacional do Viruá, Caracaraí, 1.8161, −61.1281, 13 Mar 2007, *Cid Ferreira 13,933* (INPA). **PARÁ:** Reserva Biológica. Serra dos Carajás, Parauapebas, 6 Jan 1992 (st), *Santos 563* (MO); Ramal da Comunidade São Paulo, Juruti, −2.152, −56.092, Nov 2007 (fl bud), *Amaral 3144* (INPA); Trail to Campo do Bentivi, Quatipurú, −0.8969, −47.0053, I Jan 1963 (fr), *Rodrigues 5178* (IFAM, INPA); BR22 Capanema – Maranhão, Km 64, Piritoro, 4 Nov 1965 (fr), *Prance & Pennington 1965* (COL, F, IAN, K, NY, U). **MARANHÃO:** Fazenda Madera Riberão, Urupuchete, 7°11′S, 47°25′W, 12 Jul 1993 (fl bud), *Ratter* et al., *6810* (INPA, K, NY); Morro de Baleia, c 2 km toward Carolina from Pedra Caída, 7°03′S, 47°27′W, 1 Jul 1993 (K); São Luíz I., Feb-Mar 1939 (fr), *Fróes 11,742* (MO, NY); Parque Estadual do Bacanga, São Luís, −2.599, −44.276, 6 Jun 2012, *Rodrigues 504* (SLUI); Fazenda Madeira, Riberão Urupuchete, 17 km from Carolina, −7.1833, −47.4167, 12 Jul 1963, *Irwin 6010* (UB). **RONDÔNIA.** Reserve of Samuel Hydroelectric Dam, Porto Velho,

−8.7619, −38.2833, 13 Jun 1986 (fr), *Cid-Ferreira 7030* (INPA). **PIAUÍ**. PN da Serra das Confusões, Caracol, −9.9221, −43.485, Sep 2007 (st), *Santos 1452* (UEC, UEFS). **TOCANTINS**: W of Filadélfia, Serra Jaccouba, 2 Aug 1964 (fl), *Prance & Silva 58,544* (COL, F, K, NY, UB) **MATO GROSSO**: Mun. Gaúcha do Norte, Road to Pontual, 13°06′S, 53°23′W, 16 Aug 2000 (imm fr), *Ivanuskas 4367* (ESA, RB); Reserva do Cabaçal, Chapada dos Pareceis, 15°59′24" S, 58°22′06″W, 657 m, 2 Sep 2015 (fl), *Andrade-Lima 79–8074* (INPA, RB, UEFS); Posto Indígena Cap. Vasconcelos, Rio Tuatuari 19 May 1958 (fl), *Andrade-Lima 58–3155* (IPA, K, RB); Km 264 Xavantina-Cachimbo, 12°49′S, 51°46′W, 15 Nov 1967 (imm fr), *Philcox* et al. *3060* (K, MO, NY, RB); Mun. Novo Mundo, Parque Estadual Cristalino, 9°29′S, 55°48′W, 7 Feb 2008 (st), *Sasaki 2212* (K); margin of Garapu airstrip, Canarana, −13.55, −52.268, 30 Sep 1964 (fl), *Prance & Silva 59,189* (F, NY, S, UB); Rio Alegre, Vila Bela de Santissima Trinidade, −15,008, −59.950, 17 Nov 1996 (fr), *Hatschbach 65,644* (ASU, MBM, MO, NY): Gaúcha do Norte, −13.2422, −53.0797, 7 Oct 1999, *Ivanuskas 4125* (ESA); 125 km S. of Sinop, 10 km from Primavera, Nobres, −12.9166, −55.8833, 18 Sep 1985 (fl), *Cid-Ferreira 6124* (INPA); Mun. Nova Araguarina, Fazenda Nova Canaã, 9 Oct 2004, *Fernandes-Bulhão & Stefanello CFB527* (RB); Nova Ubiratã, 12°47′52" S, 54°45′13″W, 6 May 1997 (fr), *Nave 1433* (RB). **MATO GROSSO DO SUL**: Mun. Nobres, 18 km S. of Rio Celeste, 1.5–2 km E of BR163, 12°27′S, 55°40′W, 19 Sep 1985 (fl), *Thomas* et al. *3877A* (F, MO, NY). **DISTRITO FEDERAL**: Road to Belo Horizonte, 30 km S of Brasília, 15 Jul 1966 (fr), *Irwin* et al. *18,227* (MBM, MO, NY); Brasília Botanical Garden, 1025–1150 m, 15°52′S, 47°51′W, 16 Sep 1993 (fl), *Ramos* et al. *99* (HEPH, RB). **RIO GRANDE DO NORTE**: Campo Experimental de Produção do Jiquí-EMPARN, Parnamirim, −5.929, −35.187, 30 Jun 199, *Cestano 174* (IPA); RPPN Mata Estrela, Tabuleiro, Baia Formosa, −6.374, −35.023, 30 Jul 2005 (fl bud), *Lourenço 102* (JPB): Santuário Ecológico de Pipa, Tibau do Sul, −6.1867, −35.0919, 14 Sep 1999 (fr), *Almeida 51* (PEUFR). **PARAIBA**: João Pessoa, −7.115, −34.863, 1 Jan 2000, *Barbosa 1365* (JBP); Cabadelo, −6.981, −34.833, 1 Jan 2000, *Pontes 325* (JPB); Mata dos Postes, Caaporã, −7.461, −34.969, 16 Oct 2012 (fl), *Gadelha Neto 3411* (JPB, MO, NY, RB); Rio Cabelo, Mangabeira, João Pessoa, −7.115, −34.8631, 9 Dec 2010 (fr), P*ereira & Chagas 110* (JPB, MBM, NY, RB). **PERNAMBUCO**: Ponte dos Carvalhos, Cabo de Santo Agostino, −8.286, −35.035, 6 Jun 1997 (fr), *Sacramento 105* (PEUFR); Mun. Ipojuca, RPPN Nossa Sra de Oiteiro de Maracaipe, 8°31′11.5" S, 35°01′12″W, 27 Oct 2008 (fl), *Oliveira 74* (RB); Restinga de Ariquindá, Tamandaré, −8.714, −35.095, 14 Jan 2003 (fr), *Lira* et al. *474* (IPA, PEUFR); Ponta da Pedras, Goiana, −7.560, −35.002, 26 Nov 2013 (fr), *Silva 422* (HST, HUEFS, PEUFR, RB); Mata de Dois Irmãos, Módulo 2, Recife, −8.053, −34.881, 20 Sep 1989, *Guedes 2126* (ALCB, PEUFR); Suape, Cabo de Santo Agostino, 1 Jan 1976 (st), *Andrade-Lima 76–9145* (IPA); Ilha de Itamaracá, Dec 1847, (BM, IPA); Iguarassu, 1 Nov 1887, *Ramage s.n.* (BM). **ALAGOAS**: Povoado Olho da Água, Junqueiro, −9.925, −36.475, *Santos 25* (HUEFS, MAC), Junqueiro, Olho daÁgua, 9°54′14"S, 36°25′30″W, 17 Jul 2005 (fr), *Santos 156* (RB); Peroba, Maragogi, −9.0227,

−35.3736, 7 Jul 1993 (fr), *Bayma 71* (MAC). **SERGIPE**: RPPN Mata do Crasto, Santa Luzia do Itanhy, −11.385, −37.425, Feb 2011 (fl, fr), *Farinaccio 780* (ASE, JPB); Povoado Convento, Indiaroba, −11.4812, −37.4017, 22 May 2013, *Matos 276* (ASE); 2.5 km from Distrito de Crasto, road to Sant Luzia do Itanhy, −11.3508, −37.4483, *Amorim 1482* (ASE, CEPEC, NY, RB, US); 14 Apr 1997 (fr), *Landim 1206* (ASE, CEPEC); Japoatã, 10.346176, −36.798852, 24 Apr 2014 (fr), *Pereira 68* (RB). **BAHIA**: Mun. Una, road to Distrito de Pedras, 13 Apr 1993 (fr), *Amorim* et al. *1184* (CEPEC, K, MO, NY, RB); Fazenda Bolandeira, 6 km N of Ilha de Comandatuba, Una, −15.305, −39.021, 27 Apr 2004, *Amorim 4048* (CEPEC, NY, RB); Estrada Serrra Grande, Itacaré, −14.277, −38.996, 1 Sep. 1993 (fl); *Amorim 1319* (CEPEC, MBM, NY); Mun. Ilhéus, 16.6 km S of Serra Grande, just N of Tulha, 14°41′S, 39°09′W, 8 May 1992 (fr), *Thomas 9224* (CEPEC, MO, NY, RB); Litoral Norte, Sauípe, Mata de São João, −12.561, −38.283, *Guedes 7030* (ALCB, CEPEC); Mun. Conde. BA099. Km127, Mata do Bu, 12°01′21" S, 37°41″35″W, 14 Jun 2000 (fr), *Alves 2014* (CEPEC); Prado, PN de Descobrimento, 17°11′S, 39°20′W, 10 Nov 2009 (fl), *Matos 1964* (CEPEC). **ESPÍRITO SANTO**: Itaúnas, Conceição da Barra, −18.593, −39.732, *Folli 5389* (CEPEC, CVRD); Reserva Biológica de Comboio, Estrada Porteira, Linhares, −19.3911, −40.0722, *Folli 1121* (CEPEC, CVRD, RB); Mun. Aracruz, Barra do Riacho, 15 Jan 1990 (fl), *Vinha 1013* (RB). **SÃO PAULO**: Santos, −23.9704, −46.3623, 18 Oct 1875 (fr), *Mosen 3477* (S). **RIO DE JANEIRO**: Gavea, Aug 1916, *Frazão RB8118* (K, RB, US); Porto Estrella, −22.7192, −43.21605, 23 Aug 1925, *Ducke & Kuhlmann RB19165* (RB, US). **BOLIVIA. BENI**: Vaca Diez, Serranía San Simón, 205 m, 14°25′S, 62°03'.W, 19 Jul 1993 (fr), *Quevado* et al. *993* (MO, NY, USZ). **COCHABAMBA**: Carrasco, km 240 Santa Cruz-Villa Tunari, 17°00′S, 64°37′W, 10 Jul 1989 (fr), *Smith* et al. *13,860* (LPB, MO, USZ); Chapare, Campo Surubí, 16°30′S, 65°30′W, 15 Nov 1998 (fr), *de la Barra* et al. *847* (BOL, MO); Carrasco, Entre Rios Distr. 17°23′55"S, 64°32′36″W, 2 Mar 2006 (fr), *Colque & Mendoza 443* (fr), (RB); Estac. Experimental, Valle del Sajita, 280 m, 17°0′S, 64°46′W, 2004 (fl), *Mendoza & Arroyo 1573* (NY). **LA PAZ**: Abel Iturralde, Parque Nacional Madidi, 950 m, 13°53′S, 68°09′W, 17 Nov 2001 (imm fr), *Macia* et al. *6443* (LPB, MA); Franz Tamayo, Calabatea, N of road Apolo-Charazani, 15°00′S, 68°26′W, 30 Apr 2005 (fr), *Fuentes* et al. *7637* (LPB, MO, NY, USZ); Franz Tamayo, Madidi Region, Santo Domingas, Río Turiapo, 1433 m, 14°46′S, 68°37′W, 23 Aug 2010 (fr), *Fuentes* et al. *17,102* (LPB, MO). **PANDO**: Limoero, 250 m, 11°12′S, 69°20′W, 28 Aug 1989 (fl), *Prescott 181* (MO); 54 km SW of Cobija, Triunfo, 250 m, 11°10′S, 69°07′W, 17 Jul 1988, *Pennington* et al. *5* (F, K, MO, NY); Manuripi, 11°44′S, 67°59′W, 200 m, 14 May 1994 (fl), *Jardim 743* (F, MO, NY, USZ). **SANTA CRUZ**: Reserva Ecológica El Refugio, 30 km S. of Florida, 200 m, 14°43′S, 61°09′W, 17 Oct 1994 (fl), *Killeen* et al. *6929* (F, K, L, MO, NY, USZ); Velasco, San José de Campamento, 230 m, 15°09′S, 60°59′W, 30 Aug 1996 (fl, imm fr), *Guillén & Marmañas 4622* (MO, USZ); Velasco, Parque Nacional Noel Kempff-Mercado, Lago Caimán, 220 m, 13°36′S, 60°56′W, 20 Jan 1997 (imm fr), *Garvizu & Fuentes 366* (MO, USZ); Ichilo, Dec 1995, 17° S, 64°20′W, 24 Dec 1995 (fr), *Nee 46,502*

(MBM, NY); Parque Nacional Noel Kempff Machado, Campamento La Torre, 250 m, 13°39′14" S, 60°49′50″W, 21 Nov 1993 (fr), *Killeen* et al. *6158* (K, MO, NY, USZ); Velasco, Parque Nacional Noel Kempff Mercado, 13°54′22" S, 60°48′52" S, 150 m, 12 May 1994 (fr), *Mostacedo* et al. *1660* (MO, NY, USZ).

This species is easily confused with *Sacoglottis guianensis* and *S. amazonica* but differs mainly in the globose fruit and in the thinner more flexible and prominently nerved leaves. Cuatrecasas (1961) divided this species into various varieties and forms based on the presence or absence of pubescence on the petals and the leaves. These taxa have no geographical distinction and based on more recently collected material there is considerable variation in these characters and so they are not recognized here.

The fruits are dispersed by bats and birds and the flowers pollinated by bees (Kuhlmann 2012).

9. **Sacoglottis ovicarpa** Cuatr., Trop. Woods 96: 39. 1950; Contr. U.S. Natl. Herb. 35(2): 168. 1961. Type. Colombia. Valle del Cauca: Bahía de Buenaventura, Quebrada de Aguadulce, 24 Feb 1946 (fr), *J. Cuatrecasas 19,998* (holotype, F V0060791; isotypes, COL 0000018153 (fruit), COL 000001814, COL 000001815, G 00368590, US 1206766, VALLE, WIS).

Large trees, the young branches glabrous or slightly puberulous. Leaves with petioles 4–10 mm long, terete, robust, thickened toward base; laminas elliptic or ovate-elliptic, 9–20 × 3.5–11.7 cm, rigid chartaceous or coriaceous, glabrous above, glabrous or sparsely puberulous beneath, rounded or obtuse at base, abruptly cuspidate at apex; margins slightly crenate almost entire, with one pairs of gland at base of lower surface; midrib prominulous above, prominent beneath; primary veins 8–11 pairs, prominulous on both surfaces; reticulate venations conspicuous, prominulous. Inflorescences axillary, much shorter than leaves, cymose-paniculate, dichotomous, the peduncle obsolete, branchlets slightly puberulous, reddish, bracts ovate-triangular, acute, caducous. Sepals suborbicular, ca 1.5 mm long, glabrous, the margins minutely and irregularly ciliate. Petals 5, oblong, 3 mm long, 1 mm broad, glabrous. Stamens 10, united on lower half, glabrous, filaments compressed-flattened, alternately 1.5 and 2 mm long, anthers ovoid, ca 0.6–0.7 mm long, connective thick, rather obtuse, thecae elliptic, ca 0.3 mm long. Ovary ovoid, glabrous; style ca 1 mm long, glabrous, stigma capitate. Drupe ovoid, or ellipsoid-ovoid, 5–5.5 × 2.5–4.5 cm; exocarp fleshy, hard and coriacous when dry, 6–7 mm thick; endocarp woody, smooth, irregularly 5-septate, with large, rounded abundant resinous cavities; seeds 1–2.

Distribution and habitat. Primary forest of Panama and Pacific coastal Colombia (Fig. 10.20).

Phenology. Collected in fruit from January to July and in flower from March to November.

Local names and uses. Panama: *Conocillo*. Colombia: *Corosillo, chanu, chanusillo*.

Conservation status. Least concern (LC).

Illustrations. Cuatrecasas (1961) figs. 34 h, 35 d–e.

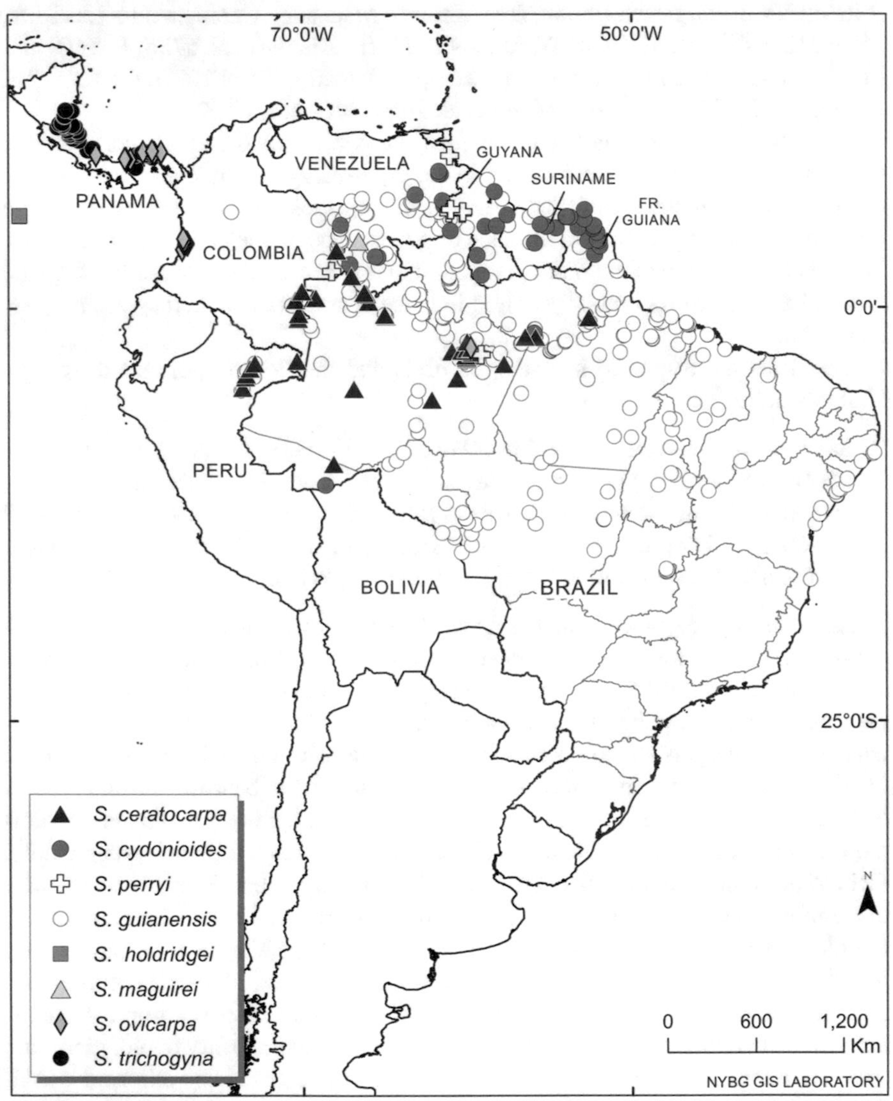

Fig. 10.20 Distribution of eight species of *Sacoglottis*

Additional specimens examined. PANAMA. BOCAS DEL TORO: 6 May 2007, *Daguerre* et al. *219* (MO, PMA). **COLÓN**: Santa Rita Ridge, 350 m, 9°23′N, 79°39′30″W, 26 Jan 1986 (fr), *de Nevers 6874* (MO); Donoso, Teck Cominco Petaquik, 130 m, 8°49′58″N, 80°38′40″W, 20 Jun 2008 (fl bud), *McPherson 20,561* (MO). **PANAMÁ**: Chepo, 9.29722, −78.94306, *Kennedy* et al. *2437* (MO, U). **SAN BLAS**: El Lano-Carti road Km 19.1, 350 m, 9°20′N, 78°58′W, 14 Jun 1985 (fr), *de Nevers 5844* (MO); Río Diablo, Nargau, 40 m, 9°23′N, 78°55′W, 11 Feb 1985 (fr),

Herrera et al. *1726* (BM, MO); Sendero Nusagandi, El Llano-Carti road Km 18, 350 m, 9°20′N, 78°58′W, 3 Jul 1994 (fr), *Galdames* et al. *1296* (F); road from El Llano to Carti-Tupile, 300–500 m, 30 Mar 1973 (fl), *Liesner 1306* (MO); in front of Isla Narganá, ribera de Río Diabo, 9°22′N, 78°34′W, 9 Aug 1994 (fl), *Galdames 1462* (BM, MO). **COLOMBIA. VALLE DEL CAUCA**: Bahía de Buenaventura, Quebrada de San Joaquín, 22 Feb 1946, *Cuatrecasas 19,927* (COL, F V0090160, VALLE, paratypes); Río Cajambre, Barco, 5–80 m, 28 Apr 1944 (fr), *Cuatrecasas 17,226* (COL, F V0090162, VALLE, paratypes); Buenaventura, Bajo Calima, 100 m, 3°55′N, 77°00′W, 25 Nov 1986 (fl bud), *Monsalve 1321* (MO); Buenaventura, Bahia de Malaga, near mouth Quebrada La Sierpe, 0–20 m, 4°10′N, 77°15′W, 17 Feb 1983 (fr), *Gentry* et al. *40,437* (COL, MO, NY); Buenaventura, 3°56′N, 77°10′W, 230 m, Nov-Dec 1979 (fl, imm fr), *Van Rooden 435* (COL, MO, NY).

10. **Sacoglottis perryi** K. Wurdack & C. E. Zartman, Phytokeys 124: 103. 2019. Type. Guyana: Cuyuni-Mazaruni Region, Below escarpmentl of Kamakusa Mt., Powis Creek, 5°48′26.7″N, 60°14′6.9″W, 662 m, 20 May 2012 (fl), *K. Redden 7264* (holotype, BRG; isotypes K, NY, US 3694797). Fig. 10.16.

Trees 4–15 m, the young branches sparsely hirtellous, glabrescent. Leaves with petioles 5–10 mm long, sparsely pubescent, often slightly winged by confluent lamina; laminas elliptic, chartaceous, 4.5–8 × 2.5–3.2 cm, with sparse appressed hairs or glabrous beneath, rounded to subcuneate at base, markedly cuspidate at apex, the acumen 1–1.8 cm long, narrow, often curved, margins slightly crenate, setae present, sometimes with a pair of glands at base of lower surface; midrib prominulous above, prominent beneath, ending in a minute apiculate, glandular tooth; primary veins rather inconspicuous on both surfaces, 8–12 pairs; reticulate venation slightly prominulous, but inconspicuous. Inflorescences of axillary glomerules with 3–8 flowers, with very short peduncles to 1 mm long; bracts 1.8 × 1–1.5 mm, margin entire, early caducous. Flowers subsessile. Calyx cupular. Sepals 5, rounded, tightly clasping, 1–1.5 mm long, sparsely puberulent, margins short-hirtellous ciliate. Petals oblong-lanceolate, ca. 4 mm long, glabrous, quincuncial, reflexed at anthesis. Stamens 10, filaments flattened, connate for lower third, alternating 5 long, antesepalous and 5 short, antepetalous, anthers dorsifixed, oblong, glabrous, connective thickened, linear, thecae 2, lateral, narrowly elliptic. Disc 0.5–0.6 mm high, cupular, margin rounded with irregular lobes. Ovary globose, 1–1.2 mm high, 5 locular, densely pubescent; style slightly exceeding the stamens, stigma 5-capitate. Drupe oblong-ellipsoid 2–4 × 1.4–3 cm, smooth and glabrous on exterior, rounded at both ends; endocarp obscurely bullate on surface, with 3 rather inconspicuous longitudinal furrows, with numetrous lacunae; seeds 1–3 per fruit, 15 mm long.

Distribution and habitat. Highlands in Colombia, Guyana, and adjacent Brazil in white sand habitats beside streams (Fig. 10.20).
Phenology. Flowering in April and May, fruiting in October.
Local names and uses. Venezuela: *Nabaru kuaha* (Warao).
Conservation status. Least concern (LC).

Additional specimens examined. COLOMBIA. GUIANIA: Maimachi, Serranía del Naiquén, trail to Cerro Minas, 455 m, 2°12′N, 68°13′W, 9 Apr 1993 (fl bud), *Madriñán & Barbosa 977* (K). **AMAZONAS**: Río Cahuinarí, 22 km W of mouth, 300 m, 7 Sep 1988 (fl), *Sanchez S. 971A* (COAH). **VENEZUELA. DELTA AMACURO**: Dep. Antonio Diaz, between Jubusujuru & Nabasanuka, 9°17′N, 61°6′W, 2 Feb 1997 (fr), *Take 5* (INPA). **GUYANA**. Cuyuni-Mazaruni, Pakaraima Mts, Imbaimadai Creek, W of Imbaimadai, 503 m, 5°42′N, 60°18′W, 22 Jun 1986 (old fl), *Pipoly 7990* (INPA, K, MO, NY, US); 7 km N of Pariuma, 980–1060 m, 5°54′N, 61°2′W, 30 May 1990 (fl) *McDowell 2993* (K, MO, NY, US); Cuyuni-Mazaruni, Pakaraima Mts, Imbaimadai Creek, 1 km W of Imbaimadai, 16 May 1992 (fl), *Hoffman 1600* (COL, F, INPA, K, US); From Utshe River to Great Falls, Kamarang River, 850–975 m, 5°43′N, 61° 7′W, 26 May 1990 (fr), *McDowell 2920* (K, MO, NY, US); 8.6 km NE of Imbaimadai on Partang River, 600 m, 5°46′N, 60°15′W, 20 May 1992 (fl), *Hoffman 1755* (BM, INPA, K, MO, US); 8.6 km NE of Imbaimadai on Partang River, 625 m, 5°46′N, 60°15′W, 30 May 1992 (ff), *Hoffman 1745* (COL, K, US); Pakaraima Mts, Mt. Aymatoi, 1150 m, 5°55′N, 61° W, 16 Oct 1981 (fr), *Maas* et al. *5753* (K, MO, US). **BRAZIL. AMAZONAS**: Rio Urubu near ferry of Manaus-Itacoatiara road, 3 Apr 1967 (fr), *Prance* et al. *4736* (F, INPA, K, MG, NY, US).

Close to *Sacoglottis ceratocarpa* but differs in the smaller leaves and fruit and the inconspicuous primary veins and venation. Similar to some forms of *Sacoglottis mattogrossensis*, but differs in the smaller leaves, the consistently long cuspidate apex, the oblong rather than round fruits and the glomerulate inflorescence. In *S. mattogrossensis,* the inflorescence is sometimes clustered but has many more flowers and is more usually borne on long peduncles.

11. **Sacoglottis trichogyna** Cuatr., Cíencia (México) 27: 171. 1972; Zamora, Manual de Plantas de Costa Rica 6:12. 2007. Type. Costa Rica. Heredia: between San Miguel de Sarapiqui and Pital, 400 m, 22 Feb 1968 (fl), *L. Holdridge 5216* (holotype, US 00101215; isotypes, INB 0004140278, US 00810990).

Trees 15–35 m tall, the young branches glabrous, lenticellate. Leaves with petioles 8–11 mm long, plane above, terete, swollen at base, glabrous; laminas oblong-elliptic to narrowly elliptic, coriaceous, glabrous, 6–15 × 2.5–7 cm, cuneate or obtuse at base, apex acuminate, the acumen 2–16 mm long; margins subentire or shallowly crenate, with small gland near the base; midrib plane above, prominent beneath, puberulous when young soon glabrous; primary veins 8–10 pairs, prominent on both surfaces; reticulate venation prominulous. Inflorescences axillary, paniculate, much shorter than the leaves, 1–3 cm long, branches minutely hirsute; bracts minute, soon deciduous, triangular, 0.7–1 mm long; pedicels thick, glabrous, 1 mm long. Sepals 5, ca 1 mm long, orbicular, eglandular, margins minutely ciliolate. Petals 5, fleshy, glabrous, linear-oblong, acute at apex, 5–6 mm long, 1.2–1.6 mm broad. Stamens 10, glabrous, filaments thick toward base, thin above, connate at base for 0.6–0.9 mm, 5 opposite sepals 4–4.3 mm long, alternating with 5 opposite petals 3–3.2 mm long, anthers ovate-oblong, 1–1.3 mm long; connective thick, subacute, thecae ellipsoid, inserted laterally toward base, ca. 0.5 mm long. Disc surrounding ovary, annular, glabrous, minutely denticulate, 0.8 mm high. Ovary ovoid, 5- angled, 5-locular, villous-hispid, 1.5 mm high; style thick, 5-angular, 4 mm long; glabrous except for a few sparse hairs, stigma capitate, 5-lobed. Drupe oblong-subovoid, 3–5 × 2–2.8 cm, narrowed toward apex, exocarp 2–3 mm thick, glabrous, minutely granular; endocarp thick, woody, attenuate at both ends, sparsely bullate, with 5 prominent costae, joining at apex and base, 5 foramina near apex, with resinous cavities; seeds 1 or 2, oblong, 1 cm long.

Distribution and habitat. Lowland rainforests of Central America, also common in secondary forest, Nicaragua to Panama (Fig. 10.20).
Phenology. Flowering from January to April, fruiting from June to August.
Local names and uses. Nicaragua: *Rosita*; Costa Rica: *Campano, chiricano, colmena, danto plombillo, erizo, terciopelo, titor;* Panama: *Conocillo, corocito.*
Conservation status. Least concern (LC).
Illustrations. Zamora (2007), p. 13.

Selected specimens examined. **NICARAGUA. ATLÁNTICO SUR**: Between Puerto Viejo and Manzanillo, 9°37′N, 82°41′W, 100 m, 18 Jan 1992 (fr), *Hammel 18,390* (MO); Caño Serrano, 40 km S. of Nueva Guinea, 1°36′N, 84°20′W, 80–100 m, 29 Oct 1984 (fl), *Sandino 5008* (MO); Río Ramón, Salto Orapéndula, 11°57′N, 84°17′W, 15–25 m, 29 17 May 1978 (fr), *Stevens 8991* (MO). **RÍO SAN JUAN**: Boca de Sábalo, 11°02′N, 84°28′W, 70–100 m, 14 Mar 1985, *Moreno 26,822* (MO); Río Indio, 2 km above Caño Negro, 11°02′N, 83°53′W, 10 m, 20 May 2002 (imm fr), *Rueda* et al. *16,996* (MO). **ZELAYA**: N of Talolinga, 19 Aug 1983 (fl), *Sandino 4510* (BM). **COSTA RICA. ALAJUELA**: Llanua de San Carlos, Boca Tapada, 10°42′N, 84°10′W, 20 m, 22 Jun 1993 (fr), *Chavarria* et al. *742* (CR, MO). **HEREDIA**: Estación Biológica La Selva, confluence of Río Sarapiqui and Río Puerto Viejo, 10°25′N, 84°00′W, 50–80 m, 3 May 1987 (fr), *Grayum* et al. *8299* (MO); Sarapiqui, 10°33′N, 84°03′W, 200 m, 11 Jan 1997, *Hammel 20,631* (INB, MNCR, MO); Parque Nacional Braulio Carillo, Estación Magassay, 200 m, 10°24′18″N, 84°03′30″W, 5 Jul 1990 (fr), *Acevedo 115* (CR, K): **LIMÓN**: Cerro Coronel, E of Laguna Danto, 10°41′N, 83°38′W, 20–170 m, 16 Jan 1986 (fr), *Stevens 23,812* (MO); Pococi, 10°11′S, 83°51′W, 500 m, 12 Oct 1999, *Acosta 69* (MO); SW ridge of Cerro Coronel, S of Río Colorado, 10°40′30″N, 83°39′30″W, 10–80 m, 17–18 Sep 1986 (fr), *Davidse & Herrera 31,480* (MO, PTBG). **PANAMA. COLÓN**: Santa Rita Ridge, 9°19′N, 79°47′W, 50 m, 10 Aug 1971 *Lao* et al. *5* (F, MO, PMA, SCZ); Salud, 120 m, *Lao & Holdridge 193* (MO, PMA). **PANAMÁ**: El Llano Carti road, 10.8 km from Pan American Hwy, 9°16′N, 78°55′W, 350 m, 27 Dec 1974 (fr), *Mori & Kallunki 4114* (LD, MO, RB); Alto de Paore, Parque Nacional Chagres, 783.6 m, N1025076.65, E672241.83, 15 Apr 2004 (fl), *Pérez 1248* (BM, NY). **SAN BLAS**: El Llano Carti road Km 26.5, 9°25′N, 78°58′W, 10 Apr 1985 (fl bud), de *Nevers* et al. *5291* (MO); El Llano Carti road 14.5 km from Pan American Hwy, 9°20′N, 78°58′W, 350 m, 17 Jun 1986 (fr), *McPherson 9511* (MO).

Hartshorn (2010) noted that the seeds of this species take 18–24 months to germinate. He hypothesized that this is linked to seed dispersal by flotation.

10.7 Schistostemon

Schistostemon (Urb.) Cuatr., Contr. U.S. Natl. Herb. 35(2): 146. 1961.*Saccoglottis* subgen. *Schistostemon* Urb., Fl. Bras. (Martius) 12 (2): 443, 445.1877; *Saccoglottis* Sect. *Schistostemon* (Urb.) Reiche, Nat. Pflanzenfam. (Engler & Prantl) 3 (4): 37. 1890; *Sacoglottis* Sect. *Schistostemon* (Urb.) Winkl. In Engl. & Prantl, Nat. Pflanzenfam. 19a: 128. 1931. Type. *Schistostemon dichotomum* (Urb.) Cuatr.

Humirium Benth., London J. Bot. 2: 374.1843, pro parte.

Trees. Leaves alternate, petiolate or sessile, coriaceous, lamina margin usually conspicuously crenate-dentate. Inflorescences axillary or subterminal, paniculate with trichotomous or dichotomous branching; bracts persistent. Sepals 5, suborbicular, imbricate, united at base. Petals 5, free, thick, aestivation quincuncial, cochlear or contorted. Stamens 20, glabrous, unequal, of 3 types, 5 opposite antesepalous longer, trifurcate at apex and triantheriferous, 5 antepetalous shorter, entire, monantheriferous, 10 intermediate shorter, monantheriferous; filaments complanate and thickish, united into tube for half length; anthers ovate or ovate-lanceolate, connective thick, lanceolate or rarely obtuse, dehiscence by valves on the ventral side that open outward, thecae 2, unilocular, ellipsoid or oblong, attached on lower side. Disc cupular, dentate or rarely of 10 free scales. Ovary 5-locular, the locules uniovulate, glabrous or pubescent; carpels opposite the sepals; style thick, short, to 1 mm long, stigma capitate and 5-lobed. Drupe ellipsoid, smooth with thick fleshy or coriaceous exocarp; endocarp woody, bullate, with 10 narrow slightly apparent furrows filled with many globose resinous cavities; on germination, the embryos pushing away longitudinal broad valves; seeds 1 or 2.

Distibution. Nine species of northern South America mainly in Colombia, Venezuela, the Guianas and northern Amazonia with a single species occurring in the Atlantic rainforest.

Herrera et al. (2010) questioned the status of *Schistostemon* because there are no differences in fruit between it and *Sacoglottis.* They proposed returning *Schistostemon* to subgeneric status in *Sacoglottis.* However, the 10 undivided stamens of *Sacoglottis* compared to the 20 stamens, 5 being longer and episepalous and trifurcate at the apex bearing three anthers is to me an adequate character to maintain *Schistostemon* as a genus.

Key to Species of Schistostemon

1. Young branchlets minutely pilose.

2. Leaves oblong (4 times long as broad), crenate, minutely papillose sparsely pilose beneath..6. **S. oblongifolium**
2. Leaves ovate, lanceolate or attenuate-elliptic (less than 3 times as long as broad), subentire, glabrous, not papillose.

3. Connective of anthers ovate attenuate, acute; calyx pilose, the margin long-ciliate..7. **S. reticulatum**
3. Connective of anthers obtuse; calyx glabrous except for minutely ciliate margin.

4. Leaf nerves loosely reticulate above; petals 3–3.3 mm long...4. **S. fernandezii**
4. Leaf nerves inconspicuous; petals 4–4.5 mm long...1. **S. auyantepuiense**

1. Young branchlets glabrous.

5. Leaves subsessile, petioles 0–1 mm, laminas rounded or retuse at apex, margins subentire..8. **S. retusum**
5. Petioles 2–14 mm long; laminas acute or acuminate at apex, margins crenate.

6. Inflorescence 0.5–2 cm long, peduncles 0.4–0.5 cm long; bracts caducous; drupe 5–6 × 4.5–5.3 cm..............................9. **S. sylvaticum**
6. Inflorescence 1.5–6 cm long, peduncle 1–6 cm long; bracts persistent; drupe 2.6–4 × 2–2.8 cm (drupe unknown in *S. dichotoma* and *S. auyantepuiensis*).

7. Petals glabrous; disc of free or united scales.

8. Inflorescence 3.5–6 cm long, peduncles 3.5–6 cm long; disc of free scales..3. **S. dichotomum**
8. Inflorescence 1–3 cm long, peduncles 1–3 cm long; disc of united scales..5. **S. macrophyllum**

7. Petals pubescent at least on dorsal surface; disc annular, dentate.

9. Leaf laminas 7–19 × 3–7.5 cm; peduncles 1.5–3 cm...2. **S. densiflorum**
9. Leaf laminas 4–9.5 × 2.5–4.5 cm; peduncles 0.2–1 cm..1. **S. auyantepuiense**

1. **Schistostemon auyantepuiensis** Cuatr., Contr. U.S. Natl. Herb. 35(2): 151. 1961; Cuatrecasas & Huber, Fl. Venez. Guayana (Steyermark et al.) 5: 637. 1999. Type. Venezuela. Bolívar: Cerro Auyantepuí around Guayaraca camp, 1100 m, *Vareschi & E. Foldats 4673* (holotype, VEN 40611; isotypes, NY 03376852, US 101250).

Small shrubs, the young branches glabrous. Leaves with thick petioles 2–5 mm long; laminas ovate to ovate-elliptic, 4–9.5 × 2.5–4.5 cm, rigid-coriaceous, glabrous, obtusely cuneate at base, acuminate or acute at apex or rarely obtuse, margin slightly crenate, with sparse, minute glands, base with two conspicuous glands on lower surface; midrib prominent on both surfaces; primary veins 8–10 pairs, prominent on both surfaces, reticulate network of veins prominulous. Inflorescences axillary or subterminal, cymose-paniculate; peduncle 2–10 mm long, shortly hirtellous-pubescent, branches dichotomous or alternate, minutely pilose or glabrous; bracts persistent, triangular, acute, 1–1.5 mm long, margin ciliolate, dorsal surface papillose and sparsely pilose; pedicels thick, 0.5 mm long, pilose-hispid. Sepals 5, thick, rounded, 1.5 mm long, slightly connate at base margins ciliate minutely papillose and glabrous on dorsal surface. Petals 5, oblong, acute, thick, 4–4.5 × 1.5 mm, papillose and minutely pubescent on dorsal surface. Stamens 20, filaments thick, connate at base into short tube, 5 longer ones 3 mm opposite sepals, 5 medium length, 2.5 mm, ten smaller ones 2.2 mm; anthers ovate, 0.5–0.7 mm, glabrous, connective thick obtuse, thecae 2, oblong-elliptic. Disc cylindrical, membranaceous, dentate, 0.8 mm high. Ovary pyriform, glabrous, 5-locular, locule uniovulate; style 0.5 mm long, stigma capitate, 5-lobed. Drupe unknown.

Distribution and habitat. Endemic to Auyantepui in Venezuela (Fig. 10.22).
Phenology. Collected in flower in November.
Conservation status. Data deficient (DD).
Illustration. Cuatrecasas (1961), pl. 15.

Additional specimen examined. VENEZUELA. BOLÍVAR: Guayaraca, S. base of Auyan-tepuí, 950 m, 5°44′N, 62°32′W, 27 Nov 1982, *Davidse & Huber 22,804* (MO, US).

2. **Schistostemon densiflorum** (Benth.) Cuatr., Contr. U.S. Natl. Herb. 35(2): 160. 1961; *Humirium densiflorum* Benth., London J. Bot. 2: 347. 1843; *Sacoglottis densiflora* (Benth.) Urb., Fl. Bras. (Martius) 12 (2): 445. 1877. Type. Guyana. Quitaro River, *R. Schomburgk 543* (lectotype **here designated**, K 000407316; isolectotypes, BM 000796061, BR 0000005119004, E 00326635, F V0060769, F V0060770, F V0060771, FI-W, G 00368225, K 000407317, L 2120736, NY 00388390, OXF, P 01903254, P 01903255, U 0002493, US 00101202; B, lost, photo F). Fig. 10.21.

   Sacoglottis kaboeriensis Bakh., In Pulle, Recuil. Trav. Bot. Néerl. 37: 292.1940; Pulle, Fl. Suriname 3 (1): 420. 1941. Type. Suriname, Kaboerie, Corantyne River, J.W.G. tree No 138, 26 Jun 1916, *Boschwezen 2068* (holotype, U 0002491; isotype, US 101222).

Trees, the young branches glabrous, lenticellate. Leaves with petioles 5–12 mm long, semiterete, thickened at base; laminas ovate to ovate-oblong, 7–19 × 3–7.5 cm, rigid coriaceous, glabrous, rounded at base, acute or shortly acuminate at apex, margins slightly crenate, setae present, sometimes with a pair of glands at base of lower surface; midrib prominulous above, prominent and thick beneath, glabrous;

Schistostemon densiflorum

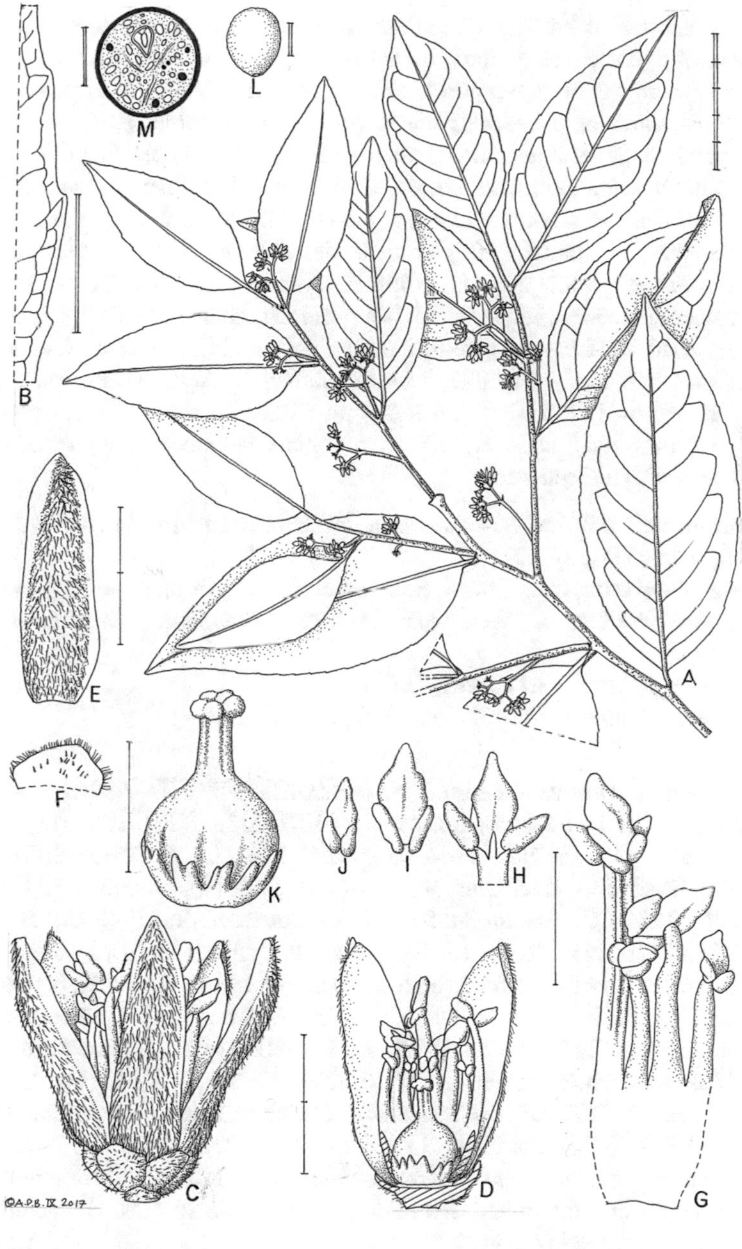

Schistostemon densiflorum del. Andrew Brown September 2017

Fig. 10.21 *Schistostemon densiflorum*. (**a**) Habit. (**b**) Abaxial leaf margin. (**c**) Flower. (**d**) Flower with 3 petals and sepals and part of staminal tube removed. (**e**) Petal. (**f**) Sepal. (**g**) Part of staminal tube, showing variable filament length. (**h**) Outer surface of long stamen, showing trifurcate tip. (**i**) Outer surface of anther from medium stamen. (**j**) Outer surface of anther from short stamen. (**k**) Pistil. (**l**) Fruit. (**m**) Fruit section. Single bar = 1 mm, graduated single bar = 2 mm, double bar = 1 cm, graduated double bar = 5 cm. (A–K from *Fanshawe 724* [FD 3460]; L from *Persaud 102*; M, from *Hohenkerk 404A*)

primary veins ca. 10 pairs, prominulous on both surfaces, reticulation of veins prominent. Inflorescences axillary, cymose-paniculate, dichotomous branching at base, alternate above, branches pubescent, peduncle 1.5–3 cm long, robust, slightly compressed, minutely pubescent; bracts persistent, clasping, 0.5–1,5 mm long, ovate-lanceolate, puberulous with ciliate margins; pedicels short, 0.1–0.3 mm long. Sepals ovate-orbicular, connate at base, 0.7–0.8 mm long, hispidulous, ciliate on margin. Petals linear, attenuate at apex, 3.5–4.5 mm long, hispidulous pubescent with reflexed hairs. Stamens 20, glabrous, filaments united into a tube for lower half, 2–3 mm long, 5 longer short trifurcate at apex, trianthiferous, others simple acute monoanthiferous, anthers ovoid, 0.7 mm long, connective fleshy, lanceolate, thecae ellipsoid. Disc membraneous, annular, 0.4 mm high, dentate. Ovary ovoid, glabrous, 1.2 mm high, 5-locular, locules uniovulate; style 1 mm long, stigma 5-lobed. Drupe ovoid-ellipsoid, 2.6 × 2.3 mm, exocarp smooth, coriaceous, glabrous, 1.5–2 mm thick; endocarp ellipsoid, slightly bullate, woody, resinous lacunose, usually with a single seed.

Distribution and habitat. River margins in Delta Amacuro Venezuela, Guyana, and Suriname (Fig. 10.22).

Phenology. Flowering from June to September, fruiting from September to May.

Local names and uses. Venezuela: *Ponsigué montañero*. Guyana: *Bukuria* (Arawak).

Conservation status. Least concern (LC).

Illustrations. Cuatrecasas (1961), pl. 19, figs. 31d, f, 33 e–g.

Additional specimens examined. VENEZUELA. **DELTA AMACURO**: E of Río Grande, ENE of El Palmar, 8.092183, −61.575613, 18 Feb 1964 (fr), *Berti 85* (COL, K, NY, U), 19 Nov 1964 (fl), *Berti 520* (INPA, IPA, MBM, MO). **GUYANA**. Essequibo River near Bartica, 4 Oct 1929 (fl), *Sandwith 374* (K, NY, RB, U), Sep 1886 (fl), *Jenman 2489* (BM, K); Bartica, 1 Jun 1942 (fl), *Fanshawe 724* (FD 3460), (K, NY); Ruri, near Blue Mountain, Aug 1924 (fl), *Persaud 102* (F, K, NY); Berbice County, Courantyne River between Mapenna and Hubudikuru, 13 May 1918 (fr), *Hohenkerk FD404A* (K); Potaro-Siparuni Region, Micobe road by Potaro River, 5°20′29″N, 58°58′51″W, 14 Sep 2006 (fr), *Redden 4236* (K); U Takutu-U, Essequibo, River, 94 m, 3°33′02″N, 58°14′12″W, 11 Oct 2008 (fl), *Wurdack* et al. *4775* (K, US). **SURINAME**. Corantyn River, Kabourie, Feb 1920, *Boschwezen 4033* (U), 1 Oct 1920, *Boschwezen 4960* (U).

Very similar to *Schistostemon macrophyllum* (Benth.) Cuatr. differing only in the pubescent petals and the united disc. Perhaps these two species, characteristic of riverine habitats, should be united.

3. **Schistostemon dichotomum** (Urb.) Cuatr., Contr. U.S. Natl. Herb. 35(2): 159. 1961; *Sacoglottis dichotoma* Urb., Fl. Bras. (Martius) 12 (2): 446. 1877. Type. Suriname, River Lava, Oct 1861, *A. Kappler 2144* (holotype B lost; lectotype, **here designated,** U 0198211; isotypes, L 0018067, P 01903253, photo F).

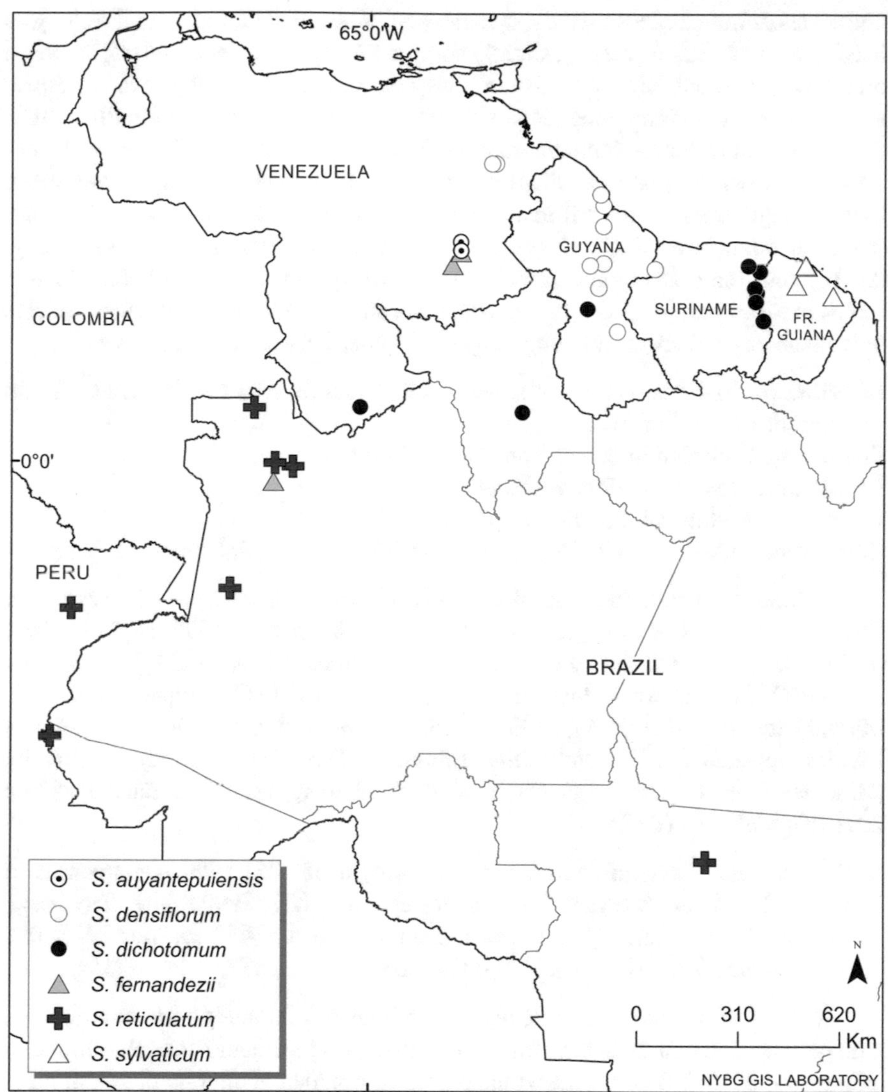

Fig. 10.22 Distribution of six species of *Schistostemon*

Trees, the young branches glabrous, lenticellate. Leaves with petioles 6–12 mm long, semiterete, thickened at base, flat above; laminas oblong-ovate or ovate-elliptic, 10–16 × 4–8 cm, rigid, coriaceous, rounded or slightly cuneate-decurrent at base, attenuate at apex and slightly cuspidate, margins slightly crenate, with a pair of basal glands on lower surface; midrib plane above, prominent and striolate beneath; primary veins 8–10 pairs, prominent on both surfaces, with minute impressed glands near margin; reticulate venation prominent and conspicuous on both sides. Inflorescences axillary, cymose-paniculate, dichotomous, half as long as leaves, peduncle 3.5–6 cm long, slightly compressed, minutely puberulous; branches minutely pubescent; bracts persistent, clasping, 0.5–1.5 mm long, ovate, obtuse; pedicels thick, to 0.5 mm long. Sepals suborbicular, glabrous except for ciliate margin, 1 mm long. Petals submembranaceous, oblong, acute, 3–3.5 mm long, glabrous. Stamens 20, glabrous, filaments complanate, connate at base, 2.2–3.3 mm long, 5 longer ones shortly trifurcate and triantheriferous, anthers ovate-rhomboid, 0.9 mm, connective thick, angulate, ovate-lanceolate, thecae ellipsoid-oblong, 0.4–0.5 mm long. Disc cupular, surrounding ovary, formed of 10 bidentate free scales, 0.5–0.6 mm long. Ovary globose, glabrous, 0.8 mm high, 5-locular, locules uniovulate; style thick, 1 mm long, stigma capitate, 5-lobed. Drupe not seen.

Distribution and habitat. Only known from riverside forest in the Guianas and adjacent Brazil (Fig. 10.22).
Phenology. Collected in flower from July to October.
Local names and uses. None recorded.
Conservation status. Least concern (LC).
Illustrations. Cuatrecasas (1961) figs. 31 d–f, 33 e–g. 1961; Sabatier (1987), fig. 12.

Additional specimens examined. GUYANA. Potaro-Siparuni, Iwocrama Reserve, 4–5 km N of Surama, 4.16667 N, 50.05 W, 17 May 1995, *Hoffman 4564* (F, MO, US). **SURINAME**. Lawa River, near Cottica, 01 Oct 1903, *Versteeg 265* (U). **FRENCH GUIANA.** Maroni River, *Rech s.n.* 1892 (P); Crique Beiman, afl, Maroni Riviére, 4.61, −54.33, 21 Feb 1984, *Sabatier 989* (CAY, MO, NY); Maroni Riviére, between Belligui and Langa Tabiki, 23 Feb 1984, *Sabatier 1013* (US). **BRAZIL. RORAIMA**: Parque Nacional do Viruá, margin of Rio Barauana, 29 Sep 2011 (fl), *Melo 898* (INPA).

4. **Schistostemon fernandezii** Cuatr., Phytologia 68: 261.1990; Cuatrecasas & Huber, Fl. Venez. Guayana (Steyermark et al.) 5: 637. 1999. Type. Venezuela. Bolívar: Gran Sabana, S of Urimán, Laguna Capaura, 440 m, Mar 1986, (fl), *A. Fernández 2276* (holotype, US 00289003, isotypes, MY, VEN 287154).

Low trees or sprawling shrubs, the young branches minutely hirtellous. Leaves with petioles 3–6 mm long, base thickened into a pulvinus; laminas ovate-elliptic or elliptic, 5–10 × 3–5.8 mm, coriaceous, glabrous, rounded or slightly obtuse at base, apex attenuate-cuspidate, the acumen 0.5–1 mm long, margins entire or inconspicuously crenate, small setae present, with a pair of basal glands on lower surface; midrib plane above, prominent beneath; primary veins 7–8 pairs, inconspicuous on both surfaces, venation prominently reticulate on both surfaces. Inflorescences axillary, 1.5–3 cm long, cymose-paniculate; peduncle thick, to 6 mm long often shorter, branching simple or dichotomous, branches densely hirtellous; bracts persistent, deltoid, semiamplectant, acute, 0.5–1 mm long. Sepals 5, orbicular, 1–1.2 × 1.3–1.5 mm, slightly connate at base, glabrous except for ciliate margins, surface resinous. Petals 5, oblong, acute, 3–3.3 × 1.2 mm, hirtellous on abaxial surface. Stamens 20, filaments connate at base into a tube 1 mm high, 5 opposite to sepals longer, 3 mm long, apex shortly trifurcate with 3 anthers, central anther

0.8 mm long, connective thick, obtuse, thecae 2, elliptic, basally attached, lateral anthers smaller, 0.3 mm, sterile, 5 filaments opposite petals slightly shorter, 2.5 mm long, bearing one fertile anther similar to the large anther, alternating with 10 smaller filaments 2 mm long, each with one fertile anther. Disc cylindrical, 0.8 mm high, dentate with 20 teeth. Ovary globose, 1.5 mm high, 5-locular, glabrous; style short, robust, 0.3 mm long, stigma capitate, 5-lobed, glutinous. Drupe 2.4–2.8 × 1.9–2 cm, ellipsoid, exocarp surface smooth, 1 mm thick; endocarp woody, 2.1–2.6 mm thick, 5–grooved on exterior, with 5 foramina near apex, with large resiniferous cavities.

Distribution and habitat. Known from two collections from lowland forest in Bolívar, Venezuela (Fig. 10.22).
Phenology. Collected in flower in March and in fruit in May.
Local names and uses. None recorded.
Conservation status. Near Threatened (NT), B1a + 2a.
Illustration. Cuatrecasas & Huber (1999), fig. 544.

Additional specimen examined. **VENEZUELA. BOLÍVAR**: Guayaraca, above Valle de Camarata, 100 m, 8 May 1964 (fr), *Steyermark 94,197* (NY, US).

5. **Schistostemon macrophyllum** (Benth.) Cuatr., Contr. U.S. Natl. Herb. 35(2): 157. 1961; *Humirium macrophyllum* Benth., Hooker's J. Bot. Kew Gard. Misc. 5: 102. 1853; *Sacoglottis macrophylla* (Benth.) Urb., Fl. Bras. (Martius) 12 (2): 446.1877. Type. Brazil. Amazonas: Barra do Rio Negro (Manaus), *R. Spruce 1714* (lectotype **designated here**, K 000407318; isotypes, BM 000796062, K 000407319, M, P 01903252, TCD 0003829). Fig. 9.4E.

   Sacoglottis duckei Huber, Bol. Paraense Hist. Nat. 5: 4–3. 1909. Type. Brazil. Amazonas: Manaus, 23 Jun 1905 (fl), *A. Ducke RB 21024* (holotype, MG 007174; isotypes, BM 000796064, F V0060787).

Trees, the young branches glabrous, lenticellate. Leaves with petioles 8–18 mm long, terete, flattened above, thickened at base, glabrous; laminas ovate-oblong, 8–20 × 3.5–7.5 cm, coriaceous, glabrous, rounded at base, acuminate at apex, margins slightly crenate, with widely spaced glands and a pair of basal glands on lower surface; midrib prominulous and flattened above, prominent and glabrous beneath; primary veins 10–14 pairs, prominulous on both surfaces, reticulate venation conspicuously prominulous, with small punctiform glands along margin. Inflorescences axillary, cymose-paniculate, much shorter than leaves, peduncle stout, compressed, short-pubescent, 1–3 cm long, bifurcate, branches pubescent; bracts persistent, coriaceous, ovate-triangular, obtuse, clasping, glabrous except for ciliate margin; pedicels thick, short, 0.3 mm long. Sepals orbicular, 1.5 mm long, glabrous except for ciliate margins. Petals thick, oblong, glabrous, obtuse to subacute, 4.5–5 × 1.5–1.8 mm. Stamens 20, glabrous, filaments compressed, lower half connate into tube, 2.5–4 mm long, 5 longer ones trifurcate and trianthiferous sterile filaments often present, anthers ovoid or rhomboid, 1 mm long, connective thick, angulate, subdeltoid-lanceolate, thecae oblong-elliptic, 0.5 mm long. Disc annular-cupular, surrounding ovary, formed of 10 bidentate connate scales, 0.6–0.7 mm

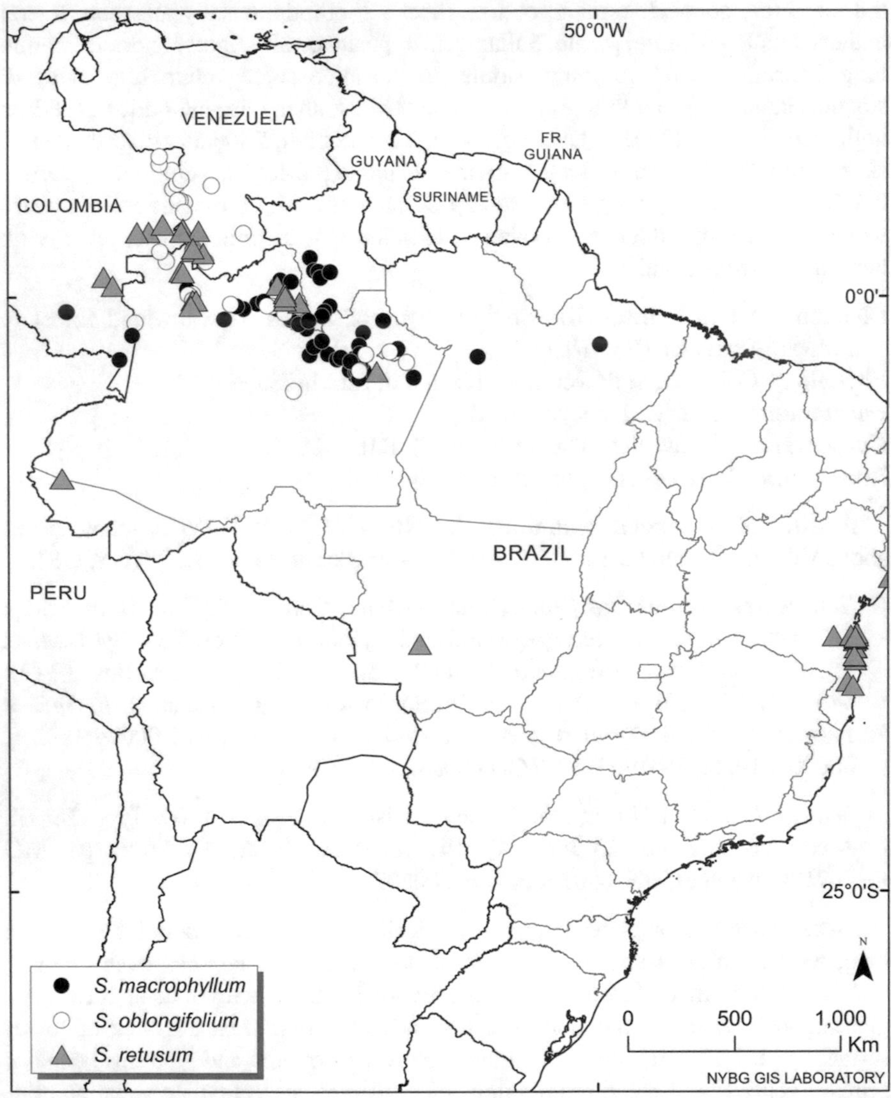

Fig. 10.23 Distribution of three species of *Schistostemon*

high. Ovary subglobose, 1.2 mm high, glabrous, 5-locular, locules uniovulate; ovules oblong, 1 mm long; style columnar, glabrous, 0.8 mm long, stigma capitate, 5-lobed. Drupe subovoid, or oblong-ovoid, 3.5–4 × 2–2.8 cm, rounded at base, abruptly attenuate at apex, exocarp smooth, thin, glabrous; endocarp woody, smooth or slightly bullate, with many spherical resinous cavities.

Distribution and habitat. A common tree in flooded river margins and igapó in central Amazonian Brazil and southern Colombia (Fig. 10.23).

Phenology. Flowering from April to August, fruiting from September to November.

Local names and uses. *Umirí-rana*. A decoction of the leaves is taken to relieve heavy colds, bronchial constipation and tuberculosis (Schultes, 1979).

Conservation status. Least Concern (LC).

Illustrations. Cuatrecasas, (1961), pl. 18, figs. 31a–c, 33a. 1961; Holanda et al. (2015a, b) fig. xxx; Huber (1909), pl. 15, fig. 37.

Selected specimens examined. COLOMBIA. AMAZONAS: Mun. Leticia: Río Caquetá, Cerro Yupati, 1°21′S, 69°31′W, 7 Nov 1994 (fr), *Cárdenas 5564* (COAH). **GUAINÍA**: Mun. Taraira, Est. Biol. Caparú, 1°4′36.1"S, 69°30′54.1″W, 1 Jan 2009 (fr), *Cano 51* (COAH). **BRAZIL. AMAZONAS**: Igarapé Mindú, São Gabriel da Cachoeira, Jul 2007, *Stropp & Assunção 336* (EAFM); Manaus, Rio Tarumã, 24 Jul 1936 (fl), *Ducke 255* (A, F, IAN, K, MG, MO, NY, S, US), 6 Aug 1949 (fl), *Fróes 24,916* (IAC, IAN); Cachoeira Grande, Tarumã, 7 May 1933, *Ducke RB 23816* (INPA, K, RB, US, U), 26 Sep 1955, *Rodrigues INPA 2044* (INPA, MG); Rio Negro near Tarumã, 13 Oct 1966 (fl), *Prance* et al. *2646* (F, INPA, K, NY); Manaus, Ponta Negra, 16 Mar 1967 (fl, fr), *Prance* et al. *4663* (F, INPA, K, NY, S); Tarumã-mirim, 3°02′S, 60°17′W, 31 Jan 1992 (fr) *Ferreira 136* (INPA, K, NY); Tarumã Grande, 3°02′S, 60°08′W, 1 Nov 1977 (imm fr), *Keel & Coelho 244* (MO, NY, RB); Baixo Rio Negro, Tanacoera, 26 Apr 1911, *Ducke MG11550* (MG, RB16680); 1 km N of junction of Rio Negro and Igarapé Tarumã, 23 Nov 1977 (fr), *Keel & Guedes 311* (INPA, NY, RB); Cacau-Pirêra, Iranduba, −3.284, −60.186, 3 Oct 1961 (fl), *Rodrigues & Lima 2559* (INPA); Mun. Barcelos, Rio Aracá near Rio Jauari, 0°30′N, 68°30′W, 1 Jul 1985 (fr), *Sette Silva 196* (FEMACT, INPA, K, NY); Rio Negro between Rio Cuiní & Rio Ararirá, 0°30′S, 63°30′W, *Maas 6633* (INPA, K, NY, RB); Barcelos, 26 Sep-14 Oct 1947 (fl) *Schultes & López 8881* (F, IAN, K, MO, US); Rio Aracá, 100 m, 8 Aug 1996 (fl), *Acevedo-Rodrigues 8114* (INPA, NY, US); mouth of Rio Negro, Igarapé da Colonia Loge, 21 Jul 1874 (fl), *Traill 81* (K); Rio Uneiuxi, 100–200 km above mouth, −0.931, −65.756, 21 Oct 1971 (fr), *Prance* et al. *15,509* (INPA, K, MO, NY); Rio Negro, between mouth Rio Caurés and Barcelos, 12 Oct 1971 (imm fr), *Prance* et al.*15123* (INPA, NY); Rio Negro, Boca do Ararirá, Barcelos, 1 Jan 2011 (fl, imm fr), *Cid Ferreira* et al. *9272* (F, HFSL, INPA, RB); Reserva Biológica Parque do Jaú, Feb 2008, *Zartman 7326* (INPA); Rio Jauari below Igarapé Pretinho, 0°42′N, 63°22′W, 1 Jul 1985, *Silva 205* (FEMACT, INPA, LBT, MO, NY, SP, US). **RORAIMA:** Parque Nacional da Serra da Mocidade, Rio Capivara, 1.175, −62.553, 6 Dec 2013 (fl, fr), *Perdiz* et al. *2072* (MIRR, UFRR); Rio Capivara, 1.09, −61.927, 6 Dec 2013 (fl), *Schutz Rodrigues* et al. *2295* (MIRR, UFRR); Parque Nacional Serra da Mocidade, Igarapé do Vale do Cumaru, 11 Dec 2013 (fr), *Perdiz 2236* (UFRR); Parque Nacional Serra da Mocidade, Rio Capivara, 55 m, 1°03′33″N, 61°44′48″W, 25 Mar 2012 (fr), *Forzza 6983* (K); Estação Ecológico de Niquiá, Rio Agua Boa do Univiní, 500 m, 26 Mar 2012 (fr), *Forzza 7029* (K); Parque Nacional Viruá, above Estirão Jurema, Caracaraí, 29 Jul 2011 (fl), *Zartman 8496* (INPA); Parque Nacional Viruá, Rio Inuá, 30 Jul 2012 (fl), *Holanda 472* (INPA). **PARÁ**: Alter do Chão, Lago Verde, 25 Dec 1991 (fr), *Ferreira 45*

(INPA, K, NY); Road Oeras do Pará-Cametá, Km 14, 12.00306, −49.8544, 15 Aug 2000 (fl), *Cid Ferreira 12,057* (INPA).

6. **Schistostemon oblongifolium** (Benth.) Cuatr., Contr. U.S. Natl. Herb. 35(2): 148.1961; Cuatrecasas & Huber, Fl. Venez. Guayana (Steyermark et al.) 5: 637. 1999; *Humirium oblongifolium* Benth., Hooker's J. Bot. Kew Gard. Misc. 5: 103. 1853; *Sacoglottis oblongifolia* (Benth.) Urb., Fl. Bras. (Martius) 12 (2): 447. 1877. Type. Brazil. Amazonas: between Barcelos and San Isabel, Dec 1851 (fl), *R. Spruce 1969* (**lectotype designated here**, K 000407324; isotypes, BM 000796048, GH, FI-W, K 000407323, M, NY 00388395, OXF, P 04693169, TCD 0003831).

Small trees or shrubs, the young branches minutely hirtellous pubescent, lenticellate. Leaves with petioles 5–7 mm long, puberulous, thickened at base; laminas oblong-elliptic to oblong-lanceolate, 7–18 × 3–5.5 cm, rigid coriaceous, rounded or obtuse at base, slightly decurrent onto petiole, acute or bluntly acuminate at apex, the acumen 4–7 mm long margins slightly crenate, glabrous above, with diffuse appressed hairs or a ferrugineous pulverulent indumentum beneath or glabrous, with 1–2 basal glands on lower surface; midrib plane above, glabrous on upper part, puberulous toward base, strongly prominent, sparse-puberulous beneath; primary veins 9–12 pairs, prominulous above, prominulous or obscure beneath, reticulate venation obscurely prominulous, minutely papillose. Inflorescences axillary, short-cymose-paniculate, glomerulate, usually dichotomous, peduncle robust, 4–10 mm long, hirtellous, branches hirtellous; bracts persistent, ovate, obtuse, clasping, 0.5–1.5 mm long, pubescent, densely ciliate; pedicels 0–0.5 mm long, Sepals 1 mm long, rounded, minutely pubescent, densely ciliate. Petals thick, oblong, subacute, appressed pubescent, 4.5 × 1.5 mm. Stamens 20, glabrous, filaments 2.7–3.5 mm long, connate at base to form a tube to mid length, 5 longer ones trifurcate, trianthiferous, 5 medium and 10 smaller ones alternating, each with one anther, sterile filaments often present, anthers ovate-lanceolate, 0.9 mm long, connective thick, angulate, lanceolate, thecae oblong, 0.4 mm. Disc thick, cupuliform with dentate margin. Ovary ovoid, glabrous, 1.4 mm high, 5-locular, locules uniovulate, ovules 0.6 mm long; style thick, 0.5–0.6 mm long, stigma capitate, 5-lobed. Drupe oblong, fusiform at apex, especially when young, attenuate at base, 4–4.5 × 1.3–1.5 cm broad, attenuate at both ends, exocarp glabrous, smooth, thick, fibrous; endocarp woody weakly fusiform, smooth or slightly bullate, obtuse at base, apiculate at apex.

Distribution and habitat. Riverine forest in Colombia, southern Venezuela and northwestern Amazonian Brazil (Fig. 10.23).

Phenology. Flowering from October to March, fruiting from February to August.

Local names and uses. Venezuela*: Merecure de ardita, yurí de caracol.*

Conservation status. Least concern (LC).

Illustrations. Cuatrecasas (1961) figs. 30 h-I, 31 g-i; Cuatrecasas and Huber (1999), fig. 545.

Selected specimens examined. **COLOMBIA. GUAINÍA**: Panapaná, Río Cuyarí, 1°54′48.97″N, 68°28.7′9464″W, 26 Apr 2014 (fr), *Castro 18,072* (COAH).

VENEZUELA. AMAZONAS: Río Negro near San Carlos, 1853–54 (fl), *Spruce 3073* (BM, GH, K, OXF, NY, P, S); between San Carlos and mouth of Río Casiquiare, 25 Jun 1984 (fl), *Davidse & Miller 26,643* (F, INPA, K, MO, US); Río Autana, 4°44′29′N, 67°44′10″W, 26 Feb 2000 (imm fr), *Castillo 7143* (MO); Río Negro, Isla Mayabo, 1°56′N, 67°03′W, 119 m, 21 Apr 1981 (fr), *Clark & Maquirino 7902* (MO, NY); 12 km above San Fernando de Atabapo, 110 m, 14 Jan 1988 (fl bud) *Aymard 6470* (MO, US); Río Sipapo, Santa Lúcia, Mun. Autana, 18 Feb –4 Mar 1986 (fl bud, fr), *Stergios & Aymard 9215* (MO, NY); Caño Jiboa, N side Río Orinoco, 100 m, 3° 58′N, 67°25′W, 25 May 1993 (fr), *Berry* et al. *5507* (MO); Junction Ríos Casiquiare & Guainía, 120 m, 1°59′N, 67°04′W, 5 Feb 1980 (fr), *Liesner & Clark 9072* (MO, NY, US); Río Sipapo, between Boca del Cuao & Piedra Chamii, 13 May 1998 (imm fr), *Castillo 5787* (MO). **BRAZIL. AMAZONAS**: Rio Negro, São Gabriel, 28 Oct 1932 (fl), *Ducke RB 23817* (INPA, K, P, RB, S, U, US), 1 May 1947, *Pires 588* (IAN, NY); Camanaus, Rio Negro, 17 Apr 1947 (fr), *Pires 365* (COL, IAN); Igarapé Cucucuhy, São Gabriel, 26 Nov 1945 (fl), *Fróes 21,437* (F, IAC, K, NY, US); Rio Negro, near Serra Jacamin, 18 Jan 1978 (fr), *Steward 411* (F, INPA, NY); Rio Negro, Jerusalem, 1 Oct 1955 (fl), *Fróes 21,090* (F, IAN, K, NY, US); mouth of Rio Içana, 5 Mar 1944 (fl), *Baldwin 3187* (IAN, US); Rio Uatumá, mouth Rio Pitinga, 25 Aug 1979 (fl, imm fr), *Cid* et al. *776* (F, INPA, MO, NY, RB); Rio Negro between Rio Quinini & Moreira, 15 Jan 1975 (fl), *Prance* et al. *15,175* (F, INPA, NY); Rio Negro, Lago de Marajó. Manaus, 2 Jan 1956, *Coelho INPA3255* (INPA); Mauá road, Manaus, 22 Mar 1971 (fr), *Prance* et al. *11,541* (INPA, MO, NY); Rio Negro between mouth of Rio Ariraháara & Paraná São José, Ilha Florianópolis, –0.384026, –63.72257, 29 Apr 1973, *Silva 1135* (INPA); Rio Negro, Ilha Tamanduá, 0°06′N, 67°16′W, 20 Oct 1987 (fl), *Maas 6804* (INPA, K, NY, RB); Manaus Distrito Agropecuário, Reserva 1501, km 41, 2°24′28"S, 59°43′40″W, 5 Dec 1988 (fr), *Boom* et al. *8742* (NY).

This species has buoyant fruits that are dispersed by water. The Kew herbarium has two specimens of the type number. I have chosen as the lectotype the specimen (K000407324) with original drawings and notes by the author on it.

7. **Schistostemon reticulatum** (Ducke) Cuatr., Contr. U.S. Natl. Herb. 35(2): 153. 1961; *Sacoglottis reticulata* Ducke, Arq. Inst. Biol. Veg. 1: 206. 1935.Type. Brazil. Amazonas, São Paulo de Olivença, 3 Oct 1931 (fl), *A. Ducke RB 23819* (holotype, RB 00539072, isotypes, K 000407325, P, U, US 00101217).

Schistostemon reticulatum var. *froesii* Cuatr., Contr. U.S. Natl. Herb. 35(2): 153. 1961. Type. Brazil. Amazonas: Foz de Cairay, Serra Tunuhy, 500 m, 13 Nov 1945 (fl), *R. L. Fróes 21,370* (holotype, NY 00388414; isotypes, F V0060294, IAN 016816, K 000407321, US 00101249).

Trees, the young branches minutely puberulous. Leaves with short, thick, flattened petioles 5–8 mm long, pilose or glabrescent; laminas, ovate or elliptic, 4–12.5 × 2–7.5 cm, rigid coriaceous, glabrous, obtuse to somewhat rounded at base, obtusely acuminate at apex, margin slightly crenate, 1–2 basal glands on lower surface; midrib prominulous above, prominent and thick beneath, glabrous or sparsely puberulous; primary veins 6–9 pairs, prominulous above, prominent beneath, reticulate venation prominulous above, prominent beneath. Inflorescences axillary, cymose-paniculate, dichotomous at base, branches minutely hirtellous; peduncle striate, short-pubescent; bracts persistent, clasping, ovate-triangular, 0.5–1 mm long, obtuse or acute; pedicels 0.5 mm long, hispidulous. Sepals 1.4 mm long, ovate, obtuse or roundish, acute, puberulous or hispidulous, margin ciliate, apex with small callous gland. Petals oblong, acute, 4–4.5 × 1–1.5 mm, appressed pubescent. Stamens 20, glabrous, filaments 2.2–3 mm long, connate at base into a tube, 5 longer trifurcate and triantheriferous, anthers ovate, 0.8 mm long, connective ovate, acuminate, thecae elliptic-oblong, 0.3–0.4 mm long. Ovary subglobose, glabrous, 0.5 mm high; style 0.4 mm long, thick, stigma capitate, 5-lobed. Drupe round, 1.5–7 cm diam; exocarp smooth, glabrous; endocarp hard, fibrous, 3 mm thick, inner layer 11 mm thick.

Habitat and distribution. Forests on *terra firme* of western Amazonia in Peru and Brazil (Fig. 10.22).
Phenology. Collected in flower in November and in fruit in April and November.
Local names and uses. Venezuela: *Guaquito*.
Conservation status. Least concern (LC).
Illustrations. Cuatrecasas (1961) figs. 30 g, j–k, 32c, 33 h.

Additional specimens examined. PERU. LORETO: Mishuyacu near Iquitos, July 1930, *Klug 1564* (F, US). **BRAZIL. AMAZONAS**: Mun. São Gabriel da Cachoeira, Rio Tuari, Piraiauara, 13 Nov. 1987 (fr), *Lima 3180* (F, INPA, MO, NY, RB). **ACRE**: Māncio Lima, 7°30′S, 73°44′W, 29 Apr 1971 (fr), *Prance* et al. *12,600* (F, INPA, MO, NY, US).

Cuatrecasas (1961) described variety *froesii* of this species. The small differences in leaf size are due to the drier habitat of this collection, and it does not seem necessary to maintain the variety.

8. **Schistostemon retusum** (Ducke) Cuatr., Contr. U.S. Natl. Herb. 35(2): 156. 1961; Cuatrecasas & Huber, Fl. Venez. Guayana (Steyermark et al.) 5: 638. 1999; *Sacoglottis retusum* Ducke, Arq. Inst. Biol. Veg. 4: 26, 29. 1938. Type. Brazil. Amazonas. Rio Curicuriari, above Cajú Cataract, 22 Feb 1936 (fl), *A. Ducke RB 30131* (holotype, RB 00539072, isotypes, K 000407320, US 101216). Fig. 9.4A.

Small- to medium-sized trees, the young branches glabrous. Leaves sessile or subsessile, with very short vaginate petiole; laminas broadly elliptic or suborbicular-elliptic, 6–15 × 4.5–10 cm, thick-coriaceous, rigid, rounded or obtuse at base, rounded or retuse at apex, margins entire, revolute, 1 pair of basal glands on lower surface; midrib prominulous above, prominent, thick, glabrous beneath; primary veins 8–9 pairs prominulous above, prominent beneath, reticulate veins prominulous. Inflorescences axillary, cymose-paniculate, much shorter than leaves, branches hirtellous-pubescent, branches short, alternate or rarely dichotomous; peduncle erect, striolate pubescent-hirtellous; bracts persistent, ovate or ovate-oblong, obtuse, clasping, puberulous, margins ciliate, 0.5–2 mm long; pedicels short, 0.2–0.3 mm long. Sepals thick, rounded, 1–1.2 × 1.2–1.5 mm, puberulous, margins ciliate with minute gland at apex. Petals thick, oblong, 4.5–5 × 1.5 mm, acute at apex, puberulous. Stamens 20, uniseriate, glabrous, connate at base into a tube, 5 episepalous longer, 2.5 mm, shortly trifurcate and trianthiferous, medial anther larger and fertile, lateral anthers sterile, 5 epipetalous filaments 2 mm long, and 10 alternating shorter ones, 1.5 mm, entire, monantheriferous, anthers dorsifixed, glabrous, connective thick, ovoid, acute at apex, interior carinate, thecae basal, ellipsoid, unilocular. Disc cupular, membranous, 0.8 mm high, surrounding ovary, margin dentate. Ovary globose, 3–4.5 mm thick, ovoid, apex acute, glabrous; style 0.4 mm long, stigma capitate, 5-lobed. Drupe globose, 3–6 cm diam, exocarp smooth, coriaceous, 3–5 mm thick; endocarp globose, hard, woody, almost smooth, copiously filled with resinous cavities, 2.6–4.5 cm diam, usually 2-seeded.

Distribution and habitat. Western Amazonia in Colombia, Venezuela and Brazil, and disjunct in Bahia, common in forest on sandy soils (Fig. 10.23).
Phenology. Flowering from November to February, fruiting from May to October.
Local names and uses. *Zaizīna* (Witito). Fruit edible.
Conservation status. Least concern (LC).
Illustrations. Ducke (1938), fig. 2d; Cuatrecasas (1961), pl. 17, figs. 30a–f, 31j, 32d; Cuatrecasas and Huber (1999), fig. 543.

Additional specimens examined. COLOMBIA. AMAZONAS: Río Igara-Paraná, tributary Río Putumayo, Corrego La Chorrera, 17 Jun 1974 (fl), *Gasche & Desplats 212* (COL). **VAUPÉS**: Cerro de Circasia, 200–400 m, 10 Oct 1939, *Cuatrecasas 7203* (COL, US); Mun. Mitú, Comunidad Puerto Nariño, Caño Gallineta, 0°29′3″N, 70°27′39″W, 8 Jun 2009 (old fl), *Cárdenas 23,419* (COAH, NY). **GUAINÍA**: Río Inírida, Raudal Guacamayo, 180 m, 4 Feb 1953, *Fernández 2142* (COL, US); Mun. Inírida, Río Inírida, Caño Nabuquen, 3°02′86.5″N, 68°20′52.7 W, 3 Jan 2007 (fr), *Cárdenas 20,386* (COAH). **VENEZUELA. AMAZONAS**: Yavita-Maroa km 9.3, 2° 51′40″N, 67°29′24″W, 22 Feb 1998, *Berry* et al. *6739* (MO); San Carlos de Rio Negro-Solano 4 km from San Carlos, 120 m, 1°55′30″N, 67°02′04″W, 24 Mar 2000, *Berry* et al. *7091* (MO); 4–7 km N of San Carlos, 75 m, 1°56′N, 67°02′W, 19 May 1979, *Liesner 7547* (MO); Mamurividi, Río Pacimoni, 1°50′N, 66°38′W, 27 Jun 1984 (fr), *Davidse & Miller 26,712* (MO). **BRAZIL. AMAZONAS**: Camanaus, Rio Negro, 31 Oct 1971 (fr), *Prance* et al. *15,914* (F, INPA, NY); Rio Içana, Santana, 18 Nov 1945 (fr), *Fróes 21,411* (F, IAN, K, NY, US, VEN), 2 May 1948 (fr), *Black 48–2514* (IAN, NY, U, US, VEN); Rio Preto, 5 Nov 1947 (fl), *Fróes 22,747* (IAC, IAN, RB); Yútica, Rio Vaupés, 15 Nov 1952 (fr), *Romero-Castañeda 3542* (COL). **ACRE**: Cruzeiro do Sul, Estrada Alemanha, 6 May 1971 (fr), *Maas* et al. *P12746* (F, INPA, K, MO, NY); Mun. Barcelos, Rio Aracá, Bacuquara, 30–45 m, 0°9′6″N, 63°10′41″W, 16 Apr 2014 (fr), *Forzza 7947* (INPA, NY, RB); Rio Aracá, comunidade de Bacuquara, 0°09′16" S, 63°10′41″W, 30–50 m, 16 Apr 2014 (fr), *Amorim 8601* (CEPEC, INPA, RB). **MATO GROSSO**: Estrada Tangará, da Serra, Chapadão dos Parecis, −14. 6196, −57.4858, 30 Sep 2011, *Silva 409* (TANG). **BAHIA**: Mun. Itacaré, 6 km W of Itacaré, road to Serra Grande, 14°15′S, 39°16′W (fl), 2 May 1993, *Thomas 9788*

(CEPEC, MO, NY); Mun. Una, Reserva Biológica Mico-leão, Km 46 Ilheus-Una, 20 Apr 1996 (fr), *Jardim* et al. *792* (CEPEC, MO, NY); Res. Biológica Mico-leão, Km 46 Ilheus-Una, 15°09′S, 39°05′W, 28 Nov 1993 (fl), *Amorim* et al. *1563* (CEPEC, NY); Marau-Ubaitaba Km 8, *Carvalho 6777* (CEPEC); road Pontal-Olivença, Km 10, Ilhéus, *Silva 1415* (CEPEC, HUEFS); Mun. Itacaré, Km 20–23 Serra Grande Itacaré, 4 Jul 1996 (fl), *Carvalho & Sant'Ana 6239* (F, CEPEC, NY); Mun. Una, Maruim, 33 km SW of Olivença road to Buerarema, 29 Apr 1996 (fr), *Mori* et al. *13,825* (CEPEC, NY); Mun. Itacaré, 14 km N of Serra Grande, 14°22′S, 39°4′W, 15 Nov 1992 (fr), *Thomas 9466* (CEPEC, NY); Mun. Santa Cruz de Cabrália, Est. Pau-Brasil, 16 km W of Porto Seguro, 2 Jul 1978, *Mori 10,215* (CEPEC, K, NY).

This species has a disjunct distribution between western Amazonia and Bahia. I have examined the specimens carefully and think that Cuatrecasas (1961) was correct in recognizing this unusual distribution.

9. **Schistostemon sylvaticum** Sabatier, Proc. Kon. Ned. Akad. Wetensch. C 90 (2): 206. 1987. Type. French Guiana. PK14, Piste de St. Elie, near Sinnamary, 8 Apr 1984 (fl), *D. Sabatier 842* (holotype, formerly at CAY, P 00077220; isotypes, CAY, U0008171).

Large trees, the young branches glabrous. Leaves with petioles 5–7 mm long, weakly winged, robust, thickened at base, semiterete; laminas oblong-elliptic to obovate-elliptic, subcoriaceous, glabrous, rounded to obtuse at base; cuspidate acuminate at apex, margins slightly crenate to subentire, 1–2 basal glands on lower surface; midrib slightly impressed above, prominent beneath; primary veins 13–15 pairs, inconspicuous and prominulous on both surfaces Inflorescences axillary, paniculate, very short, 0.5–2 cm long; peduncle 0.4–0.5 cm long; bracts caducous. Sepals 5, rounded, 2.5 mm long, margins minutely ciliate. Petals oblong, acute, thick, 4.5–5.4 × 1.5–1.9 mm, strigose on dorsal surface. Stamens 20, filaments connate into a tube for lower half, 5 longer, 4.5 mm long, trifurcate, triantheriferous, 5 medium ones 3.6 mm, 10 shorter ones 3.1 mm long; anthers glabrous, lanceolate, ca 0.55–0.9 mm long, connective, thick, lanceolate, apex acuminate, thecae 2, oblong, basal. Disc membranous, annular, 0.9 mm high, margin dentate. Ovary globose, glabrous, 5-locular, 1.3 mm high; style 1.2 mm long, stigma capitate, 5-lobed. Drupe ovoid, 5–6 × 4.5–5.3 cm, exocarp fleshy, smooth, 0.5–0.7 mm thick; endocarp woody, 4.2–4.9 cm long, surface with resinous cavities, dehiscing by lateral opercula.

Distribution and habitat. Only known from the lowland rainforest of French Guiana (Fig. 10.22).

Phenology. Collected in flower in April and May and in fruit from March to May and October.

Local names and uses. None recorded.

Conservation status. Endangered (EN), B1ab(iii,v). Only known from a restricted range outside of reserves.

Illustrations. Sabatier (1987), figs. 1–9.

Specimens studied. FRENCH GUIANA. PK14, Piste de St. Elie, near Sinnamary, 5.365524, −53.031706, 22 Feb 1984 (st), *Sabatier 815* (CAY, U), 30 Mar 1984 (st), *Sabatier 836a* (CAY, U), 30 Jan 1981, seedling, *Sabatier 836b*, May 1981 (fr), *Sabatier 38–4* (CAY); Piste de St. Elie, 14 Mar 1991 (fr), *Loubry 1074* (NY, US); Concession Cyprio, Route de l'Est, 16 Apr 1977 (st), *Lescure 668* (CAY); R. Counana, 4°32′N, 52°17′W, 6 Oct 2000 (fr), *Sabatier & Prévost 4672* (CAY, NY).

This species has the largest fruit in the genus, to 6 × 5.3 cm. The holotype of this species, *Sabatier 842*, was transferred from CAY to P along with many other holotypes as noted in Cremers (2001).

10.8 Vantanea

Vantanea Aubl., Hist. Pl. Guiane 1: 572. 1775; Benth., Hooker's J. Bot. Kew Gard. Misc. 5: 98. 1853; Urb., Fl. Bras. (Martius) 12 (2): 450. 1877; Zamora, Manual de Plantas de Costa Rica 6: 115. 2007. Type. *Vantanea guianensis* Aubl.

Lemniscia Schreb., Gen. Pl. ed. 8: 1: 358. 1789. Type. *Vantanea guianensis* Aubl.

Lemnescia Willd., Sp. Pl. 2: 1172. 1800, orth. var.

Helleria Nees & Martius, Nova Acta Phys.-Med. Acad. Caes. Leop.-Carol. Nat. Cur. 12: 38. 1824; Martius, Nov. Gen. et Sp. 2: 147. 1827.Type. *Helleria obovata* Nees & Mart.

Houmiri Sect. *Vantanea* Baill., Adansonia 10: 370. 1870; Hist. Pl. (Baillon) 5: 48. 1873–1874.

Houmiri Sect *Vantaneoides* Baill., Adansonia 10: 370. 1870; Hist. Pl. (Baillon) 5: 48. 1873–1874.

Trees. Leaves alternate, petiolate or sessile, chartaceous or coriaceous, lamina margins entire. Inflorescences axillary or terminal, paniculate, usually dichotomous, or with alternate branching; bracts deciduous. Sepals 5, suborbicular, united at base. Petals 5, free, thick, aestivation contorted. Stamens 15–200, usually in 3 or 4 rows, of varied lengths, filaments thin, glabrous, connate at base into a tube surrounding the ovary; anthers ovate-lanceolate, attached near to base, glabrous, connective thick, ovate-oblong, acuminate, acute or subobtuse, thecae 2, bilocular, ellipsoid, attached at lower side, each cell dehiscent by a longitudinal slit. Disc cupular, dentate or fimbriate, surrounding ovary. Ovary glabrous or pubescent, 5-locular, 2 ovules per locule, the lower one with long funicle; carpels opposite the sepals; style erect, stigma more or less thickened, 5-lobed. Drupe medium to large, smooth, ovoid or ellipsoid, exocarp fleshy, subcoriaceous when dry, thick or thin; endocarp woody, without resinous cavities, dehiscent at germination by longitudinal, linear or oblong valves or opercula; seed usually 1.

Distribution. Twenty-one species from Nicaragua to southern Brazil.

Key to Species of Vantanea

1. Stamens 15–40; disc glabrous.

2. Petioles 8–16 mm long; stamens 30–40; leaves thin coriaceous, obovate; ovary striate, tomentose……………………..……………1. **V. bahiaensis**
2. Petioles 2–9 mm long; stamens 15–26; leaves rigidly coriaceous, elliptic; ovary not striate, glabrous or puberulent.
   3. Leaf base obtuse; midrib impressed above; fruit almost globose, exocarp glabrous………………………………...………………13. **V. morii**
   3. Leaf base cuneate, midrib plane above; fruit ellipsoid.
      4. Ovary densely puberulent; exocarp pubescent; Inflorescence terminal or subterminal; stamens 15–18………..……………… 6. **V. depleta**
      4. Ovary glabrous; exocarp glabrous; inflorescence axillary; stamens 20……………...……………………………21. **V. spiritu-sancti**

1. Stamens 50–230; disc pubescent or glabrous.
   5. Disc tomentose or puberulous.
      6. Stamens 80–120; leaves coriaceous, acute at apex; drupe 2.5–2.8 x 2.2–2.5 cm, endocarp rugose with 5–6 opercular val ves……………………………………………………………………18. **V. parviflora**
      6. Stamens 120–230; leaves rigid coriacous; blunt or retuse at apex; drupe smooth, ca 5–7 x 4–5 cm, or rough, ca 2.5 cm diam and covered with numerous white lenticels, with 5–7 opercular valves.
         7. Petioles ca. 4–6 mm long; endocarp glabrous with numerous white lenticels; drupe c.2.5 cm diam………..…………9. **V. maculicarpa**
         7. Petioles 5–10 mm long; endocarp velutinous sericeous, not lenticellate; drupe 4.5–7 cm long.
            8. Leaves 3.5–8.5 × 1.5–4.5 cm; petals 7–11 × 1–3 mm..…….4. **V. deniseae**
            8. Leaves 8–15 × 4–7 cm; petals 9–13 × 3.5 mm...……………5. **V. ovicarpa**
   5. Disc glabrous except for ciliate margins in some cases.
      9. Ovary tomentose to hirsute; petals glabrous or pubescent.
         10. Leaf laminas small, 3.5–7(−9.5) × 1.3–2.5(−4) cm, oblanceolate or lanceolate-elliptic, apex cuspidate; endocarp rugulose...3. **V. compacta**

10. Leaf laminas 4–22 × 2.5–11 cm, ovate or obovate or elliptic, apex often rounded or retuse, sometimes acuminate; endocarp smooth.

11. Petals pubescent or tomentose on exterior.

12. Petioles 2–4 mm long; ovary oblong, glabrous on lower third, villose above; petals 5.5–8 × 2.5–3 mm.

13. Ovary oblong, villous on upper portion, disc membraneous; tubular, 1–3 mm high; drupe with smooth pericarp; petals 5.5–6 × 2.5 mm.................10. **V. magdalenensis**

13. Ovary ovoid, hirsute; disc cupular, less than 1 mm high; drupe with tuberculate, warty pericarp; petals 8 × 2–3 mm ....22. **V. tuberculata**

12. Petioles 4–20 mm long; ovary ovoid appressed villous or hirsute; petals 7–13 mm long.

14. Petioles 10–20 mm long; leaves 11–22 × 6.5–11 cm; drupe 5–6× 3.3–5 mm, 8 valved.................. 20. **V. spichigeri**

14. Petioles 4–12 mm long; leaves 4–12 × 2.4–6 cm; drupe 2.3–3 × 1–1.2 cm, 5-valved.

15. Petals ca. 7 mm long; stamens 50–60, 5–7 mm long; endocarp glabrous................................2. **V. barbourii**

15. Petals 9–13 mm long; stamens 65–120; drupe velvety pubescent

16. Leaf apex retuse or rounded, not conduplicate; stamens 100–120.................................14. **V. obovata**

16. Leaf apex apiculate, lamina conduplicate; stamens ca 65...15. **V. aracaensis**

11. Petals glabrous.

17. Petioles 1–5 mm long; ovary densely long-hirsute-lanate; petals ca. 9 mm long; drupe ovoid-ellipsoid, 2.2–3.5 × 1.2–2 cm; endocarp ellipsoid, attenuate at both ends, 3–3.3 × 1.5–1.7 cm.................16. **V. occidentalis**

17. Petioles 10–18 mm long; ovary short-velvety-tomentose; petals 6–8 mm long; drupe oblong-ellipsoid, 5–5.3 × 2.8–3 cm; endocarp ellipsoid-oblong, obtuse at both ends, 4.6 × 2.5 cm .... 17. **V. paraensis**

9. Ovary glabrous; petals glabrous.

18. Inflorescence glabrous; flowers 25–40 mm long; petals red, 25–40 mm long; stamens 25–32 mm long..............7. **V. guianensis**

18. Inflorescence more or less densely tomentulose-hirtellous; flowers smaller; petals white or greenish–white, less than 15 mm long.; stamens not exceeding 10 mm.

19. Flowers small; petals 4–5 mm long; calyx 0.6–0.7 mm high, papillose on exterior, margin ciliate; leaf laminas thin, flexible, lanceolate or lanceolate-elliptic, 5–12 × 2–5 cm, apex cuspidate-acuminate; drupe 2–2.5 × 1.4–2.2 cm; endocarp strongly corrugate-sulcate, 2 × 1.6 cm; drupes as below............11. **V. micrantha**
19. Flowers larger; petals 10–14 mm long; calyx 1.2–2 mm high; leaf laminas rigid, coriaceous, apex rounded or acuminate, not cuspidate.

    20. Leaf laminas 10–20 × 5–10 cm, apex acuminate; petioles 8–16 mm; calyx pubescent-hirtellous; disc 1.8 mm high, laciniate; anther connective elongate, subacute; drupe large, ellipsoid, 5–10 × 3.5–4.5 cm; endocarp anfractuose-rug ose..8. **V. macrocarpa**
    20. Leaf laminas smaller, 3–8 × 1.2–4.5 cm, rigid, apex rounded to retuse; petioles 1–4 mm; calyx glabrous; disc 1 mm high, denticulate; drupe as below.

        21. Petals 12–14 mm long; stamens 9–12 mm long, tube 3–4 mm high; anther connective short, thick, obtuse; drupe smooth, 3.5–5 × 2.2–3 cm; endocarp densely anfractuose-rugose, 3–4 × 2–3 cm; petioles 1–2 mm long; primary veins 13–15 pairs.....................................**12. V. minor**
        21. Petals ca 11 mm long; stamens 7–10 mm long, tube 1–1.5 mm high; anther connective elongate, subobtuse; drupe unknown; petioles 3–4 mm long; primary veins c. 10 pairs..**19. V. peruviana**

1. **Vantanea bahiaensis** Cuatr., Phytologia 68: 263.1990. Type. Brazil. Bahia: Mun. Belmonte, Itapebí, 1 Aug 1981 (fl, imm fr), *A. M. de Carvalho & J. Gatti 484* (holotype, NY 00023367; isotypes, CEPEC, RB 00539073, US 00288994).

Trees 10 m tall, the young branches glabrous. Leaves with petioles 8–16 mm long, glabrous, thickened at base; laminas obovate, 4–9 × 3.4–6.4 cm, thin-coriaceous, glabrous, cuneate at base, rounded to obtuse at apex, margins entire, 1 pair of basal glands on lower surface; midrib prominulous and flattened above, prominent beneath, primary veins 10–12 pairs, plane above, prominulous beneath. Inflorescences terminal or subterminal, cymose-paniculate, corymbose, 6–10 × 6–10 cm, branching monochasial or dichasial, branches minutely velutinous hirtellous; peduncles thick, 1–2 cm long, minutely puberulous; pedicels thick, 0.5–1.5 mm long densely hirtellous; bracts deciduous, oblong, 0.6 mm long, acute at apex. Flowers 10–12 mm diameter, white. Sepals 1–1.2 × 1.5–2 mm, connate at base, rounded at apex, 1–3 glands on dorsal surface, almost glabrous, margins ciliate. Petals 4–5.8 × 2.1–2.3 mm, thick, oblong, apex acute, minutely, sparsely appressed-pubescent and with minute thick resiniferous hairs. Stamens (26-) 30–40, filaments 3.5–3.8 mm long alternating with shorter ones 2.5–3.3 mm long, connate at base into a tube 0.7–1 mm high, anthers 0.8 mm long, attached at base, connective ovate, acute, thecae 2, elliptic, bilocular, 0.5–0.6 mm long, laterally attached, longitudinally dehiscent. Disc cupular, surrounding ovary, 1 mm high, dentate, glabrous. Ovary globose, 1.8–2 mm diam, striate, densely velutinous-tomentose, 5-locular, locules biovulate; ovules 0.6–0.7 mm long; style short, 1–1.5 mm long, hirtellous near base, glabrous above, stigma capitate, slightly 5-lobed. Drupe ellipsoid, rounded at both ends, 2.2 × 1.8–1.9 cm, exocarp thin, minutely velutinous, rugulose, resinous; endocarp elliptic, rugulose, woody, with 5 or 6 longitudinal plates for dehiscence, 1.7–1.8 × 0.3–0.5 cm.

Distribution and habitat. Atlantic coastal forest and restinga of Bahia and Espírito Santo, Brazil (Fig. 10.28).
Phenology. Flowering mainly from April to July, fruiting from July to October.
Local names and uses. *Murtin.*
Conservation status. Least Concern (LC).

Selected specimens examined. BRAZIL. BAHIA: Mun. Itabuna, Km 80 Betanha-Canavieiras, 13 Jul 1964 (fr), *Silva 58,410* (K, NY, US); Mun. Ilhéus, Km 22 S. Luzia-Canavieiras, 27 Apr 1972 (fl), *Santos 2290* (CEPEC, MO, RB, US); Mun. Canavieiras, 22 km W. of city, 13 Jul 1978 (fr) *Santos & Silva 3287* (CEPEC, K, NY, RB, US); Campo Lució, Ramal da Fez 4 Jun 1981 (fl bud), *Hage & Santos 906* (CEPEC, US); Camacan road, 11 Apr 1965 (fl), *Belém & Magalhães 748* (NY, UB, US); Mun. Maraú, road Ubaita-Mara, 13 Jun 1979, *Mori* et al. *11,983* (CEPEC, K, NY, RB, US); Mun. Belmonte, Rio Ubú, toward Itapebí, 18 May 1979 (fl, imm fr), *Mattos Silva* et al. *403* (K, NY, US); Belmonte, 6 Jul 1966, *Belém 2496* (NY, UB, US); Itapeibí-Belmonte, ramal Mogiquiçaba, 9 Jul 1980 (fr), *Silva & Brito 960* (CEPEC, RB, US); Mun. Prado, 10 km N of Prado, road to Curumuxatiba, 17°17′S, 39°18′W, 19 Oct 1993 (fl, imm fr), *Thomas 9978* (CEPEC, MBM, K, MO, NY); 4.5 km N of Curumuxatiba, 21 Oct 1993 (fr), *Thomas 10,083* (CEPEC, K, MBM, MO, NY, RB); 3 km N of Prado, −17.31, −39.22, 17 Jun 2005 (fl, fr), *Stapf 462* (HUEFS, NY); Serra Chapadinha, near Rio Mucugezinho, Lençois, −12.462567, −41.390413, 1 Dec 1979 (fl), *Guedes* et al. *5456* (ALCB, CEPEC); Mun. Una, Reserva Biológica Una, 15°09′S, 38°05′W, 21 Aug 2009 (fr), *Mattos-Silva 5085* (CEPEC). **ESPÍRITO SANTO**: Linhares, −19.39, −40.07, 6 Apr 2006, *Paciencia 2335* (CVRD, ESA), Linares, 16 Oct 1992, *Folli 1683* (CVRD, MO, RB); Meaípe, Guarapari, −20.658, −40.511, *Lima 2918* (K, MBM, NY, RB); Itaúnas, Conceição da Barra, −18.5933, −39.7322, *Lima 2976* (K, MBM, RB); Reserva Natural da Vale do Rio Doce, Nativo da Gavea, Km 5, Linhares, −19.13, −39.96, 23 Jun 2009 (fr), *Meireles 613* (NY, RB); Mun. Vila Velha, Restinga de Lagoa do Milho, 20 Jul 1973 (fr), *Araújo 320* (K, RB).

2. **V. barbourii** Standl., Trop. Woods 75: 5. 1943; Cuatrecasas, Contr. U.S. Natl. Herb. 35(2): 57. 1961; Zamora, Manual de Plantas de Costa Rica 6: 14. 2007. Type. Costa Rica, San José: 1 mile N of San Isidro del General, 2500 feet, 11 Jun 1943, *W. A. Dayton & W. R. Barbour 3129* (holotype, US 00101248, isotypes, A

00043832, F V0060795, F 00043832, MO 158115; NY 00388415, WIS 00000514).

Large trees, the young branches angular, glabrous, inconspicuously lenticellate. Leaves with petioles 6–10 mm long, flat above, slightly winged, thickened at base; laminas elliptic, or oblong-elliptic, 2.5–15 × 2.4–6.5 cm, coriaceous, glabrous, cuneate at base, obtuse or emarginate at apex, margins entire, 1 pair of basal glands on lower surface; midrib plane above, prominent beneath; primary veins 7–9 pairs, plane above, prominent beneath, venation prominulous and conspicuous. Inflorescences axillary or subterminal, cymose-paniculate, as long as the leaves, peduncle glabrous, branches dichotomous, short-pubescent; peduncles 0.5–2 cm long, puberulous; pedicels thick, 1–2 mm long, minutely puberulous; bracts caducous. Calyx 1–1.5 mm high, slightly puberulous, deeply lobed. Sepals rounded, ciliate. Petals oblong, subacute, attenuate toward apex, 7 × 2.5 mm, adpressed pubescent on exterior. Stamens 50–60, glabrous, filaments 5–7 mm long, connate at base, anthers ovate-lanceolate, 0.8 mm long, with 4 small ellipsoid lobes, connective thick, acuminate-lanceolate, twice as long as thecae, thecae 2, 0.3 mm long, basal, elliptic. Disc thick, glabrous, short-denticulate. Ovary ovoid, 1.5–2 mm high, densely tomentose-hirsute, 5-locular, locules biovulate; style erect, glabrous, 4 mm long. Drupe ovoid-oblong, 2.8–3 × 1.8–2 cm, smooth, obtuse at base, attenuate at apex; endocarp ellipsoid-ovoid, 2.7 × 1.6 cm, obtuse at both ends, with 5 valves ca 1.8 cm long, 4–5 mm wide.

Distribution and habitat. Lowland rainforest of Nicaragua and Costa Rica (Fig. 10.27).

Phenology. Flowering from January to June, fruiting from September to October.

Local names and uses. *Campano, caracolillo, chiricano, chiricano alegre, níspero, ira chiricana, chirricano triste*.

Conservation status. Least concern (LC).

Illustrations. Allen (1956), fig. 20: Cuatrecasas (1961), fig. 9 f–h; Zamora (2007), 14.

Additional specimens examined. NICARAGUA. ATLÁNTICA SUR: La Bodega, Río Kuhra, S. of Bluefields, 11°54′N, 83°55′W, 10–20 m, 15 Feb 1993, *Zamora & Castrillo 1963* (CR, MO). **COSTA RICA**. **CARTAGO**: 2 km S of Río Hermoso, 8 km from San Isidro del General, 2100 feet, 6 Sep 1943, *Barbour 1018* (F, WIS, US). **PUNTARENAS**: Reserva Indigena. Guaimí, Cantón de Osa, NE junction of Ríos Pavón & Rincón, 8°36′N, 83°31′W, 50–200 m, 20 Oct 1990 (fr from ground), *Hammel* et al. *17,917* (MO); Osa, 27 Aug 1993 (fr), *Zamora* et al. *1862* (CR, K, MO, NY); Llorona-San Pedrillo trail, 8°35′N, 83°43′W, 100 m, 10 Mar 1978 (fl), *Hartshorn 2139* (MO); Golfito, Jiménez, 8°24′N, 83°20′W, 100 m, 20 Sep 1991 (fr), *Aguilar 453* (MO); Osa, Ganado-Guerra, 8°43′N, 83°09′W, 1–300 m, 29 Jan 1992 (fl bud), *Hammel* et al. *18,171* (CR, K, MO); Osa Rancho Queimado, 8°41′N, 83°33′W, 0 m, 30 Mar 1991 *Saborio* et al. *130* (INB, MO); Reserva Forestal Golfo Dulce, 300 m, 8°43′N, 83°34′W, 4 Apr 1996 (fl), *Aguilar 4537* (INB, K, MO, NY).

3. **Vantanea compacta** (Schnizl.) Cuatr., Contr. U.S. Natl. Herb. 35(2): 65. 1961; *Humirium compactum* Schnizl., Iconogr. Fam. Regn. Veg. [Abbild. Nat. Fam.] 3: Ordo 222, pl. 222. 1843–1870; *Vantanea panniculata* Urb., Fl. Bras. (Martius) 12(2): 450, pl. 96, 1877. synypes: Brazil: "In silvis prob. Bahiensis prope Jacobinam: *J. S. Blanchet n. 3305, 3362, 3837* (BM, P)." Vantanea contractum Moric. ex Urb. [based on the unpublished name *Humirium contractum* Moric. msc.], Bot. Jahrb. Syst. 15, Beibl. 34: 3. 1892. Type. Brazil. Bahia: Jacobinha, Igreja Velha, 1842 (fl), *J. S. Blanchet 3362* (holotype, lectotypified by Cuatrecasas 1961, p. 67, P; isolectotypes, BM, BR 0000005119004, BR 0000005111985, BR0000005111626, F V0060766, F V0060767, FI-W, G, GH00043822, K 000407353, K 000407354, MO 1836401, NY 00445960, OXF, P 01903288, P 01903289, P 01903290, WU).

   Vantanea compacta var. *grandiflora* Urb., Bot. Jahrb. Syst. Beibl. 34: 3. 1893; *Vantanea compacta* (Schnizl.) Cuatr. subsp. *compacta* var. *grandiflora* (Urb.) Cuatr., Contr. U.S. Natl. Herb. 35(2): 67. 1961. Type. Brazil, Rio de Janeiro, Alto Macahé, 5 Mar 1988 (fl), *J. F. M. Glaziou 16,723* (holotype B, lost; lectotype, US 00101246[1]; isolectotypes, BR 0000005118984, E 00326637, K 000407352, NY 00446024, P 01903285, P 01903286).

Large trees, the young branches glabrous or slightly puberulous, shiny, densely lenticellate. Leaves with petioles 3–6 mm long, thickened at base; laminas oblanceolate or lanceolate-elliptic, 3.5–9.5 × 1.3–4 cm, coriaceous, glabrous or sparsely appressed puberulous, cuneate at base, acute or acuminate or long-cuspidate at apex, margins entire, callous-thickened and slightly revolute, 0–4 pairs of basal glands on lower surface; midrib plane above, prominent beneath; primary veins 8–9 pairs, slightly prominulous on both surfaces. Inflorescences axillary, cymose-paniculate, equalling or shorter than leaves; branches angled, minutely hirtellous-pubescent; peduncle thin, 0.5–1.5 cm long, sparsely short-puberulous; pedicels 0.5–1.5 mm long, thick, minutely hirtellous; bracts deciduous. Calyx cupular, 1 mm high, slightly puberulous. Sepals, subrotundate, 0.3–0.5 mm long, margins minutely ciliate. Petals linear-oblong, white, subobtuse, glabrous, 4–5 × 1.5–1.8 mm. Stamens 50–60, filaments slender, 1.5–7 mm long, connate at base into a tube 0.6 mm high, anthers ellipsoid, 0.7–0.9 mm long, with 4 oblong, apiculate lobes 0.5–0.6 mm long, connective with acute tip 0.3–0.4 mm long, thecae 2, lanceolate-ellipsoid inserted along most of anther. Disc annular, surrounding ovary, glabrous, 1 mm high, strongly denticulate. Ovary ovoid, 1.5 mm high, minutely velvety-sericeous, 5 locular, locules biovulate; style 2–3 mm long, erect, pubescent on lower half, stigma obtuse. Drupe oblong-obovate-ellipsoid, rounded at apex, attenuate at base, 2.4–2.8 × 1.5–1.8 cm; exocarp smooth, 1–1.5 mm thick, coriaceous; endocarp

[1] <Footnote ID="Fn1"><Para ID="Par770">Cuatrecasas (<CitationRef aid:cstyle="CitationRef" CitationID="CR28">1961</CitationRef>) selected Glaziou 16,723 as the lectotype of var. *<Emphasis aid:cstyle="Italic" Type="Italic">grandiflora</Emphasis>* and cited the US and P sheets as isotypes. Since the Berlin holotype was lost, it is also necessary to indicate a lectotype specimen, so I have chosen the US sheet cited by Cuatrecasas.</Para></Footnote>

woody, oblong ellipsoid, obtuse at apex, slightly narrowed at base, 2.5 × 1.4 cm, surface rugulose, with 5 slight furrows toward base and 5 longitudinal, oblong valves 1.8–2 cm long.

Key to Varieties of Vantanea Compacta

1. Drupe oblong-obovate-ellipsoid, 2.4–2.8 × 1.5–1.8 cm.. ....**3a. V. compacta var. compacta**
1. Drupe globose-ovoid, 1.6–1.8 × 1.4–1.6 cm.. ....**3b. V. compacta var. microcarpa**

3a. **Vantanea compacta** (Schnizl.) Cuatr. **var. compacta**, Contr. U.S. Natl. Herb. 35(2): 65. 1961.

Drupe oblong-obovate-ellipsoid, 2.4–2.8 × 1.5–1.8 cm.

Distribution and habitat. Forests of eastern and southern Brazil (Fig. 10.28).
Phenology. Flowering from December to April, fruiting from May to October.
Local names and uses. *Aroeirana, guaraparím.*
Conservation status. Least concern (LC).
Illustrations. Schnizlein. (1843–1870) pl. 222; Cuatrecasas (1961), figs. 7a, b, 10c–e.

Selected specimens examined. BRAZIL. CEARÁ: Guaramiranga, 8 Aug 1908 (fr), *Ducke 1513* (INPA, MG). **BAHIA:** Abaíra, Catolés de Cima, 13°17′S, 41°52′W, 1150 m, 23 Oct 1992 (fr), *Ganev 1329* (HUEFS, K, NY, USP); Rio de Contas, road to Fazenda Marion, 900 m, 13°37′14" S, 41°45′51″W, 5 Feb 2004 (fl), *Harley 54,757* (HUEFS, K); Jacobina, 11°12′S, 40°30′W, 800–1000 m, 30 Dec 2004 (fl), *Forzza 3884* (K, NY, RB); Diamantina, Morro de Chapada, 8 Sep 1990 (fr), *Lima* et al. *3914* (CEPEC, K, RB); Chapada Diamantina, track to Fazenda Volta da Pedra, km 511, Itagu, Andarai, −12.88, −42.13, 9 Nov 1997 (fl), *Guedes 5517* (ALCB, MBM); Piatã, road to Inúbia, 1100 m, 15 Feb 1987 (fl), *Harley 24,305* (F, K, NY); Mun. Abaíra, Samabaia-Serrinha, 4 km N of Castolés, 22 May 1992 (fr), *Ganev 348* (HUEFS, K, NY); Summit of Morro do Chapeu, 8 km SW of Morro do Chapeu, road to Utinga, −12.58, −41.2, 3 Mar 1997 (fl), *Harley* et al. *19,327* (CEPEC, IPA, NY, UEC). **ESPÍRITO SANTO**: Reserva Florestal, Linares, near Estrada 142, 29 Mar 1973 (fl), *Spada 233* (CEPEC). **RIO DE JANEIRO**: Mun. Novo Friburgo, Reserva Ecológica Municipal Macaé de Cima, near Hotel Garlip, 22°33′S, 42°30′34″W, 11 Sep 1989 (fl), *Lima 3678* (NY, RB); Petrópolis, 5 Mar 1880 (fl), *Glaziou 11,829* (K, P, NY, R); Corocovado, 4 Feb 1880, *Glaziou 11,828* (BM, F, GH, IAN, K, NY, P); Alto Macahé, 11 Mar 1891 (fl), *Glaziou 18,181* (K, NY, P, RB, US), 7 Apr 1891 (fl), *Glaziou 18,182* (F, NY, P, RB, US); Estrada do Redentor, Corcovado, 16 Sep 1986 (fl), *Kuhlmann RB47414* (K, RB); PN Serra das Orgãos, Rio Paquenque, trail to Rancho Frio, Mun. Teresópolis, 1385 m, −22.4122, −43.9656, 10 Oct 2005 (fr), *Wesenberg 1012* (CEPEC, ESA, NY, RB). **SÃO PAULO**: Jardim Botánico, São Paulo, 10 Oct 1933 (fl), *Hoehne, 29,281* (A, GH, NY, P, RB, S); Parque da Avenida, São Paulo, 20 Feb 1934 (fl), *Lemos s.n.* (K, NY, P, US, USP); Mun. Cananéia, Ilha do Cardoso, 10 Apr 1986 (fr), *Barros & Martuscelli 1272* (ESA, F, K, SP); Km 31 road SP226, Cananéia, −24.88, −47.93,

27 Apr 2004 (fr), *Urbanetz 381* (UEC); Parque Estadual Carlos Botelho, São Miguel Arcanjo, −24.06, −47.96, 5 May 2005, *Baitello* et al. *1799* (FUEL, MBM, SPSF). **PARANÁ**: Paranaguá, Pico Torto, 14 Mar 1969 (fl), *Hatschbach 21,265* (F, FUEL, K, MBM, MO, NY, RB, UFPR, UPBC); Mun. Morretas, 23 Jan 1968 (fl) *Hatschbach 20,893* (K, MBM, UPBC); Serra de Araraquara, Morro do Cauvi, Guaratuba, −25.88, −48.57, 25 Mar 1965 (fl), *Hatschbach 12,500* (F, K, MBM, NY, UEPG, UPBC, UFPR); Fazenda Primavera, Mun. Adrianópolis, 21 Feb 2000 (fl), *Silva & Abe 3164* (ESA, HUCS, HUEFS, K, MBM, MO, RB, SPSF, UFPR, US); Rio Guraqueçaba, −25.30, −48.99, 23 Jan 1991 (fl), *Hatschbach 54,916* (ASL, FUEL, HUEFS, IPA, MBM, MO, NY, RB, SPSF, UEC). **SANTA CATARINA**: Mata da Azambuja, 23 Feb 1950, *Klein 37b* (S, US); Horto Florestal Instituto Nacional do Pinho, Ibirama, 1 Mar 1954 (fl), *Reitz & Klein 1589* (IAP, NY, S, U, US); Morro da Fazenda, Itajaí, 14 May 1954, *Reitz & Klein 1730* (US), 18 Mar 1954 (fl), *Reitz & Klein 1744* (IAP, MBM, NY, S, U, US); Saco Grande, Ilha de Santa Clara, Florianópolis, −27.59, −48.54, *Klein & Bresolin 7298* (FLOR, MBM, UFRGS); Reserva Volta Velha, Itapoá, −26.11, −48.61, 17 Feb 1993 (fl), *Negrelle 740* (UPCB); Piloes, Palhoça, 400 m, −27.64, −48.66, 5 Apr 1956 (fl, imm fr), *Reitz & Klein 3024* (F, IAP, K, NY, UPCB). **RIO GRANDE DO SUL**: Parque Zoológico do Rio Grande do Sul, Porto Alegre, −30.03, −51.23, 21 May 2002 (fr), *Faustino 32* (HUEFS, K).

3b. **Vantanea compacta** var. **microcarpa** Cuatr., Contr. U.S. Natl. Herb. 35(2): 67. 1961. Type. Bolivia. Chailla (=Challa?), May 1866 (fr) *R. W. Pearce s.n.* (holotype, K 000407348; isotypes, K 000407347, US 00101247).

Trees 10–20 m, the young branches glabrous. Drupe globose-ovoid, 1.6–1.8 × 1.4–1.6 cm, rugulose, rounded at base and acute at apex; endocarp rugulose, rounded at base, obtuse at apex, with 5 obovate-oblong longitudinal valves, ca. 1.4 cm long, in section with 5–10 cavities containing one or two seeds.

Distribution and habitat. Highland forests of eastern Bolivia (Fig. 10.28).
Conservation status. Vulnerable (VU), B1.
Illustration. Cuatrecasas (1961), pl. 2.

Additional specimens examined. BOLIVIA. YUNGAS: Moro, 500–600 feet, Jan1866 (fl), *Pearce s.n.* (BM, K, MO, paratypes). **LA PAZ**: Madidi, Apolo, Machariapo, 1560 m, −14.62083, −68.37027, 2 Sep 2004 (imm fr), *Fuentes & Aldana 6506* (LBP, MO, NY); Franz Tamayo, Madidi, Apolo, Machariapo, −14.42944, −68.36611, 7 Dec 2002, *Miranda* et al. *682* (LPB, MA, MO); Parque Nacional Madidi, Río Tuichi, Centro Cuatro Vientos, 14°35′39" S, 68°42′03″W, 1380 m, 27 Nov 2005 (fl bud), *Arauj-M 2394* (LPB, MO, RB, USZ), 27 Nov 2005 (fr) *2397* (LPB, MO, RB, USZ).

4. **Vantanea deniseae** W.A. Rodrigues, Acta Amazonica 12 (2): 297. 1982. Type. Brazil. Amazonas: Manaus-Itacoatiara Km 99, 17 Sep 1965 (fl, fr), *W. A. Rodrigues & A. Loureiro 7150* (holotype, INPA 15812; isotypes, INPA 15812, MG 100800).

Large trees up to 25 m tall, the young branches glabrous, longitudinally striate. Leaves with petioles 5–10 mm long, thickened at base; laminas obovate or oblanceolate, 3.5–8.5 × 1.5–4.5 cm, rigid-coriaceous, glabrous, cuneate at base, obtuse or emarginate at apex, margins entire; midrib prominulous above, prominent beneath; primary veins 5–10 pairs, slightly impressed above, prominent beneath, venation obscure above, reticulate and prominulous beneath. Inflorescences terminal or subterminal, cymose-paniculate, subcorymbose, shorter than the leaves, branches dichotomous, hirtellous-pubescent; pedicels thick, 1–3 mm long; bracts triangular, ca. 2 mm long, deciduous, hirtellous-pubescent. Calyx cupuliform, 1–1.5 mm high, 5-lobed. Sepals roundish, hirtellous pubescent on exterior, margins ciliate. Petals linear-lanceolate, 7–11 × 1–3 mm, thick, glabrous except on margins. Stamens ca. 200, unequal inserted in 2–3 rings, filaments thin, glabrous, connate at base, 5–10 mm long, anthers 1 mm long, connective ovoid-lanceolate, thecae 2, elliptic inserted at base. Disc annular, tomentellous, 1 mm high, margins entire. Ovary ovoid, 2.5–3 mm high, velutinous-tomentose, 5–7-locular, locules biovulate; style erect, 6–8 mm long, glabrous, except around base, stigma obtuse. Drupe ovoid, 4.5–7 × 3–4.5 cm, rounded at base abruptly pointed at apex, densely minutely velutinous-sericeous; exocarp coriaceous when dry, 3–5 mm thick; endocarp 5.5–6 cm long, ca 3.8 cm broad, woody, rounded at base, abruptly pointed at apex, smooth on exterior, with 6–8 longitundinal slits or opercula for germination, 4–5 cm long, 0.6–0.8 cm wide.

Distribution and habitat. Forest on *terra firme* in the vicinity of Manaus, Brazil (Fig. 10.25).

Phenology. Flowering between August and September, fruiting from November to December.

Local names and uses. None recorded.

Conservation status. Least concern (LC).

Illustrations. Rodrigues (1982), fig. 2.

Additional specimens examined. BRAZIL. AMAZONAS: Manaus-Itacoatiara Km 118–135, 25 Aug 1975 (fl), *Monteiro s.n.* (INPA 50891-A, paratype); Km 57, 14 Sep 1976 (fl), *Damião & Mota 617* (INPA, paratype); Distrito Agropecuario 2.5 km N of Km 34, Estrada ZF-3, 11 Dec 1981 (fr), *Costich & Cardoso 1071* (INPA, paratype); Distrito Agropecuaria 90 km N of Manaus, Fazenda Esteio, −2.38, −59.85, *Sette Silva 1301.421.2* (INPA, NY); Dist. Agropecuario, Reserva 15,018, −2.41, −59.74, Dec 1988 (fr), *Boom 8780* (INPA, MO, NY); Tonantins, −2.8731, −67.8022, 18 Nov 1986, *Cid Ferreira 8459* (INPA).

5. **Vantanea ovicarpa** Sabatier, Brittonia 54: 233. 2002. Type. French Guiana. Pic Matécho, 22.5 km NE of Eaux Claires, 515 m, 3°44.5′N, 53°02.2′W, 7 Sep 2000 (fl), *S.A. Mori & N. P. Smith 25,047* (holotype, NY 718019; isotypes, CAY, MG 175242, MPU, NY 007180019, NY 718020, P 00710334, US 00810841).

Large trees up to 25 m tall, the young branches glabrous, longitudinally fissured, finely lenticellate. Leaves with petioles 8–10 mm long, sulcate, thickened at base;

laminas obovate or oblanceolate, 8–15 × 4–7 cm, rigid-coriaceous, glabrous, sparsely punctate-glandulose, cuneate at base, rounded and emarginate at apex, margins entire; midrib prominulous above, prominent beneath; primary veins 7–10 pairs, slightly impressed above, prominent beneath, venation obscure above, reticulate and prominulous beneath. Inflorescences terminal or subterminal, cymose-paniculate, subcorymbose, shorter than the leaves, minutely hirtellous-pubescent; pedicels thick; bracts triangular, ca. 2 mm long, deciduous, hirtellous-pubescent; pedicels thick; Calyx cupuliform, 1.5–2 mm high, 5-lobed. Sepals roundish, ca 2 mm long, minutely hirtellous pubescent on exterior, margins ciliate. Petals white, linear-lanceolate, 9–13 × 3.5 mm, thick, sparsely appressed puberulous, margins ciliate. Stamens 120–200, unequal inserted in 2–3 rings, filaments thin, glabrous, connate at base, 6–12 mm long, anthers 1.1 mm long, connective ovoid-lanceolate, thick, thecae 2, bilocular, elliptic, inserted at base. Disc cupular, dentate, densely tomentellous, 1.2 mm high, margin entire. Ovary ovoid-turbinate, 2.5–3 mm high, densely crisp-pubescent, 6–7 locular, locules biovulate; style erect, 6–7 mm long, glabrous, except pilose around base, stigma obtuse. Drupe ovoid, 5–7 × 3–4.5 cm, rounded at base abruptly pointed at apex, densely minutely velutinous-sericeous; exocarp coriaceous when dry, ca 5 mm thick; endocarp 5–6 cm long, 3–4.5 cm broad, woody, smooth, rounded at base, abruptly pointed at apex, smooth on exterior, with 6–7 longitudinal valves or opercula for germination, 4–5 cm long, 0.6–0.8 cm wide.

Distribution and habitat. Forest on *terra firme* in French Guiana and adjacent northern Amazonian Brazil (Fig. 10.27).
Phenology. Flowering in September, fruiting between December and January.
Local names and uses. None recorded.
Conservation status. Least concern (LC).
Illustrations. Sabatier (2002), fig. 1.

Additional specimens examined. FRENCH GUIANA. Atachi Bacca, Mt. Inini region, 780 m, 3.55, −53.9166, 20 Jan 1989 (old fr), *Granville* et al. *10,792* (CAY, MO, NY, US, paratypes); Pic Matécho, NE Saül, 590 m, 3.7449, −53.0402, Feb 1980 (fr from ground), *Granville 1753* (CAY, paratype); Pic Matécho, 22.5 km NE of Les Eaux Claires, 500 m, 3°45′N, 53°2′W, 14 Sep 2000 (fl), *Mori & Smith 25,159* (CAY, K, NY, paratypes). **BRAZIL. AMAZONAS**: Manaus, Distrito Agropecuário, Reserva 1501, 2°24′26" S, 59°43′40″W, 9 Aug 1989 (fl), *Mori* et al. *20,717* (INPA, NY); Distrito Agropecuário, Reserva ZF3, BR 174, Km 64, 2°25′52" S, 59°47′38″W, 9 Sep 2006 (fl), *Mello 58* (INPA); ARIE-PDBFF, Estrada ZF3, Km 37, −2.4416, −59.7855, 10 Oct 2014 (fl bud), *Correa 431* (INPA); Manaus-Itacoatiara road, Km 69–70, −2.65696, −59.64833, 5 Sep 1973 (fl), *Prance* et al. *17,533* (INPA, MG, MO, NY); Rio Abacaxis, 1 Jan 1989 (fl), *Todzia 2341* (INPA, NY). **PARÁ**: Oriximinã, Rio Trombetas, Cachoeira Porteira, −1.015, −51.018, 20 Aug 1986 (fl), *Cid Ferreira 7867* (F, INPA, MIRR, MO, NY).

This species is close to *Vantanea deniseae*, and both species are characterized by the exceptionally large fruit and the densely pubescent discs. *Vantanea. ovicarpa* consistently has much larger leaves than *V. deniseae*. Sabatier (2002) noted in his description of this species that tooth marks on the endocarps suggest that rodents prey upon and/or disperse the seeds.

6. **Vantanea depleta** McPherson, Ann. Missouri Bot. Gard. 75: 1148. 1988. Type. Panama. Panamá: Cerro Jefe, 650 m, 2 May 1987 (fl), *G. McPherson & Stockwell 10,892* (holotype, PMA 775; isotypes, F V0060796, MO 155975, US 385521).

Trees up to 40 m tall, the young branches often strongly angled, glabrous, with elongate lenticels. Leaves with petioles 3–9 mm long, swollen at base, glabrous; laminas elliptic or elliptic-ovate, 5.5–16 × 3–8 cm, coriaceous, glabrous, cuneate at base often somewhat reflexed abaxially, apex obtuse acute or bluntly acuminate, margins entire, 1 pair of basal glands on lower surface; midrib prominent above, prominulous beneath; primary veins 6–11 pairs, slightly prominulous on both surfaces, glabrous, with 1 or 2 small sunken laminar glands associated with the veins. Inflorescences terminal or subterminal, cymose-paniculate, shorter than the leaves, branches puberulent; peduncles 0.5–2 cm long, puberulous; pedicels 1.5–3 mm long, less puberulent than branches; bracts deciduous. Sepals semicircular, 1 mm long, 1.5–2 mm broad, obtuse, puberulent, mostly bearing one centrally raised crateriform gland. Petals narrowly triangular-ovate, 5 mm long, 2 mm wide, glabrous on both surfaces, white. Stamens 15–28, filaments 3–5 mm long, of alternating lengths, fused basally into a tube for 1 mm, glabrous, anthers ca 0.7 mm long, thecae bilocular, about as long as the distal prolongation of the connective. Disc 1 mm high, glabrous, sharply dentate. Ovary ovoid, 2 mm long, densely puberulent; locules biovulate, style 3 mm long, geniculate. Drupe ovoid, 2.5–3.3 × 1.5–1.7 cm, puberulent, rounded at base, acute at apex; endocarp smooth with 5 broad ribs alternating with 5 oblong lingulate valves, 2.4–3.3 cm long, 1.5–1.8 cm diam.

Distribution and habitat. Forests of Nicaragua, Costa Rica and Panama from sea level to 1200 m (Fig. 10.27).
Phenology. Flowering from February to June, fruiting from May to July.
Local names and uses. None recorded.
Conservation status. Least Concern (LC); Vulnerable (VU), A1a. (M. Mitré, pers. comm. 1998).
Illustrations. McPherson (1988), fig. 1.

Selected specimens examined. NICARAGUA. ATLÁNICO SUR: Caño El Tigrillo (La Picada), 50 m, 11.65, −84.03333, 8 Nov 1982 (imm fr), *Laguna 159* (MO). **COSTA RICA. HEREDIA**: Sarapiqui, N. of Chilamate, 7 km W. of Puerto Viejo, 10°28′12″N, 84°03′36″W, 30 m, 21 Jun 1982 (fl), *Hammel & Trainer 12,954* (MO). **PANAMA. CANAL AREA**: Fort Sherman, S. of Chagres, 9°17′N, 79°59′W, 150 m, 17 Jul 1997 (fr), *Foster* et al. *15,737* (F); Barro Colorado I, Tree B535, 9°09′20″N, 79°51′10″W, 10–170 m, 17 Sep 1976 (fr), *Garwood & Foster 440* (MO, W, paratypes). **COCLÉ**: Road La Pintada-Cochlesito, 8°41′06″N, 80°27′15″W, 600 m, 7 Feb 1983 (fl), *Hamilton & Davidse 2817* (MO). **COLÓN**: Donosos, 8°49′51″N, 80°40′55″W, 250 m, 8 Mar 2008 (fr), *McPherson 20,471* (MO). **PANAMÁ**: Cerro Jefe, 9°15′N, 79°30′W, 650 m, 29 May 1987 (fr), *McPherson 11,008* (MO, paratype). **SAN BLAS**: between Río Irgandi and Río Cartí Senni,

9°25′N, 78°51′W, 20 Dec 1985 (old fr), *de Nevers & Herrera 6597* (MO, paratype). **VERAGUAS**: Cerro Tute, 8°28′N, 81°05′W, 1200 m, 28 Sep 1972 (st), *Lao & Gentry 530* (MO, NY, paratypes); Santa Fé, Carretera Guabal, 689 m, 8°30′N, 81°05′W, 24 Apr 2012 (fl), *Ortiz 557* (SCZ).

7. **Vantanea guianensis** Aubl., Hist. Pl. Guian. 1: 572. 1775; Urban, Fl. Bras. (Martius) 12 (2): 452. 1877; Cuatr., Contr. U.S. Natl. Herb. 35(2): 71. 1961; Cuatrecasas & Huber, Fl. Venez. Guayana (Steyermark et al.) 5: 640. 1999. *Lemniscia guianensis* (Aubl.) Gmel., Syst. nat. (Gmelin): 817. 1791, Raeuschel, Nomencl. Bot. ed. 3: 156. 1797; *Lemnescia floribunda* Willd., Sp. pl. 2: 1172.1800; *Lemniscia floribunda* Spreng., Syst. Veg. 2: 600.1825. Type. French Guiana, Comté de Gêne. Lectotype of Cuatrecasas (1961: 73), *J. C. B. F. Aublet s.n.* (lectotype, S-R9886; isolectotype, BM 000795993).

Large trees, the young branches glabrous, subterete, lenticellate. Leaves with petioles 6–12 mm long, flat and sulcate above, glabrous, swollen at base; laminas elliptic or oblong-ellipitic, 6–18 × 2.7–7 cm, chartaceous or thin-coriaceous, glabrous, obtuse-cuneate at base, decurrent onto petiole, attenuate, shortly obtusely acuminate at apex, margins entire; midrib prominent on both surfaces; primary veins 13–15 pairs, prominulous on both surfaces, reticulate venation inconspicuous. Inflorescences axillary and terminal, cymose-paniculate, subcorymbiform, shorter than the leaves; branches glabrous, peduncle stout, short, 0.5–1 cm long, subterete, glabrous; pedicels 4–6 mm long, thin, glabrous; bracts deciduous, ovate, 1 mm long. Calyx thick, deep-cupular, 4 mm long, glabrous, slightly dentate with obtuse teeth, glandular on exterior. Petals red or purple, linear, acute, glabrous 3–4 cm long, 2.5 mm broad. Stamens 60–80, glabrous, filaments 2.5–3.2 cm long, connate at base into a tube 5–7 mm long, anthers 1 mm long, oblong-ellipsoid, lobes 4, oblong, 0.8 mm long, connective thick with a short acute tip, thecae 2, elliptic, ca 0.4 mm long. Disc tubular, glabrous, 1.5 mm high, surrounding ovary, margin smooth. Ovary ovoid, 2 mm high, glabrous, 5-locular, locules biovulate; style glabrous, 3–3.5 cm long, stigma obtuse. Drupe ellipsoid-ovoid, smooth, 6 cm long, 4.5 cm broad; exocarp fleshy, 3–5 mm thick, coriaceous when dry; endocarp 5 × 3.8 cm, hard, woody, broadly rounded at base, abrubtly and obtusely apiculate at apex, deeply anfractuose-rugose-cerebriform, with 5 longitudinal deep, narrow furrows, with a hidden longitudinal narrow operculum hidden at base; seeds oblong, 3.2 cm long, 0.3 cm thick.

Distribution and habitat. Widely distributed in the riverine and *terra firme* forests of eastern Venezuela, Colombia, the Guianas and northern Brazil and west to Peru and Ecuador (Fig. 10.25).

Phenology. Flowering from March to September, fruiting from November to January.

Local names and uses. French Guiana: *Iouantan*, which is the derivation of Aublet's name *Vantanea*. Brazil: *uchirana*; Peru: *manchari caspi*; Colombia: *largatocapi*.

Conservation status. Least concern (LC).

Illustrations. Cuatrecasas (1961), figs. 14 a, b, 15; Spichiger et al. (1990), fig. 4; Mori et al. (2002), fig. 147; Kubitzki (2013), fig. 50.

Selected specimens examined. COLOMBIA. AMAZONAS: Amacayacu, −3.7, −70.2500, 25 Mar 1992 (st), *Rudas* et al. *3681* (MO). **VENEZUELA. AMAZONAS**: Río Casiquiare, Ríos Pacimoni-Yatua, 20–25 km above Piedra Arauicana, 30 Sep 1957 (fl), *Maguire* et al. *41,640* (F, K, NY, RB, US, Usw); Cerro Neblina, Río Mawarinuma, 190 m, 0°50′N, 66°05′W, 24 Apr 1984 (fl), *Gentry & Stein 46,932* (F, K, MO). **GUYANA**. Roraima, 1842–43 (fl), *Schomburgk 982* (BM, NY, P); Corentyn River, s.d. (fl), *Rob. Schomburgk 1581* (BM, NY, US), without locality, s.d. *Schomburgk 47* (K). **FRENCH GUIANA**. Gourdonville, 2 Aug 1914, *Benoist 1530* (P); Eaux Claires, 14 Feb 1993 (st), *Pennington* et al. *13,855* (K); Saül, Monts La Fumée, 14 Aug 1987 (fl), *Mori 18,722* (NY). **ECUADOR. NAPO**: Orellana, Yasuni NP, 14–15 km from Maxus/YPF Pipeline, 0.51666, −76.5333, 200–399 m, Sep 1999 (st), *Pitman & Delinks 1951* (MO, QCA, QCNE). **SUCUMBIOS**: Reserva Cuyabeno, Lago Agrio, Laguna Canangüeno, 230 m, −0.3333, −76.216666, 18 Nov 1991 (fr), *Palacios* et al. *9132* (MO, ECUAMZ). **PERU. LORETO**: Maynas, Quebrada Sucursari, Río Napo, 150 m, −3.25, −72.9166, 5 Jul 1983 (fr), *Gentry* et al. *42,586* (MO). **BRAZIL. AMAZONAS:** Manaus, Estrada do Aleixo, 20 May 1936 (fl), *Ducke 200*, (A, F, IAN, K, MO, NY, S, US), 31 Mar 1932 (fl), *Ducke RB 23814* (K, RB, S, U, US); Manaus, Reserva Ducke, 9 Oct 1989 (fl), *Kawasaki* et al. *377* (INPA, K, MO, NY), 1 Nov 1994 (fl bud, fr), *Vicentini 761* (INPA, K, MG, NY, RB); Mun. Presidente Figueredo, Balbina, 9 Mar 1986 (fl), *Cid Ferreira 6686* (HFSL, INPA, K, NY, PNFM); Humaitá-Lábrea Km 90, 24 Nov 1966 (fl), *Prance* et al. *3283* (F, INPA, K, MG, NY); Porto Urucú, Tefé, −3.35, −64.71, 20 Jul 1991 (fl), *Tavares 523* (INPA); Coari, −4.84, −65.34, 31 Jul 2006 (fl), *Amaral 2987* (INPA). **PARÁ**: Belém-Brasília, Km 93, 5 Aug 1963 (fl), *Maguire* et al. *56,006* (F, K, MO, NY, RB), Km 137, 14 Aug 1964 (fl), *Prance & Silva 58,722* (F, K, NY, RB); Tomé Açú, Rio Acará, 5 Aug 1936 (fl), *Mexia 6049* (A, BM, F, GH, K, MO, NY, S, U, US); Moju, 9 May 1997 (fl), *Martins-da-Silva 31* (IAN, K, MFS). **AMAPÁ:** Cultivated, 1956 (fl), *Bastos 244* (F, MO, NY, RB). **MARANHÃO**: Parque Indígino Guajá, Monção, −3.12, −46.08, 22 Jun 1987, *Balée 3455* (NY).

8. **Vantanea macrocarpa** Ducke, Arq. Inst. Biol. Veg. 1: 205. 1935; Cuatrecasas, Contr. U.S. Natl. Herb. 35(2): 68. 1961. Type. Brazil. Amazonas: Manaus, 25 Apr 1932 (fl, fr), *Ducke RB 20427* (lectotype, **here designated**, RB 00539076; isotypes, F 0060797, K 000407351, NY 00388417, P 01903284, RB 00570495, RB 00570494, RB 00539076, S-R9887, U 0002496, US 00101244). Fig. 10.24.

Large trees, the young branches glabrous, densely lenticellate. Leaves with petioles 8–16 mm long, thick, canaliculate above; laminas ovate-oblong or oblong-elliptic, 10–20 × 5–10 cm, thick-coriaceous, glabrous, lustrous above, rounded or obtuse at base, decurrent onto petiole, acuminate at apex, margin entire and thickened; midrib prominulous and flattened above, prominent and glabrous beneath; primary veins 16–18 pairs, slender, prominulous on both surfaces, minor veins obscure, some minute glands more or less seriate, conspicuous. Inflorescences

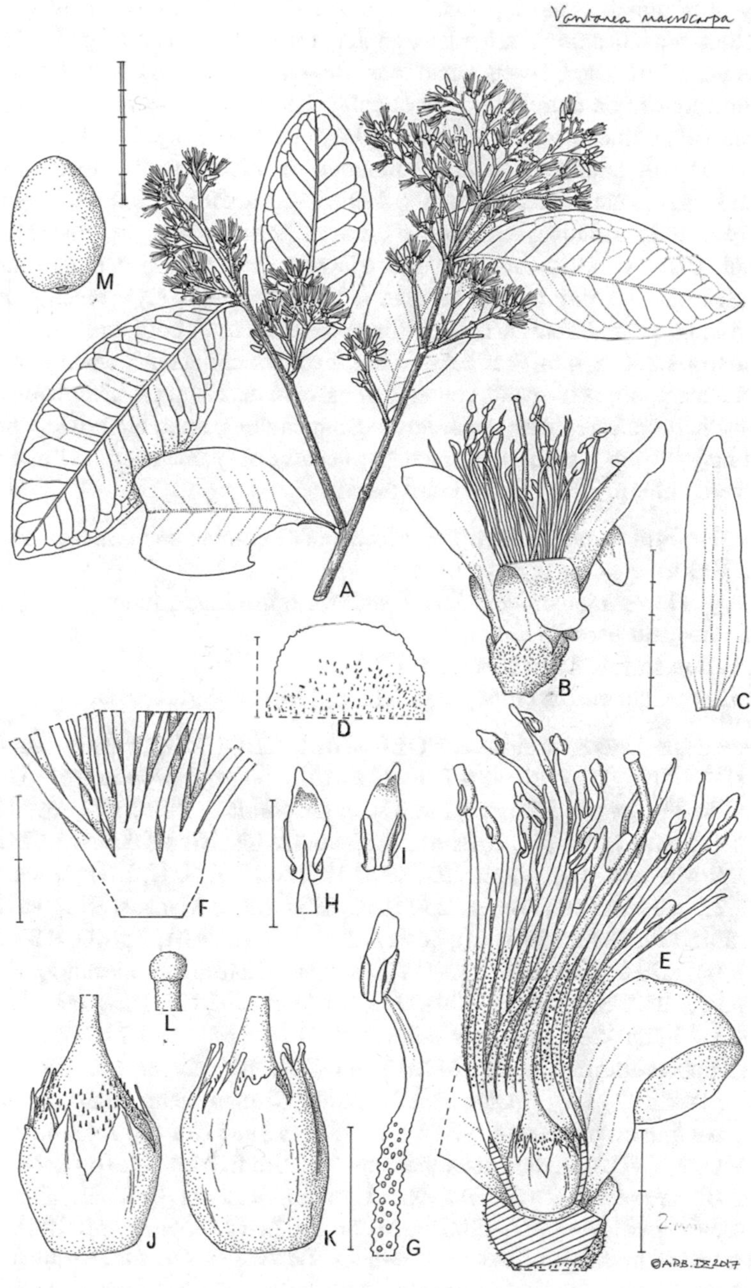

Vantanea macrocarpa del. Andrew Brown September 2017

Fig. 10.24 *Vantanea macrocarpa*. (**a**) Habit. (**b**) Flower. (**c**) Petal. (**d**) Sepal. (**e**) Flower with 2 petal and sepals, part of staminal tube removed. (**f**) Inner surface of part of stamina tube to show branching. (**g**) Upper part of stamen. (**h**) Outer face of anther and filament; inner face of anther. (**j**, **k**) Two examples of disc. (**l**) Stigma. (**m**) Fruit. Dashed bar = 500 μm, single bar = 1 mm, graduated single bar = 2 mm and 5 mm, graduated double bar = 5 cm. (A–L from *Sothers* et al. *843*; M from *Ducke RB 20427*)

axillary or terminal, cymose-paniculate, much shorter than the leaves; peduncle short, stout, branches thick, articulate, angled, minutely hirtellous; pedicels thick, 0.5 mm long, hirtellous; bracts deciduous, thickish, clasping, ovate, 1.5–2.5 mm long, tomentulose on exterior. Calyx cupular, 1.5 mm high, hirtellous. Sepals 5, roundish. Petals linear, white, subacute, glabrous, 1.cm long, 1.5–1.8 mm wide. Stamens ca 70, filaments slender, minutely papillose-verruculose, unequal, 6–8 mm long, connate at base in a tube 2 mm long, anthers oblong, 0.9 mm long, with 8 basal ellipsoid lobes, connective fleshy, elongate, subacute, thecae 2, elliptic, basal. Disc cylindric, 1.8 mm high, glabrous, the margin dentate-fimbriate. Ovary glabrous, 1.5 mm high, 5 locular, locules biovulate; style erect, glabrous, 7–8 mm long, stigma obtuse. Drupe ellipsoid, rounded at both ends or obtuse at apex, almost smooth, 4.6–10 × 3.6–4.5 cm; exocarp coriaceous when dry, 3 mm thick; endocarp ellipsoid-ovoid, rounded at base, obtuse at apex, alveolate-rugose, the rugosities connected with ten large irregular cavities; opercula broadly oblong, 3.4 cm long, 9 mm broad, inconspicuous; usually only one seed and one fertile cavity, seed oblong, 2.6 cm long 4 mm thick.

Distribution and habitat. *Terra firme* forest of Colombia and Amazonian Brazil (Fig. 10.27.)

Phenology. Flowering mainly in March and April, fruiting in June.

Local names and uses. *Uchirana.*

Conservation status. Least concern (LC).

Illustrations. Cuatrecasas (1961), figs. 11a, b, 12a, b.

Selected specimens examined. **COLOMBIA. CAQUETÁ**: Solano, Est. Puerto Abeja, 0°4′27″N, 72°27′05″W, 6 Nov 1999 (fl), *Eusse & Montes 767* (COAH). **GUIANÍA**: Panapaná, 150 m from Caño Guaviarito, 1°52′51″N, 69°0′34″W, 149.44 m, 6 May 2014 (fr), *Aymard 14,214* (COAH). **BRAZIL. AMAZONAS**: Manaus, 9 Mar 1937 (fl), *Ducke RB 30133* (INPA, K, P, S, U, US); Manaus, Rio Tarumã, 22 Feb 1949 (fl), *Ducke 2230* (IAC, IAN, MG); Manaus, Reserva Ducke, 2°53′S, 59°58′W, 28 Mar 1996 (fl), S*others 843* (HFSL, INPA, K, MG, RB), 29 Jun 1993 (fl, fr), *Ribeiro 929* (INPA, K, NY); Manaus, Distrito Agropecuario, Fazenda Porto Alegre, 2°25′S, 59°54′W, 50–120 m, 10 June 1992 (fl), *Dick 147* (INPA, K, MO, NY), 13 Jun 1989 (fr), *Sette Silva INPA/WWF3402.4075.2* (INPA, K, NY); Estrada Manaus-Itacoatiara, Km 140, 13 Jun 1972 (fl), *Monteiro 72–136* (INPA), Km 79, 2 Apr 1970 (fr), *Coelho 70-31* (INPA); Transamazon Highway, 400 km from Humaitá, Novo Aripuanã, −7.25, −60, 4 May 1985 (fl), *Cid Ferreira 6031* (F, INPA, MO, NY, RB); Parque Nacional Jaú, Rio Unini, Mun. Novo Airão, −2.62, −60.94, (fl), *Ferreira 40* (INPA); Rio Uatumã, Reserva UHE Balbina, Mun. Presidente Figureido, −2.03, −60.025, 27 Jun 3997 (fr), S*ilva 1298* (INPA).

There are four sheets of the type number RB20427 in Rio herbarium, and as Cuatrecasas did not indicate or label any as the holotype, I have selected one as the lectotype. Several collections noted the scented flowers of this species.

9. **Vantanea maculicarpa** Sabatier & Engel, Phytotaxa 338 (1): 130–134. 2018. Type. French Guiana, RN2 Cayenne-Régina, Petites Montagnes Tortues, 4°18′N, 52°15′W, *D. Sabatier 5574* (holotype, CAY 111685; isotype, P 1156374).

Trees up to 40 m high, trunk buttressed, the young branches glabrous. Leaves with petioles 4–6 mm long, sulcate, thickened at base; laminas obovate, 4–11 × 3–6 cm, rigid-coriaceous, glabrous, with several small glands scattered on lower surface, cuneate or attenuate at base, rounded and emarginate at apex; margins entire; midrib prominulous above, prominent beneath; primary veins 8–10 pairs, arcuate and united near margin. Inflorescences axillary or terminal, paniculate-cymose, ca 9 cm long, axis pubescent; pedicels ca. 0.5 mm long, pubescent; bracts deciduous. Calyx broadly cupuliform, ca. 3 mm long, 2 mm diameter, entire or slightly 5-lobed, minutely pubescent. Petals 5, white, linear, rather thick, 5–8.5 × 1–1.5 mm, densely pubescent on exterior, glabrous within. Stamens 150–230, united at base, filaments glabrous, white, of varying length within and among flowers, 1–6 mm long; anthers lanceolate, ca. 7 mm long, affixed near to base, connective long and acute (half length of anther), thecae 2, bilocular, 0.3 mm long. Disc thick, cupular, surrounding ovary, densely pubescent abaxially and on upper third of adaxial surface. Ovary 1.6 mm high, globose-ovoid, densely crisp-pubescent, 5-locular, locules biovulate; style ca. 3.6 mm long, sparsely pilose at base, stigma small, rounded. Drupe globose, ca. 2.5 cm diam wide, dark with numerous white lenticels; mesocarp ca 3 mm thick, firm; endocarp strongly rugose, with 5 linear valves ca 5 mm wide, alternating with broad ribs; 1 or 2 seeds 5 × 2.5 mm.

Distribution and habitat. Rainforests on *terra firme* in French Guiana (Fig. 10.27).
Phenology. Flowering from June to November, fruiting from April to May.
Local names and uses. None recorded.
Conservation. Rated as least Concern (LC) in the protolog of Sabatier.
Illustrations. Sabatier (2018), figs. 1, 2.

Additional specimens examined: **FRENCH GUIANA**. Petite Montagne Tortue, 4°13′N, 52°13′W, 10 Jun 2005 (fl.), *D. Sabatier & M.F. Prévost 4911* (CAY 073328, P 04782086, NY); piste de Nancibo, 4°41′N, 52°30′W, 12 Apr 1985 (fr.), *D. Sabatier 1071* (CAY 080681, CAY 080680); DZ Saut Dalles, 3°16′N, 53°49′W, 05 May 1990 (fr.), *D. Loubry 706* (CAY 166269, NY, US, CBS, MPU); Massif des Emérillons, crête à 4.5 km au Sud du Piton Baron, 3°17′N, 53°4′W, 27 Oct 2007 (imm fr.), *D. Sabatier & J.F. Molino 5342* (CAY 104064, CAY 104065); Crique Wapou, 4°25′N, 52°9′W, May 2005 (fr.), *D. Sabatier 4898* (CAY 171556, P 01156368); Nancibo, 4°40′N, 52°29′W, 15 Nov 1985 (st.), *D. Sabatier 1130* (CAY 166268); Nancibo, 4°40′N, 52°29′W, 27 Dec 1985 (st.), *D. Sabatier 1172* (CAY 166267); St-Georges-Régina, entre pk 25,4 et pk 27, 4°1′N, 51°58′W, 4 Nov 1998 (st.), *P. Grenand 3062* (CAY 000286, MPU, MO, G).

This recently described species is distinct by the lenticellate exocarp of the fruit.

10. **Vantanea magdalenensis** Cuatr., Brittonia 8: 195. 1956; Cuatrecasas, Contr. U.S. Natl. Herb. 35(2): 54. 1961. Type. Colombia, Santander: Valle del Magdalena, Cimitarra, 3 km on road to Ermitaño, 24 Jul 1954 (fl), *F. B. Lamb 133* (holotype, US 00101243; isotype, COL 000001816).

Large trees, the young branches subterete, glabrous, lenticellate. Leaves with petioles 2–4 mm long, thick, glabrous; laminas obovate, oblong-ovate or obovate-elliptic, 11–19 × 6.5–10 cm, rigid-coriaceous, glabrous, obtusely cuneate or rounded at base, obtuse at apex, margins entire, 1 pair of basal glands on lower surface; midrib prominulous above, prominent beneath; primary veins 8–10 pairs, prominent on both surfaces, minor veins prominulous, loosely reticulate. Inflorescences terminal, paniculate, shorter than the leaves, dichotomous, branches hirtellous-tomentose; pedicels thick, 1 mm long; bracts deciduous. Calyx 3 mm diam. Sepals 5, rounded, pubescent on exterior. Petals obovate-linear, thick, 5.5–6 mm long, 2.5 mm broad, tomentose on exterior. Stamens numerous, triseriate, unequal, filaments glabrous, connate in a ring for 1–2 mm, anthers ovate-rhomboid, 4-lobed, connective thick, attenuate toward apex, thecae 2, bilocular. Disc tubular, 1–3 mm high, glabrous. Ovary oblong, 3 mm high, the lower third glabrous, pubescent above, locules biovulate; styles filform, glabrous, 3 mm long. Drupe subovate-elliptic, rounded at base, slightly narrowed at apex, 3.2 cm long, 2 cm thick; exocarp 1 mm thick, minutely velutinous tomentose; endocarp woody, 2.8 × 1.8 cm, rounded at base, acute at apex, surface slightly rugulose, with 5 marked ribs and 5 oblong valves 2.5 cm long, 5 mm wide.

Distribution and habitat. Known from only two collections in the upland rainforests of the Magdalena Valley of central Colombia (Fig. 10.27).
Phenology. Collected in flower in July.
Local names and uses. *Macabalo*.
Conservation status. Endangered (EN) B1ab (iii,v).
Illustrations. Cuatrecasas (1961) fig. 8 a–c.

Additional specimens examined. COLOMBIA. ANTIOQUIA: San Francisco, road Topacio-Aquitania, 5°53′N, 74°56′W, 600–1000 m, 26 Nov 1990, *Cárdenas 3085* (JAUM, MO). **SANTANDER SUR**: Puerto Parra between Ríos Carare & Opón, 6°57′1″N, 73°55′3″W, *Cabrera 883* (COL).

11. **Vantanea micrantha** Ducke, Arq. Inst. Biol. Veg. 4: 30. 1938; Cuatrecasas, Contr. U.S. Natl. Herb. 35(2): 62. 1961. Type. Brazil. Amazonas: Manaus, Rio Tarumã, 6 Aug 1937 (fl), *Ducke RB 30135* (lectotype **here designated**, RB 00539077; isotypes, G 00368595, INPA 230795, K 000407355, K 000543566, MG 195577, NY 00388416, NY 01104810, P 01903283, S-R9888, RB 00570496, U 0002497, UFMT 11132, US 00101242).

Large trees, the young branches smooth, grey, glabrous or puberulous-glabrescent, sparsely lenticellate. Leaves with petioles 3–6 mm long, canaliculate above, glabrous; laminas elliptic–oblong or elliptic-lanceolate, 5–12 × 2–5 cm, thin-coriaceous, glabrous, broad near base and abruptly tapering onto petiole; narrowly acuminate or cuspidate at apex, margins entire; midrib prominulous or prominent above, prominent beneath; primary veins 10–12 pairs, prominulous or inconspicuous on both surfaces, veins reticulate, prominulous, a few bearing seriate glands. Inflorescences axillary and terminal, cymose-paniculate, shorter than the leaves; peduncle and branches minutely hirtellous or papillose; peduncles 4–10 mm long pedicels thick, 0.5 mm long; bracts early deciduous, ovate-lanceolate, 1.2 mm long. Calyx cupular, 0.6–0.7 mm high, slightly 5-lobed, the lobes obtuse, sparsely papillose on exterior, minutely ciliolate on margins. Petals white, glabrous, oblong, 4–5 mm long, 1.5 mm broad at base. Stamens ca. 50, glabrous, filaments slender, unequal, 2.5–4 mm long, connate at base into a short tube, anthers ovate-lanceolate, 0.6–0.7 mm long, with 4 elliptic lobes, connective fleshy, elongate, acute or subacute, thecae 2, elliptic, basal. Disc glabrous, surrounding the ovary, 1.2 mm high, upper half laciniate. Ovary ovoid, glabrous, striate, 1 mm high, locules 5, biovulate; style rigid, glabrous ca. 4 mm long. Drupe ellipsoid, 2–2.5 × 1.4–2.2 cm; exocarp thin, oleaginous; endocarp ellipsoid obtuse, woody, strongly costate-corrugate.

Distribution and habitat. Forest on *terra firme* in the vicinity of Manaus, Brazil (Fig. 10.27).
Phenology. Flowering from August to December, fruiting from September to January.
Local names and uses. None recorded.
Conservation status. Vulnerable (VU) A2ac; B1ab (iii,v).
Illustrations. Cuatrecasas (1961) figs. 5 m, n, 7 e, f, 10 f.

Selected specimens examined. BRAZIL. AMAZONAS: Manaus, Cachoeira do Tarumã, 6 Aug 1937 (fl), *Ducke RB 9053* (INPA); Manaus, Colonia João Alfredo, 7 Dec 1941 (fl), *Ducke 751* (F, GH, IAN, K, MG, MO, NY, US, paratypes); Manaus-Caracaraí, km 57, 14 Sep 1976, *Mota 608* (INPA); Reserva Ducke, Manaus, 8 Mar 1995, *Assunção 185* (INPA, MG, NY), 19 Aug 1994, *Nascimento 570* (INPA, K, MO, NY, RB, U); 5 Aug 1963 (fl), *Rodrigues 5421* (INPA); Manaus-Itacoatiara, Km 145 (st), *Rodrigues 8165* (INPA); Reserva Desenvolvimento Sustentável, Rio Madeira, Novo Aripuanã, 28 Sep 2010 (fr), *Prata 777* (INPA).

There are two sheets of the type number RB 30135 in Rio herbarium, and as Cutrecasas did not indicte or label either as the holotype, I have selected one as the lectotype.

12. **Vantanea minor** Benth., Hooker's J. Bot. Kew Gard. Misc. 5: 99. 1853; Urb., Fl. Bras. (Martius) 12 (2): 452. 1877; Cuatr., Contr. U.S. Natl. Herb. 35, 2: 63. 1961; Cuatrecasas & Huber, Fl. Venez. Guayana (Steyermark et al.) 5: 640. 1999. Type. Guyana, *Rich. Schomburgk 1552* type B lost, photo F 12589 (lectotype **here designated**, *Schomburgk 10* (K 000370548).

Medium-sized trees, the young branches, subterete, smooth, glabrous or minutely sparse-puberulous-glabrescent. Leaves subsessile with short petioles 1–2 mm long; laminas obovate or elliptic-obovate, 3–8.5 × 1.6–4.5 cm, rigid-coriaceous, glabrous, acute or obtusely cuneate at base, rounded or obtuse at apex, sometimes emarginate, margins entire; midrib slightly prominulous above, prominent beneath; primary veins 13–15 pairs, prominulous, inconspicuous on both surfaces, venation reticulate slightly prominulous, bearing a few seriate glands on both sides. Inflorescences terminal and subterminal, cymose-paniculate, the branches furrowed, minutely puberulous; peduncles 1–2.5 cm long, puberulous; pedicels thick, 1.5–2.5 mm long, minutely puberulous. Calyx cupular, 2 mm high. Sepals 5, thick, rounded, glabrous on exterior, margins ciliate. Petals linear-oblong, white, glabrous, thickish, 12–14 mm long, 2 mm broad. Stamens ca 100 or more, pluriseriate, 9–12 mm long, glabrous, connate at base in a tube 3–4 mm high, anthers 0.6 mm long, with 4 ellipsoid lobes, connective thick, obtuse, thecae 2, elliptic, basal. Disc tubular, surrounding the ovary, 1.2 mm high, glabrous, the margin dentate. Ovary ovoid, glabrous, 2 mm high, locules 5, biovulate; styles flexuous, ca. 12 mm long, glabrous, stigma obtuse. Drupe oblong-elliptic or ovoid-ellipsoid, 3.5–5 × 2.2–3 cm rounded at base, obtuse at apex; exocarp coriaceous when dry, resinous, smooth, glabrous, 2 mm thick; endocarp ellipsoid, woody, rounded at base, apiculate or obtuse at apex, 3–4 × 2–3 cm, strongly anfractuose-rugose, 5-locular with 5 oblong opercula, obtuse at apex, 2.2 cm long, 5 mm wide, alternating with 5 small foveolae at top.

Distribution and habitat. The Guayana region of Venezuela and Guyana in forests on white sands and rocky terrain at altitudes 750–1300 m (Fig. 10.27).

Phenology. Flowering from December to April, fruiting from April to November, often simultaneously in flower and fruit.

Local names and uses. None recorded.

Conservation status. Least concern (LC).

Illustrations. Cuatrecasas (1961), figs. 5 f–l, 7 g–h, 10 g; Cuatrecasas and Huber (1999), fig. 546.

Additional specimens examined. VENEZUELA. BOLÍVAR: Sabana de Icabarú, 450–850 m, 22 Dec 1955 (fl, fr), *Bernardi 2613* (NY, VEN); Gran Sabana, Río Urari, Mar 1946, *Tamayo 3123* (US); Camino del Río Tirica a Gran Sabana, 900 m, Oct 1947, *Cardona 2362* (MO, US); Río Caroni, Uaiparu, tributary of Río Icabarú, 500 m, 27 Oct 1946, *Cardona 1912* (US, VEN); Rio Cuquenan, south of Mount Roraima, Jan 1910, *Ule 8801* (K, MG); 35 km NNE of Ikabarú, 4.6333, −61.633, 16 Feb 1986 (fl, fr), *Huber 11,315* (INPA, MO, NY, US); 7 km SSE of Salto Aponguao, 5.333, −61.466, 13 Jun 1985 (fl, fr), *Huber 10,497* (INPA, NY, US); Dist. Roscio, Río Kamúran, 1000 m, 5°37′N, 61°13″W, 18 Jun 1986 (fl,fr), *Huber 10,620* (MO, NY, US); 10 km SW of Karaurin Tepuí, 1000 m, 5°19′N, 61°03′W, 29 Apr 1988 (fl), *Liesner 24,032* (MO, NY, US); Quebrada El Cajón, 26.5 km E of Icabarú, 750 m, 4°25′N, 61°32′W, 18 Dec 1978, *Steyermark* et al. *117,842* (F, MO); Ayavaparú, 10 km SW of Wadaca Pipué, 1100 m, 15 Nov 1986 (fr), *Hernández 356* (NY, US); Ilu-tepuí, 2900 feet, 24 Mar 1952 (fl), *Maguire 33,583* (NY). **GUYANA**. Makarang River, Akapai, 910 m, 13 Oct 1960 (fr), *Tillett & Tillett 45,678* (NY); Cuyuni-Mazaruni Region. Paruima-Utushi trail to Venezuela, 5°41′N, 61°07′W, 1000 m, 29 Jan 1996 (fl), *Clarke 868* (K, MO, NY); Potaro-Siparuna Region, Mt. Wokomung, 1135 m, 5°6′36″N, 59°49′14″W, 14 Jul 2003 (fl), *Clarke 10,859* (NY).

There appears to be no material extant of the type, *Richard Schomburgk 1552*, and a photo in the Chicago herbarium of the original destroyed at Berlin.There is one sheet at Kew marked *Vantanea minor, Schomburgk 10* and so I selected this as a lectotype unless further material of 1552 is found elsewhere. Johnson et al. (2001), report that the fruits of this species were invaded by the bruchid beetle *Amblycerus crassipunctatus* Ribero-Costa. About 28.5–39% of the fruits studied were damaged by these beetles. *Vantanea minor* and the family Humiriaceae were reported as a host of bruchids for the first time.

13. **Vantanea morii** Cuatr., Phytologia 68: 265. 1990. Type. Brazil. Bahia: Mun. Andarai, road Andarai-Mucugê via Igatu, 2 km S of Igatu, 800 m, 23 Dec 1979 (fl), *S. A. Mori & F. P. Benton 13,181* (holotype, US 00289005; isotypes, CEPEC, NY 00076863).

Scandent shrubs spreading over rocks, the young branches striate, glabrous. Leaves with petioles 2–5 mm long, glabrous; laminas rounded-elliptic or elliptic, 3.5–7.5 × 3.2–5 cm, rigid-coriaceous, glabrous, covered densely punctate with resinous glands and papillose beneath, obtuse or bluntly cuneate at base, apex rounded, margins entire, 2 pairs of basal glands on lower surface; midrib plane or impressed above, prominent beneath; primary veins ca 10, prominulous beneath, reticulate venation prominulous and visible. Inflorescences terminal, cymose-thyrsoid-paniculate, 4–10 cm long and wide, branches alternate, minutely hirsutulous-pubescent; peduncles 0.5–4 cm long, puberulous; pedicels thick, ca.1 mm long; bracts deciduous, 0.5–1 mm long, clasping, minutely hirsute. Sepals 5, rounded, connate at base, 0.8 × 1.2 mm, glabrous except for ciliate margin, with minute sunken glands on interior. Petals 5, oblong, 4 × 1.6 mm, obtusely attenuate at apex, thick, glabrous, white. Stamens 21–26 in single row, filaments connate at base for 0.5 mm, longer ones 2.8–3 mm long, alternating with shorter ones 2.2 mm long, anthers 0.7–0.8 mm long, connective thick, acute, thecae 2, bilocular, elliptic, 0.4–0.5 mm long, longitudinally dehiscent. Disc cupular, of ca. 22 scales surrounding ovary, 0.9–1 mm high, dentate, glabrous. Ovary glabrous, pyriform, 1.5 mm long, 5-locular, locules biovulate; style robust, 0.6–0.7 mm long, stigma capitate, slightly 5-lobed, the lobes glutinous. Drupe subglobose, 1.7–2.5 × 1.6–2.4 cm; exocarp smooth, 1 mm thick; endocarp woody 1.5 × 1.6 cm, almost globose, with 5 locules each with one seed, with 5 longitudinal valves 3.5 mm wide.

Distribution and habitat. Known only from the campo rupestre of Bahia and Pernambuco, Brazil, in brejos (Fig. 10.28).
Phenology. Collected in flower from October to February and in fruit in February.
Local names and uses. None recorded.
Conservation status. Least concern (LC).

Selected Specimens examined. BRAZIL. PERNAMBUCO: road to Catimbau, Buique, −8.62, −37.15, 8 Oct 1971 (fl), *Andrade-Lima 71–6558* (IPA, RB); Mun. Buique, Sopé da Serra, 8°37′30" S, 37°09′15″W, 19 Nov 1995 (fl), *Figueirêdo 260* (RB, UFRPE). **BAHIA**: Mun. Andaraí, road Andaraí-Mucugê, 900 m, 12 Jan 1983 (fl bud), *Mattos Silva* et al. *1613* (CEPEC, RB, US); Serra de Andaraí, Capa Bode road to Musugê, −12.92329, −41.297835, 700–1200 m, 30 Oct 1978 (fl), *Martinelli 5410* (RB); between Igatu and the Mucugê-Andaraí road, −12.90, −41.31, 31 Jan 2011, 755 m, *França 4144* (HUEFS); 5 km S of Andaraí, Rio Paraguaçu −12.833, −41.31, *Harley 18,577* (CEPEC, IPA, K, NY, UEC); Chapadinha, Itaeté, −12.55, −41.39, 21 Feb 2004, *Funch 104* (HUEFS); Parque Municipal, Mucugê, −12.98, −41.33, 22 Aug 2011, *Costa 772* (HUEFS); Mucugê. 26 Feb 2001 (fl fr), *Ribeiro-Filho 219* (HUEFS, K); Mucugê-Igatu, 4 km from BR142, 830 m, 12°53′58" S, 41°18′39″W, 16 Feb 2002 (fl), *Harley 54,508* (HUEFS, K): Mucugê, road Andaraí-Mucugê, 4–5 km from Andaraí, 8 Sep 1981 (fl), *Pirani* et al. *CFCR2066* (K); 24 km N of Seabra, Serra da Agua de Rega, 1000 m, 25 Feb 1971 (fl), *Irwin* et al. *31,059* (NY, UEC); Mun. Andaraí, 12°52′S, 41°18′W, 560 m, (fl), *Saar 5604* (CEPEC, K); Mun. Caetité, road to Licinio de Almaida, 14°09′36" S, 42°29′46″W, 10 Feb 1997 (imm fr), *Saar* et al. *PCD5380* (ALCB, CEPEC, HUEFS, K).

This species is close to *Vantanea compacta* but consistently differs in the more rounded leaves with a round or retuse apex and fewer stamens. These species are sympatric in the campos rupestres of Bahia, but *V. compacta* is more widespread.

14. **Vantanea obovata** (Nees & Mart.) Benth., Hooker's J. Bot. Kew Gard. Misc. 5: 99. 1953; Urb., Fl. Bras. (Martius) 12 (2): 451. 1877; Cuatrecasas, Contr. U.S. Natl. Herb. 35(2): 55. 1961. *Helleria obovata* Nees & Martius, Nova Acta Phys.-Med. Acad. Caes. Leop.-Carol. Nat. Cur. 12: 40, tab. 7. 1824. Type. Brazil. Minas Gerais: Tejuco, *C. E. P.* von *Martius s.n.* (holotype, M, photos NY, US; isotype, M, possible syntypes, *Wied-Neuwied s.n.* (BR 0000006587123, BR 0000006586799). Fig. 9.4D.

    Helleria ovalifolia A. Juss., In St. Hil., Fl. Bras. merid. (A. St.-Hilaire) 2: 91. 1829; *Vantanea ovalifolia* (A. Juss.) Benth., in Hooker's J. Bot. Kew Gard. Misc. 5: 99. 1853. Type. Brazil. Minas Gerais, near Bonfim, Minas Novas, Aug (fl), *A de St. Hilaire B 1705* (lectotype of Cuatrecasas (1961), P 01903280; isotypes, MPU 012220, P 01903281, P 01903282, MPU 012220).

Trees, the young branches glabrous or sparsely puberulous, granulate-lenticellate. Leaves with petioles 4–12 mm long, subterete, canaliculate above; laminas obovate or obovate-elliptic, rigid-coriaceous, glabrous, 4–12.5 × 2.3–7 cm, obtusely or acutely cuneate at base, rounded or obtuse or retuse at apex; margins entire, 1–2 pairs of basal glands on lower surface; midrib prominulous above, prominent beneath; primary veins 7–9 pairs, prominulous above, prominent and glabrous beneath, venation reticulate, prominulous, often with glandspots on middle of veins. Inflorescences terminal or subterminal, cymose-paniculate, the branches dichotomous, densely hirtellous; peduncles 1–2 cm long, puberulous; pedicels thick, 1–2 mm long, densely pubescent; bracts deciduous, ovate, 1.5 mm long. Calyx suborbicular, 2 mm high. Sepals 5, ciliate. Petals white, linear-oblong, 8–13 × 2.5–3 mm, densely hisutulous-pubescent on exterior with spreading retrorse hairs, margins glabrous. Stamens ca 100 or more, 2–3 seriate, connate at base in a tube, filaments glabrous, 1 mm high, unequal, 6–9 mm long, anthers ovate-lanceolate, 0.8–0.9 mm long, connective fleshy, lanceolate, acute, thecae elliptic, basal. Disc annular,

Fig. 10.25 Distribution of five species of *Vantanea*

surrounding ovary, glabrous, 1 mm high. Ovary subglobose, 2 mm high, densely hirsute-villous, locules 5, biovulate; style thick, erect, pilose at base only, 4–6 mm long, stigma obtuse. Drupe ovoid, subglobose, rounded at base, obtuse at apex, 2.2–3.5 cm long, 1.6–3 cm diam, exocarp densely tomentose; endocarp ovoid with 5 lingulate valves.

Distribution and habitat. Characterstic of the cerrado region and the Atlantic restingas of central and eastern Brazil, common in campo rupestre. (Fig. 10.28).

Phenology. Mainly flowering from December to April and fruiting from May to December.
Local names and uses. None recorded.
Conservation status. Least concern (LC), a common widely distributed species.
Illustrations. Cuatrecasas (1961) fig. 8d-e; Nees and Martius (1824) tab. 7.

Selected specimens examined. BRAZIL. GOIÁS: Portal da Chapada, Alto Paraíso de Goiás, −14.13, −47.51, 15 Jul 2011 (fr), *Soares-Silva* et al. *1529* (UB). **MINAS GERAIS**: Serra Mendanha, road Bom Gosto to Olivença, −18.1103, −43.5283, (fl), *Gardner 4452* (BM, F, K); Serra de Espinhaço, 9 km SW of Mandanha, Rio Jeqití, 1150 m, −18.1575, −43.56013, 14 Apr 1973 (fl), *Anderson 8837* (MO, NY, RB); road Bom Gosto to Olivença, 15 Mar 1943 (fl), *Fróes 19,933* (F, IAC, IAN, K, NY, US); Botumirim, Serra da Canastra, Rio Bananal, 16°49′S, 43°02 W, 1260 m, 22 Dec 2004 (fr), *Mello Silva 2751* (BHCB, BM, K, RB, UFMG); Mun. Grão Mogul road Cristalina-Cabras, 700 m, 21 Jul 1985 (fl), *Martinelli 11,232* (BHCB, INPA, K, MO, NY, RB); Serra do Cipó, 18 Feb 1993 (fl), *Anderson 36,233* (NY, UB); Santana de Pirapama, Serra Cipó, Fazenda Toucan Cipó, 19°0′18" S, 43°46′06″W, 16 Feb 2007 (fl), *Zappi 759* (RB). **BAHIA**: Morro de Chapéu-Jacobina Km 6, −11.55, −41.15, 20 Dec 1991 (fr), *Harley 50,165* (K, NY); Chapadinha, 1 km W of road to Lençois, −12.46, −41.43, 850 m, 12 Mar 2002 (fl), *Thomas 12,967* (CEPEC, MBM, MO, NY, RB); Road Lençois-Seabra, 20 km NW of Lençois, 900–1000 m, 12°27′46" S, 41°25′42″W, 14 Feb 1994 (fr), *Harley* et al. *CFCR14058* (ESA, HUEFS, K); Campos Casa Pedra, 27 Jul 1914, *Luetzelburg 40* (M); Mun. Andaraí, road Xiqe-Xique to Igatú, 560 m, 12°52′S, 41°18′W, 14 Dec 2015 (fl), *Santos PCD5648* (ALCB, CEPEC, HUEFS, IAC, K); Morro do Pai Inacio, Serra dos Lençois, −12.27, −41.28, 700–1000 m, 21 May 1980 (fl), *Harley* et al. *22,286* (CEPEC, IPA, K, NY, UEC); Espigão Mestre, 100 km WSW of Barreiras, −12.15, −44.99, 800 m, 7 Mar 1972 (fl), *Anderson 36,781* (MO, NY); Dist. De Catolés, road Engenho-Marques, 9 km from Catolés, −13.37, −41.8, 1100 m, 10 Apr 1992 (fl), *Ganev 90* (F, HUEFS, K, NY); Prado, Fazenda Riacho das Ostras, 17°12′54" S, 39°14′15″W, 35 m, 30 Aug 2007 (imm fr), *Rezende 1731* (BHCB, CEPEC). **ESPÍRITO SANTO**: Linhares, BR101, Km 120, 12 May 1987 (fr), *Lima 2957* (K, RB).

Many labels of collections mention the strong, sweet scent of the flowers.

15. **Vantanea aracaensis** Prance, sp. nov. Type. Brazil. Amazonas: Serra Aracá, 1200 m, 12 Feb 1984 (fl), *G. T. Prance* et al. *28,996* (holotype, INPA132663; isotypes, K000370549, MG114944, MO, NY 02270928, NY 00446002, RB 00579725, US). Fig. 10.26.

Ab *Vantanea obovata* foliis maioribus, apicibus apiculatis, staminibus 65 (haud 100–120) differt.

Shrubs, 1–4 m tall, the young branches glabrous, shiny. Leaves with petioles 7–10 mm long, glabrous, slightly winged by confluent lamina, thickened at base; laminas elliptic to oblong-elliptic, rigid-subcoriaceous, shiny above, dull beneath, 6.5–10 × 3–6.2 cm, subcuneate at base, acute to abruptly short-apiculate at apex, margins entire, 1–2 pairs of basal glands on lower surface; midrib prominulous above, prominent beneath; primary veins 7–9 pairs, slightly prominulous above,

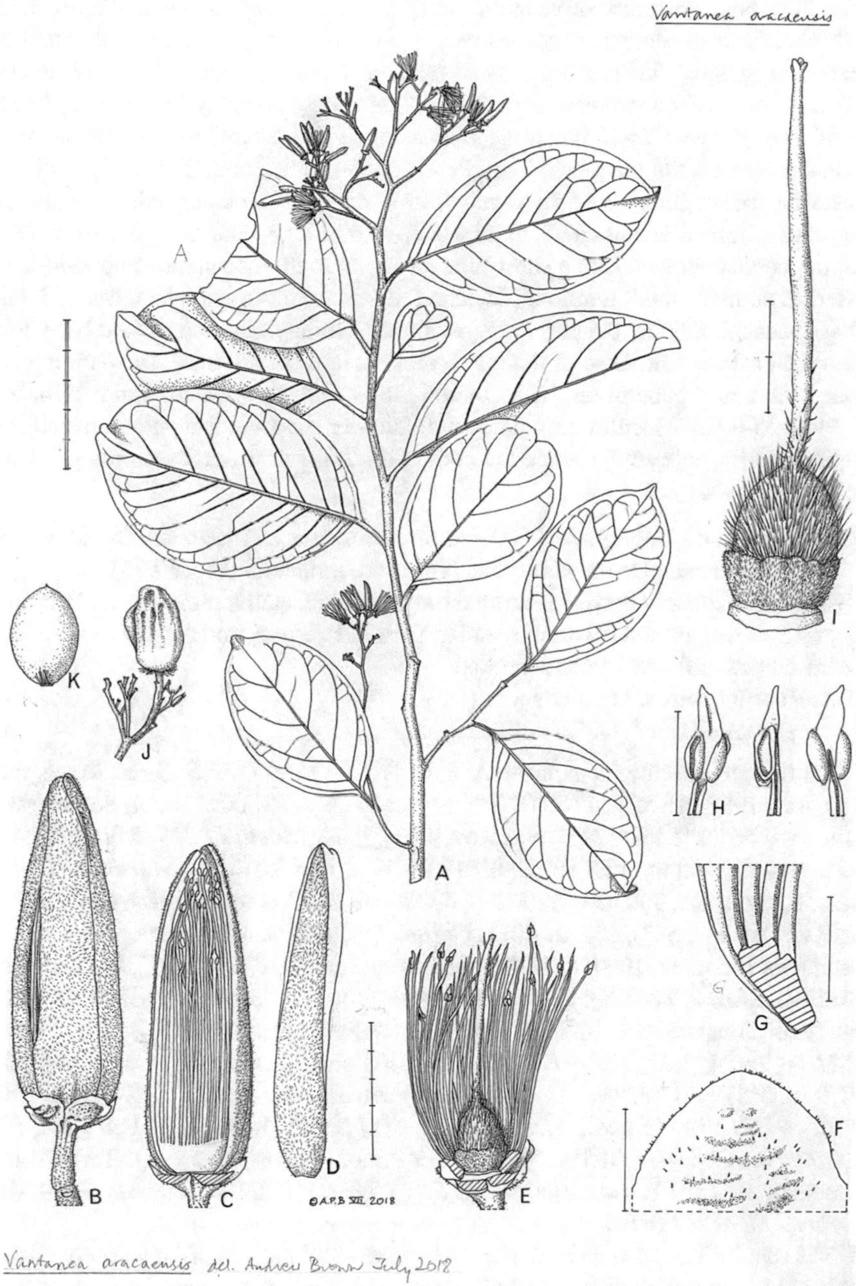

Fig. 10.26 *Vantanea aracaensis*. (**a**) Habit of flowering branch. (**b**) Side view of flower. (**c**) Flower with removal of 2 sepals and petals. (**d**) Outer surface of petal. (**e**) Flower after removal of petals and half of androecium. (**f**) Calyx lobe. (**g**) Section through base of androecium. (**h**) Anther. (**i**) Pistil. (**j**) Young fruit. (**k**) Mature fruit. Single bar = 1 mm, graduated single bar = 2 mm and 5 mm, graduated double bar = 5 cm. (A–I from *Prance* et al. *28,996*; J, K from *Tavares* et al. *29*)

prominent beneath, reticulate venation conspicuous and prominulous. Inflorescences terminal, cymose-paniculate, corymbose, to 6 cm long, slightly shorter or equalling leaves in length, dichotomous, the branches minutely puberulous; peduncles 0.5–1.5 cm long, puberulous; pedicels 1–2 mm long, sparsely hirsutulous; bracts deciduous. Flowers 10–12 mm long, white. Sepals ca 1.5 mm long, connate at base, rounded at apex, the margins ciliate. Petals 5, 9–11 mm long, 2 mm wide at base, linear-lanceolate, minutely sparse-puberulous with stiff resinous hairs on exterior, glabrous within. Stamens ca 65, filaments 7–10 mm long, short and long ones alternating, connate at base into a short tube ca 1 mm high, tube sparse-puberulous on exterior, anthers ca 0.8 mm long, attached at base, connective ovate, acute, 1 mm long, thecae 2, elliptic, 0.5 mm long, bilocular, laterally attached toward base, longitudinally dehiscent. Disc of dentate scales, united toward base, surrounding the ovary, the scales puberulous, with ciliate margins. Ovary connate, 2 mm high, hirsutulous, 5-locular locules biovulate; style shorter than stamens, glabrous, stigma capitate. Drupe ellipsoid, rounded at both ends, 2.3 × 3 cm; exocarp thin, minutely velutinous-pubescent.

Distribution and habitat. At 1000+ m on Serra Aracá, Amazonas, Brazil, and in the lowland campina on white sand below the mountain (Fig. 10.25).

Phenology. Probably flowering around the year. Collected in flower from February to April, August and November and in fruit in February and August.

Local names and uses. None recorded.

Conservation status. Least concern (LC).

Additional specimens examined. BRAZIL. AMAZONAS: Serra Aracá, plateau, Mar 1984 (fl), *Silva 7160* (INPA, MG, MO, NY, RB 00127600); Serra Aracá, 1140 m, 0°53′12” S, 63°19′29″W, 2 Nov 2011 (fl), *Martinelli 17,327* (RB00686739); Serra Aracá, 1140 m, 0°53′12” S, 63°19′29″W, 2 Nov 2011 (fl), *Martinelli 17,331* (RB); Serra Aracá, 960 m, 0°52′31” S, 63°20′40″W, 15 Aug 2011 (fr), *Quinet 2718* (INPA, RB 00726537), 15 Aug 2011, *Quinet 2732* (INPA, RB); Serra Aracá, trail to Cachoeira Eldorado, 1000 m, 16 Aug 2011 (fr), *Lima 7288* (RB00651512), 16 Aug 2011 (fl), *Lima 7282* (RB); Serra Aracá, 950–1150 m, 0°51′48” S, 63°19′59″W, 20 Apr 2014, *Amorim 8630* (CEPEC, INPA, RB); Serra Aracá, Feb 1984 (fl), *Amaral 1541* (F, INPA, MO, NY, RB 00124127); SE part of northern massif of Aracá, 1250 m, 0°51′S, 63°22′W, 14 Feb 1984 (fr), *Tavares & Silva 29* (F, INPA, K, MO, NY), 14 Feb 1984 (fl bud), *Tavares & Silva 30* (INPA, K, NY), 21 Feb 1984, (fl), *Tavares & Silva 109* (INPA, NY); Serra Aracá, *Zartman 9593* (INPA); Aracá, Acampamento do Fosso, 1000 m, 0°51′47″N, 63°19′59″W, 19 Apr 2014 (fl), *Barbosa-Silva 261* (RB).

The species *Vantanea bahiaensis, V. obovata* and *V. aracaensis* are closely related all having drupes with distinct velutinous exocarps. *Vantanea aracaensis* is distinct by its intermediate number of stamens and the shortly apiculate, conduplicate leaves. It also differs from *V. bahiaensis* in the shorter more winged petioles. The type collection of *V. aracaensis* was previously seen and identified by Cuatrecasas as *V. obovata.*

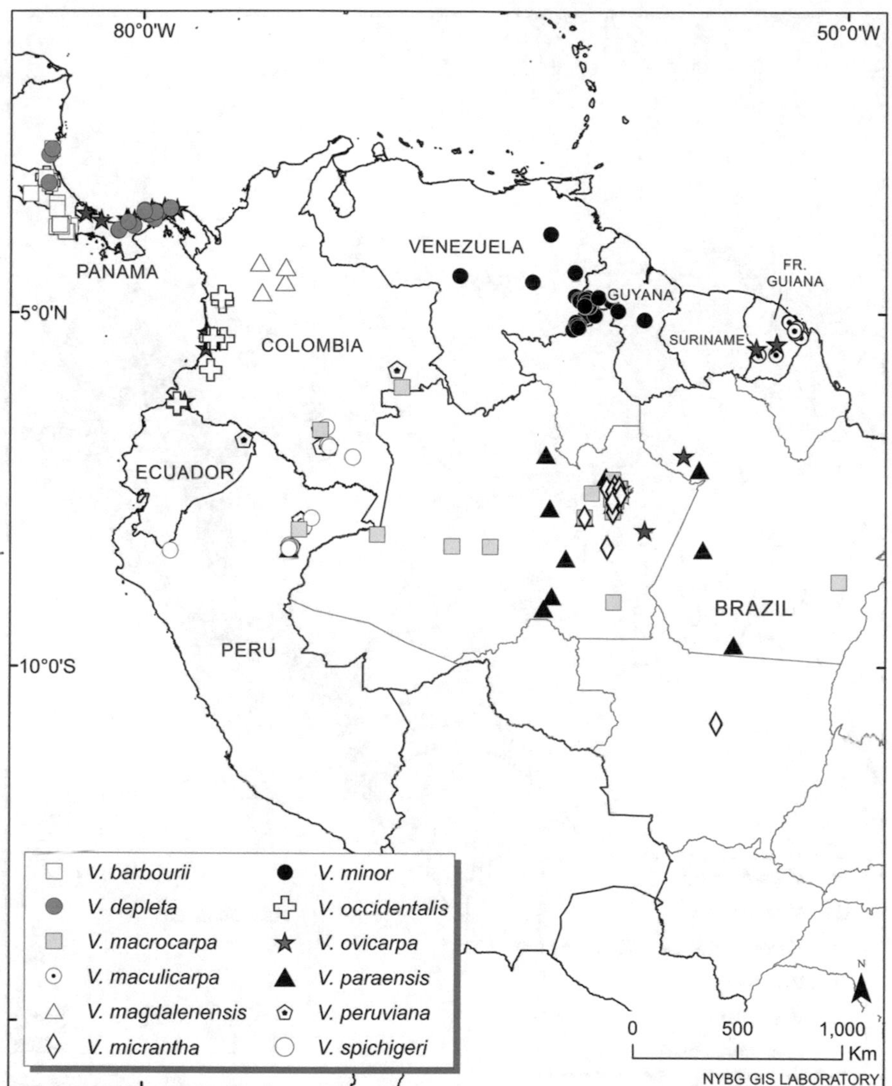

Fig. 10.27 Distribution of twelve species of *Vantanea*

16. **Vantanea occidentalis** Cuatr., Trop. Woods 96: 40. 1950; Cuatrecasas, Contr. U.S. Natl. Herb. 35(2): 59. 1961; Gentry Fl. Nicaragua 2: 1150. 2001; Zamora, Manual de Plantas de Costa Rica 6: 14. 2007. Type. Colombia. Valle, Buenaventura. 22 Feb 1946 (fl, fr), *J. Cuatrecasas 19,937* (holotype, F V0060798; isotypes, F V0060799, G, MO 1296339, US 1206550, VALLE 000282).

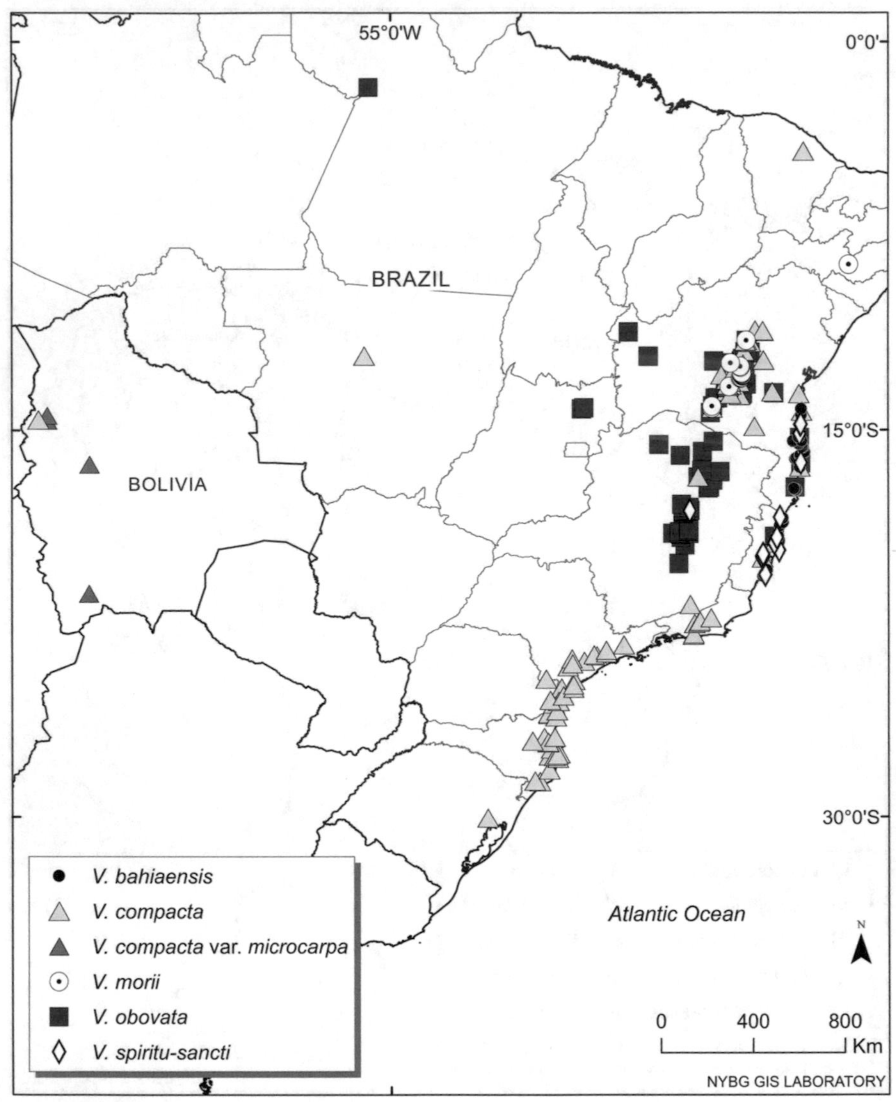

Fig. 10.28 Distribution of six taxa of *Vantanea*

Large trees up to 40 m, the young branches glabrous, shiny, angular, lenticellate. Leaves with petioles 1–10 mm long, robust, sulcate above, thickened at base; laminas elliptic, obovate–elliptic or oblong-elliptic, 7–18 × 2.5–8.5 cm, coriaceous, glabrous, cuneate at base, abruptly attenuate at apex, obtuse or short-acuminate, margins entire, midrib prominulous above, prominent beneath; primary veins 8–10 pairs, prominent on both sides, venation reticulate and slightly prominulous. Inflorescence terminal, paniculate, shorter than the upper leaves, branches dichotomous, short-pubescent; peduncles 0.4–1 cm long, puberulous; pedicels thick,

1–2.5 mm long, puberulous; bracts deciduous. Sepals orbicular, glabrous, 6–7 mm long with a single gland on exterior. Petals oblong, obtuse, 9 × 3 mm, glabrous. Stamens ca. 60–80, filaments unequal to 8 mm long, glabrous, connate at base into a short tube, anthers ovate-lanceolate, connective thick at base, acute at apex, thecae 2, 0.4 mm long. Disc cupular, 1 mm high, glabrous, minutely dentate. Ovary ellipsoid, 3–4 mm high, hirsute, striate, 5–6 locular, locules biovulate; style filiform, 5 mm long, stigma capitate. Drupe ovoid-oblong, 2.2–3.5 × 1.2–2 cm, attenuate at apex, obtuse at base; endocarp ellipsoid-ovoid, oblong, 2.7 cm long, 1.8 cm broad, obtuse at both ends, with 5 oblong valves, obtuse at apex, 1.8 cm long, 4–5 mm wide; seed oblong, 1 per fruit.

Distribution and habitat. Lowland forest from Costa Rica through Colombia to northern Ecuador (Fig. 10.27).
Phenology. Flowering from December to June, fruiting from June to October.
Local names and uses. Costa Rica: *Campano, chricano;* Colombia: *chanó, fruta de guatín.*
Conservation status. Least concern (LC).
Illustrations. Cuatrecasas (1961), figs. 1, 6j–l, 9a–b; Zamora (2007), 15.

Additional specimens examined. COSTA RICA. ALAJUELA: Llanura de San Carlos, Boca Tapada, 20 m, 10°42′N, 84°09′36″W, 23 Jun 1993 (fl), *Jiménez J.* et al. *1315* (CR, MO). **HEREDIA**: La Union de Sarapiqui, 200 m, 10°27′00″N, 84°03′36″W, 4 Jun 1986, *Hammel & de Nevers 15,335* (MO). **COLOMBIA CHOCÓ:** 4–6 km S. of Yuto on road Quibdó-Istmina, 80–100 m, 5°28′43″N, 76°37′30″W, 12 Jan 1979 (fr), *Gentry & Renteria 24,054* (COL, MO, NY); Quibdó, 40 m, Mar 1958 (fr only), *Ramos & Patiño s.n.* (US). **VALLE DEL CAUCA**: Buenaventura, 1945, *Patiño 12* (F, US, paratypes); Bajo Calima, 11 km E of Buenaventura, 50 m, 3°55′00″N, 77°00′00″W, 14 Dec 1985 (fl), *Gentry* et al., *53,294* (MO). **ECUADOR. ESMERALDAS**: San Lorenzo, Reserva Etnica Awá, 200 m, 1°8′00″N, 78°33′00″W, 21 Sep 1992 (fr), *Aulestia* et al. *603* (MO, NY).

17. **Vantanea paraensis** Ducke, Arch. Jard. Bot. Rio de Janeiro 4: 99. 1925; Cuatrecasas, Contr. U.S. Natl. Herb. 35(2): 60. 1961. Type. Brazil. Pará: Rio Tapajós, Bella Vista, 24 Sep 1922 (fl), *A. Ducke RB 17782* (holotype, RB 00539078; isotypes, K 000407358, S-R9889, U 1412496, US 00101241; B lost, photo F-12591).

    Vantanea celativenia (Standl.) Cuatr., Contr. U.S. Natl. Herb. 35(2): 61. 1961; *Licania celativenia* Standl., Publ. Field Mus., Bot. 17(3): 254. 1937. Type. Brazil. Amazonas: Mun. Humaitá, Livramento, 7 Sep 1934 (fr), *B. A. Krukoff 7182* (holotype, F V0053975; isotypes, A 00046253, BM, BR 0000005119547, K 000407356, K 000407357, MICH 1192010, MO 1836387, NY 00428363, RB 00283406, S-R7855, U 0002494, US 00130725, US 00130726).

Large trees, the young branches subterete, glabrous, sparsely lenticellate. Leaves with petioles 10–18 mm long, canaliculate above, thickened at base; laminas obovate-elliptic, 7–12 × 3–6.5 cm, thick coriaceous, glabrous, decurrent onto petiole, cuneate at base, broadly obtuse or rounded and sometimes emarginate at apex, margins entire, 1 pair of basal glands on lower surface; midrib prominulous above, prominent beneath; primary veins 7–10 pairs, slightly prominulous above, prominent beneath venation reticulate, prominulous. Inflorescences terminal or axillary in upper leaves, cymose-paniculate, corymbose, dichotomous at summit, slightly shorter than leaves, branches angled, hirtellous-puberulous; peduncle 0.5–3 cm long, ribbed, hirtellous-puberulous or glabrous; pedicels thick, subterete, minutely puberulous or glabrous; bracts soon deciduous. Calyx cupular, 0.5 mm high, 1.5 mm diam. Sepals 5, rounded, glabrous with cilate margins. Petals oblong, thickish, white, 6–8 × 2–2.5 mm, glabrous, acute. Stamens ca. 50, filaments glabrous, shortly connate at base, anthers ovate-lanceolate, 0.9 mm long, with 4 elliptic lobes, connective fleshy, lanceolate, thecae 2, 0.5 mm long, basal. Disc annular, glabrous, 1 mm high, dentate. Ovary ovoid, 2 mm high, furrowed, densely and short-velvety-tomentose, 5-locular, locules biovulate; style 3.5–5 mm long, pilose only at base, thick but attenuate toward apex, stigma obtuse. Drupe oblong-ellipsoid, 5–5.3 × 2.8–3 cm; exocarp coriaceous when dry, 2 mm thick, glabrous; endocarp woody, oblong, obtuse at base, narrowed and acute at apex, 4.6 × 2.5 cm, surface slightly rugose, with 7 valves, linear-oblong, obtuse at apex, 2.5–3.5 × 0.5 cm.

Distribution and habitat. *Terra firme* forests of Amazonian Peru and Brazil (Fig. 10.27).
Phenology. Flowering from May to October.
Local names and uses. Peru: *Manchari caspi.*
Conservation status. Least concern (LC).
Illustrations. Cuatrecasas (1961), figs. 6 a–f, 6c–f; 10a–b; Spichiger et al. (1990), fig. 5.

Additional specimens examined. PERU. LORETO: Jenaro Herrera, 120 m, −4.98333, −73.7666, 18 Aug 1988 (fl), *Loizeau 669* (G, MO). **BRAZIL. AMAZONAS**: Manaus, 12 Jul 1941 (fl), *Ducke 752* (F, IAN, MG, MO, NY, US); Rio Mindú, 22 Oct 1929 (fl), *Ducke RB 23430* (INPA, K, NY, P, RB, S, U, US), 22 Oct 1929, *Ducke s.n.* (F, NY); Lago de Badajós, Rio Capitarí, 29 Jul 1950 (fl), *Fróes 26,428* (IAN, US); Fazenda Alegre 13 Apr 1992 (fr), *Dick 115* (INPA, K, MO, NY); Km 365 Manaus-Porto Velho Road, 19 Oct 1974 (fl, imm fr), *Prance* et al. *23,036* (INPA, MO, NY). Mun. Humaitá, Tres Casas, 14 Sep 1934 (fr), *Krukoff 6371* (A, BM, F, K, MO, NY, S, U, US). **PARÁ**: Porto Trombetas, Estrada da Mina, Km 14, 25 Jul 1986, *Soares 170* (INPA).

18. **Vantanea parviflora** Lam., J. Hist. Nat. 1: 145.1792; Urb., Fl. Bras. (Martius) 12 (2): 4541877; Cuatrecasas, Contr. U.S. Natl. Herb. 35(2): 74. 1961; Cuatrecasas & Huber, Fl. Venez. Guayana (Steyermark et al.) 5: 641.1999. Type. French Guiana, *P. la Barrère s.n.* (com. *Leblond*) (holotype, P 00287895; isotypes, G 00368596, P 01903277, P 01903278).

Vantanea parviflora var. *puberuliflora* Cuatr., Contr. U.S. Natl. Herb. 35(2): 76. 1961. Type. Brazil. Amazonas: Fonteboa, Rio Solimões, 10 Mar 1929 (fl), *A. Ducke RB 23428* (holotype, US 00101240; isotypes, K, P, RB 00539079).

Vantanea cupularis Huber, Bol. Mus. Paraense Hist. Nat. 6: 83. 1910. Type. Brazil. Pará: Moêma, Belém-Bragança, 30 Jul 1908 (fl), *J. E. Huber MG 9583* (holotype, MG, isotypes, BM 000559605, K 000407349, RB 10819, U 0002498, US 00101245; B lost, photo F-12588).

Large trees, the young branches glabrous or rarely slightly puberulous, subterete, lenticellate. Leaves with petioles 5–20 mm long, canaliculate above; laminas elliptic, 4.5–10.5 × 2–5.5 cm broad, glabrous or appressed pilose, coriaceous, cuneate, or obtusely cuneate at base, obtuse, emarginate or obtusely acuminate at apex, margins entire, 2–4 pairs of basal glands on lower surface close to midrib; midrib plane or prominulous above, prominent beneath, sometimes hirtellous beneath; primary veins 9–10 pairs, prominent beneath, venation reticulate, prominulous. Inflorescences axillary or terminal, cymose panicle, corymbose, shorter than the leaves, branches pubescent-hirsutulous, 1 mm long, subobtuse minutely puberulous on exterior; pedicels 1–2 mm long, minutely pubescent; bracts deciduous. Calyx broadly cupular, 1 mm long, 2.5 mm diam, puberulous on exterior, margins ciliolate, entire or slightly undulate. Petals white, linear, 7–8 mm long, pubescent on exterior, glabrous within. Stamens 80–120, filaments glabrous, 2–3 seriate, unequal, connate at base, 5–7 mm long, anthers 0.8–1 mm long, with 4 elliptic-oblong lobes, connective with long, thick acute tip. Disc annular, surrounding ovary, 1–1.5 mm high, densely hirtellous-tomentose. Ovary ellipsoid, 2.5 mm high, villous-hispid with long hairs, locules 5. biovulate; style 3.5 mm long, with a few hairs toward base, stigma obtuse, glutinous. Drupe ellipsoid, 2.5–2.8 × 2.2–2.5 cm, velutinous, glabrescent, rounded at base, pointed at apex; endocarp woody, rugose, 2–1–2.5 cm long, 1.8–2 cm broad, with 5 longitudinal broad ribs and 5 valves.

Distribution and habitat. *Terra firme* forest of the Guianas and Amazonian Brazil, Colombia, Venezuela, and Peru (Fig. 10.25).

Phenology. Mainly flowering from May to November, fruiting in January and February, but also irregularly throughout the year (Pires-O'Brien et al. 1994).

Local names and uses. French Guiana: A*dougoue* (Paramaka); Brazil: *Paruru*, *uchirana*; Peru: *Manchari*. The Kuripako Indians along the Río Guainía rasp the bark of this species into fermenting chicha of *M. esculenta* to improve the taste and to increase the potency of the drink (Schultes 1979). The wood is used in construction and craft in Peru (Pinedo-Vásquez et al. 1990).

Conservation status. Least concern (LC).

Selected specimens examined. COLOMBIA. CAQUETÁ: Mun. Solano, Caño Peneya, 0°03′39"S, 74°23′7″W, 14 Dec 2017 (fr), *Rodriguez* et al. *3657* (COAH); Solano, Est. Purto Abeja, 0°4′27″N, 72°27′05″W, 15 Jul 1999 (fl), *Eusse & Montes 394* (COAH). **VAUPÉS**: Mun. Taraira, Estación Biológica Caparú, 1°00′S, 69°49′W, 1 Oct 1988 (fl), *Defler 179* (COAH); Pacoa region, Buenos Aires community, raudal Jirijirimo, 0°2′51″N, 70°56′15″W, 23 Feb 2018 (fr), *Castaño-A* et al. *10,536* (COAH). **VENEZUELA. AMAZONAS**: Rio Negro, San Carlos, 1.92, −67.07, 9 Dec 1947, *Schultes & López 9267* (COL, IAN, K, NY, US); San Carlos de Río Negro, 20 km S confluence of Rios Negro & Casiquiare, 1.93, −67.05, 119 m, 14

Feb 1980 (fr), *Clark 7346* (INPA, NY, US). **FRENCH GUIANA**. Road Saint Laurent-Cayene, km 13, 5.58, −53.87, 6 Jul 1953, *BAFOG 48 M* (P); road Charvain-Acarouany, Km 1, 5.59, −53.88, 24 Sep 1954, *BAFOG 220 M* (P, U); Paracou, 5°20′N, 52°55′W, 5 Oct 2000 (fl), *Sabatier 4669* (K. MO, NY); R. Sinnamary, Petit Saut, 5°03′N, 53°03′W, 23 Feb 1989 (st), *Sabatier 2384* (K, MO, US). **ECUADOR. FRANCISCO DE ORELLANA**: Río Tipituni, NE of meeting with Río Tivacuno, 76°30′W, 00°38′S, 200–300 m, 27 Jul 2003 (fl), *Villa 1796* (BM, COL, QCA). **PERU. AMAZONAS:** Bagua, Dist. Imaza, Comun Aguaruna Putuim, SW of Yamaykat, −5.41, −78.10, 700–750 m, 21 Jan 1996 (fr), *Díaz* et al. *7736A* (MO); Condorcanqui, Caterpiza, Río Santiago, 200 m, 3°55′S, 77°42′W, 20 Sep 1979, *Huashikat 2712* (MO). **HUANUCO**: Pachitea, Pucallpa, 20 km SE of Puerto Inca, 350 m, −9.48333, −74.83333, 7 Oct 1988 (fr), *Wallnöfer 12–71,088* (K). **LORETO**: Requena, Jenaro Herrera, 4°55′S, 73°45′W, I Nov 1984, *Spichiger* et al. *1739* (G, MO); Dist. Sapuena, Mun. Jenaro Herrera, 200 m, 21 Apr 1987 (fl bud), *Acevedo-Rodíiguez & Daly 1614* (MO, US). Maynas, Dep. Iquitos, Allpahuayo, 3°58′2" S, 73°25′3″W, 29 May 1990 (fl), *Vásquez 13,775* (MO); Maynas, Santa Maria, Río Nanay, 150 m, −1.42, −74.65, 8 Nov 1967 (fr), *Vidal U.* (K, MO). **PASCO**: Oxapampa, Cerro de Pasco between Shiringamazú & Rio Carachama, 300–600 m, 9°50′S, 75°0′W, 13 Jun 1986 (fl bud), *Hartshorn* et al. *2988* (F, MO). **BRAZIL. AMAZONAS**: Rio Negro, Boca do Rio Marié, São Gabriel da Cachoeira, −0.13, −67.09, 4 Dec 1987 (fr), *Lima 3375* (F, INPA, MO, NY, RB); Serra Aracá, SE of N part of Serra Norte, Mun. Barcelos, 0.85, −63.36, 1250 m, 16 Feb 1984 (fr), *Amaral 1607* (F, INPA, NY); Summit of Central Massif of Serra Aracá, 0.90, −63.37, 1000 m, 16 Jul 1985 (fl), *Prance 29,610* (F, INPA, LBT, MO, NY); Manaus, Estrada do Aleixo, −3.07, −59.95, 3 Apr 1936 (fl), *Ducke 157* (A, F, IAN, MG, MO, NY, S, US), Estrada do Aleixo, Manaus, −3.07, −59.95, 1 Jan 1980 (fl), *Berg P18806* (F, INPA, MO, NY); Rio Urubú, Ig. Sangana, −2.78, −67.78, 2 Oct 1949, *Fróes 25,459* (IAC, IAN); Rio Embira, basin of Rio Juruá, −7.5, −70.25, 21 Jun 1933, *Krukoff 4956* (A, BM, F, K, M, MO, NY, RB, S, U, US); São Paulo de Olivença, 3.44, −68.81, 19 Aug 1929, *Ducke RB 23429* (K, P, RB, S, SCZ, U, US); Mun. Humaitá, between Rio Ipixuna and Rio Livramento, −7.52, −63.03, 7 Nov 1934 (fr), *Krukoff 7120* (BM, F, IAN, K, MO, NY, RB, S, U, US); Manaus-Caracarai, km 28, 2.73, −60.04, 10 Nov 1968 (fl), *Prance 3051* (F, INPA, K, MG, NY, RB); Reserva Ducke, Manaus, −2.83, −59.97, 24 Aug 1994 (fl, fr), *Sothers 130* (BM, INPA, K, MBM, MG, MO, NY); Rio Cuieiras, below Rio Brancinho, −2.83, −60.51, 29 Sep 1991 (fl), *Prance 15,029* (INPA, K, MG, MO, NY); Rio Jandiatuba, Alto Alegre, Mun. São Paulo de Olivença, −4.77, −69.64, 24 Nov 1986 (fr), *Cid Ferreira 8574* (INPA, K, NY); Manaus, Distrito Agropecuário, Fazenda Porto Alegre, −2.39, −59.99, 20 May 1992 (fl), *Nee 42,758* (INPA, NY); trail to Cachoeira da Onça, −2.034, −60.025, 1 Jan 2001 (fr), *Cid Ferreira 12,623* (INPA). **PARÁ**: Monte Dourado, Gleba Angelim, −1.52, −52.58. 3 Sep 1987 (fl), *Pires O'Brien & Silva 1766* (K, MO, NY): Santa Rosa, −1.17, −47.12, 10 Oct 1908, *Museu Goeldi 9721* (MG, NY, S, P, US); Moêma, −1.32, −48.2, 17 Sep 1908, *Museu Goeldi 9670* (K, P, US); Rio Trombetas, between Cuminámirim and Ariramba, −1.29, −55.93, 12 Oct 1913, *Ducke RB 14962* (BM, MG, P, US); Santarém-Cuiabá, km 1221, −5.91, −55.66, 21 May 1967 (fl), *Amaral 1379* (INPA, NY, RB); Belém, Utinga,

−1.42, −48.42, 27 Aug 1941 (fl), *Ducke 781* (F, IAN, MG, MO, NY, US); Mineração Rio Norte, Porto Trombetas, 1.7, −56.45, 1 Jan 1996 (fl), *Barbosa 24* (INPA); Belém, −1.45, −48.48, 7 Nov 1962, *Cuatrecasas 26,651* (F). **AMAPÁ**: Morros de Felipe, Reserva Genética Jarí, Mazagão, −0.87, −52.37, 1 Jan 2005 (fl), *Rabelo 3538* (EAC, F, IEPA, INPA, MO, NY, UFC); Morros de Felipe, Mazagão, Rio Jarí, −0.115, −51.28, 10 Oct 1987 (fl), *Beck 31* (EAC, INPA, UFC). **ACRE**: BR364, Cruzeiro do Sul-Tarauacá Km 40, −7.63, −72.67, 13 Sep 1985 (fl), *Rosas 293* (INPA, NY, RB); Road Cruzeiro do Sul to Japim, km 18, Mâncio Lima, −7.61, −72.89, 26 Oct 1966 (fl), *Prance* et al. *2836* (F, INPA, K, MG, NY). **RONDÔNIA**: 8 km NE of Porto Velho, −8.71, −63.81, 9 Nov 1988 (fl, fr), *Prance* et al. *8282* (F, INPA, MO, NY, US); Mun. Porto Velho, UHE Jirau, −9.26305, −78.10111, 15 Aug 2010 (fl), *Pereira-Silva 15,635* (CEN, HUEFS, INPA, NY, RB). **MATO GROSSO**: Base Camp, N of Xavantina, −14.63, −52.36, 18 Dec 1967, *Philcox 3573* (K, UB). **ALAGOAS**: Usina Coruripe, −10.28944, −36.33991, 7 Feb 2003 (fr), *Machado 262* (MAC); Maceió, 4 Feb 2009 (fr), *Araújo 543* (MAC).

19. **Vantanea peruviana** Macbr., Candollea 5: 371. 1934; Cuatrecasas, Contr. U.S. Natl. Herb. 35(2): 64. 1961, as *Ventana peruviana* (orth. var.). Type. Peru. Loreto: Mishuyacu, Apr 1930, *G. Klug 1130* (holotype, F V0041232; isotype, NY 00388418, US 00101239).

Medium-sized trees, the young branches glabrous, lenticellate. Leaves with petioles 2–4 mm long, glabrous, thickened at base; laminas obovate or obovate-elliptic, 3–8 × 1.2–4.5 cm, rigid-coriaceous, glabrous, cuneate at base, rounded, truncate or retuse at apex, margins entire, 1–2 pairs of glands near to base of lower surface and scattered glands throughout, midrib prominulous above, prominent beneath; primary veins 10–12 pairs, slender obscure, venation inconspicuous, bearing a few minute glands. Inflorescences subterminal, cymose-paniculate, corymbose; branches hirtello-puberulous, peduncle thick, puberulous; pedicels 1–1.5 mm long, thick, glabrous; bracts deciduous. Calyx cupular, 1.2 mm high. Sepals 5, rounded, glabrous on exterior, margins ciliolate. Petals white, thickish, linear, 11 × 1.5–1.8 mm, acute, glabrous. Stamens ca 70, glabrous, filaments slender, 7–10 mm long, connate at base into a tube 1–1.5 mm high, the inner ones papillose, anthers ovate, 0.6 mm long, 4-lobed, connective fleshy, oblong, obtuse. Disc tubular, glabrous, dentate at margin, surrounding the ovary, 1.5 mm high. Ovary ellipsoid, glabrous, 1.5 mm long, 5-locular, locules biovulate; style erect, glabrous, 7 mm long. Drupe ellipsoid, 4–4.5 × 2.5, exocarp rough and warty, glabrous.

Distribution and habitat. Forests of Colombia and Loreto, Peru (Fig. 10.27).
Phenology. Flowering from April to October.
Local names and uses. Peru: *Manchari caspi, parinarillo*; Colombia: *Juseenee* (Mui), *gotoday*.
Conservation status. Vulnerable (VU), B1ab (iii).
Illustrations. Cuatrecasas (1961), fig. 11c–e. 1961; Spichiger et al. (1990), fig. 7.

Additional specimens examined. COLOMBIA. AMAZONAS: Río Caquetá, Villasul, −0.56666S, −72.13333, (st), *Londoño 979* (COAH, NY); Río Caquetá, in front of Isla Sumaeta, 200–300 m, 0°39′S, 72°8′W, 20 Oct 1990 (fr), *Avarez* et al. *1182* (COL). **CAQUETÁ**: Araracuara 12 Jul 1990 (st), *Wijninga 614* (COAH); Araracuara, Villa Azul, Muinane, Sep 1992 (fl), *Posada & Matapi 477* (COL), 21 Oct 1992 (imm fr), *Duque & Matapi 1785* (COL). **PERU. LORETO**: Mishuyacu, −4.01, −73.15, Jan-Feb 1930 (fl), *Klug 1091* (F, NY, US); Reserva Florestal Jenaro Herrera, 4°55′S, 73°45′W, 125 m, 14 Feb 1987 (st), *Gentry* et al. *56,473* (MO), 26 Aug 1976 (fl, fr), *Revilla 1225* (F, MO, NY), Mar 1984, *Spichiger 1744* (G, MO); Maynas, 180 m, 16 Aug 1979 (fl), *Rimachi 4577* (F, MO, NY, RB), 22 Oct 1985 (fl bud), *Rimachi 8070* (K, MO, NY).

Differs from *Vantanea parviflora* in the much shorter petioles.

20. **Vantanea spichigeri** Gentry, Candollea 45: 379. 1990. Type. Peru: Loreto: Requena, Reserva Forestal Jenaro Herrera, 4°55′S, 73°45′W, 125 m, Mar 1984 (fl), *R. E. Spichiger* et al. *1743* (holotype, MO 2071268, isotypes, G 00368593, G 00368594, MBM, NY 00074084).

Large trees, the young branches terete, glabrous, lenticellate. Leaves with long petioles 1–2 cm long; laminas obovate, 11–22 × 6.5–11 cm, rigid-coriaceous, glabrous, minutely glandular punctate beneath, confluent onto petiole, obtusely to acutely cuneate at base, rounded or slightly obtuse at apex, margins entire.1–2 pairs of basal glands on lower surface; midrib prominulous above on both surfaces; primary veins 7–10 pairs, brochidodromous, prominulous beneath, slightly prominulous above. Inflorescences terminal, cymose-paniculate, branches puberulous, the rachis glabrescent; bracts early caducous. Calyx 2 mm long, puberulous. Sepals 5, suborbicular, connate at base. Petals lanceolate-oblong, 9–11 mm long, 2–3 mm wide, minutely puberulous on exterior toward apex, margins glabrous. Stamens ca. 100, connate at base into a tube, filaments glabrous, unequal in length, 5–10 mm long, anthers narrowly ovoid, ca 0.7 mm long, connective long and narrow, thecae bilocular. Disc cupular, surrounding ovary, 1 mm high, glabrous. Ovary ovoid, ca 3 mm long, densely hirsute-villose. Drupe ovoid to ellipsoid, 5–6 cm long, 3–3.5 cm wide, rounded at base obtusely aute at apex, inconspicuously and glabrescently lepidote with flat stellate scale and scattered small glands, exocarp 1–3 mm thick; endocarp woody, ovoid, 5 × 3 cm, with 8 narrow oblong longitudinal valves or opercula.

Distribution and habitat. Rainforests of Colombia and Loreto, Peru (Fig. 10.27).

Phenology. Collected in flower from March to October and in fruit from July to October.

Local names and uses. Peru. *Manchari caspi*. Colombia: *Aagüae*, *guachekei*, *juúsenee* (Muiraña).

Conservation. Vulnerable (VU), B1ab(iii).

Illustrations. Spichiger et al. (1990), fig. 8.

Additional specimens examined. COLOMBIA. AMAZONAS: Araracuara, Peña Roja, 0°39′31"S, 72°04′38″W, 16 Feb 1998 (fr), *Londoño & Alcaraz 2537*

(COAH). **CAQUETÁ:** Mun. Solano, Río Yaruje, 0°11′N, 72°12′W, 9 Dec 1995 (st), *Stevenson* et al. *1759* COAH). **PERU. LORETO:** Reserva Forestal Jenaro Herrera, 4°55′S, 73°45′W, 125 m, *Spichiger 1742* (G, MBM, MO), *Spichiger* et al. *1746* (G, MBM, MO); 27 Jul 1986 (fr), *Vasquez* et al. *7696* (AMAZ, MO, USM); 24 Feb 1987 (st), *Gentry* et al. *56,630* (AMAZ, MO, USM); 9 Jan 1985 (fl bud) *Chota 5/93* (AMAZ, K), 1 Oct 1985 (fl imm fr), *Chota 6/21* (AMAZ, K); 22 May 2002 (fl), *Pennington* et al. *17,414* (K, MOL); 1 Oct 1986 (fl), *Valcárcel & Chota 34,090* (K); 17 Nov 1981 (fl), *Poulain 83* (NY, P); Quebrada do blanco, Río Tahuayo, −4.06, −73.14, 10 Jan 1989 (fl), *Ruíz 1351* (K); Requena, −4.98333, −73.81666, 8 Aug 1988 (fl), *van der Werff 10,022* (MO).

21. **Vantanea spiritu-sancti** (Cuatr.) K. Wurdack & C. E. Zartman, PhytoKeys 121: 102. 2019; *Humiriastrum spiritu-sancti* Cuatr., Revista Hispano-Amer. Ciencia Puras y Aplicadas 23: 137. 1947; Giordano & Bove, Rodriguesia 59: 151–154. 2008. Type. Brazil. Espírito Santo, Mun. Santa Tereza, Reserva Biológica Augusto Ruschi, Lombardia, −19.90, −67.089, 25 Jan 1954 (fl bud), *G. Dalcomo RB86212* (holotype, RB 00539060; isotypes, MBM, US 00101190).

Large trees up to 30 m tall, the young branches striate, exfoliated, glabrous. Leaves with petioles 0.6–0.8 cm long, thick; lamina obovate to elliptical-obovate, 3.5–9 × 2–4.5 cm, rigid-coriaceous, glabrous, cuneate at base, obtuse to slightly retuse at apex, margin entire, sometimes revolute, with 5–7 adaxial glands along proximal edge; midrib plane above obsolete, prominulous beneath; primary veins 6–10 pairs, prominulous on both surfaces. Inflorescences axillary, cymose-corymbose, slightly exceeding leaves, dichotomous or rarely trichotomous branching; peduncle glabrous; branches hirtellous; pedicels tomentose-hispidous; bracts deciduous, ovate-orbicular, clasping, glabrous except for ciliate margins. Sepals obtuse, 0.5–0.6 mm long, glabrous except for ciliate margins. Petals subovate, 1.2–3.5 mm long, glabrous except for ciliate margins, pale green or whitish. Stamens 20, of 3 alternating lengths, 5 long antesepalous, 10 short adjacent and 5 medium length antepetalous, filaments thick, flattened, 2–3 mm long, of 3 alternating lengths, 5 long ones antesepalous, minutely papillose, connate toward base, anthers dorsifixed, connective rostrate, thecae 2, unilocular, dehiscing longitudinally. Disc annular, denticulate, 0.4–0.5 mm high, glabrous. Ovary globose or ovoid, 5-locular, glabrous, locules biovulate; style 0.5–0.7 mm long, erect, glabrous, stigma 5-lobed, glabrous. Drupe oblong-ellipsoid, 2.5 × 1.7 cm; exocarp subcoriaceous; endocarp woody, rugose; seed oblong.

Distribution and habitat. Only known in the Atlantic coastal rainforests of Bahia, Minas Gerais and Espírito Santo, Brazil (Fig. 10.28).

Phenology. Flowering from November to February, fruiting in September.

Local names and uses. *Carne-de-vaca.*

Conservation status. Least concern (LC), listed as vulnerable (V) in the red list for the state of Espírito Santo.

Illustrations. Giordano and Bove (2008), fig. 1.

Additional specimens examined. BRAZIL. **MINAS GERAIS:** Parque Estadual Rio Preto, São Gonçalo do Rio Preto, 18°08′42" S, 43°22′17″W, 18 Nov 1999 (fl), *Lombardi 3522* (BHCB, CEPEC, HUEFS). **BAHIA**: Ilhéus, Km 10 road Pontal-Olivença, 5 Feb 1982 (fl bud), *Mattos Silva* et al. *1436* (ALCB, CEPEC, US); Reserva Biológica de Una, 15°09′S, 39°05′W, 14 Sep 1993 (fr), *Amorim* et al. *1391* (CEPEC, MBML, MO, NY, R, US); Reserva Biológica Mico-leão, road Ilheus-Una, 28 Jul 1994 (fr), *Jardim 529* (CEPEC, NY, RB); Santa Cruz Cabrália, *Belém 3322* (CEPEC, RB, UB). **ESPÍRITO SANTO**: Linhares, Reserva Florestal CVRD, 8 Nov 1977 (fl bud), *Spada 8/77* (CEPEC, CVRD, RB, US), 23 Nov 1993 (fl bud), *Folli 2095* (CEPEC, CVRD, RB), 21 Oct 2003 (fl), *Folli 4639* (CEPEC, CVRD), 2 Dec 2003 (fl bud), *Giordano* et al. *2678* (CVRD, RB); Conceição da Barra, FLONA, Rio Preto, 22 Sep 1995 (st), *Luiza s.n.* (VIC); Guarapari, Setiba, 2 Sep 1996, *Gomes 2209* (UFES, VIES).

The pollen morphology of this species is described in some detail in Giordano and Bove (2008), and they showed the unilocular thecae and uniovulate ovary cells, and as a result, Wurdack and Zartman correctly transferred this species from *Humiriastrum* to *Vantanea*. The absence of apical foramina in the fruit also confirms this transfer. The low number of stamens for the genus is no different from *Vantanea depleta*.

22. **Vantanea tuberculata** Ducke, Arq. Inst. Biol. Veg. 4: 30. 1928; Cuatr., Contr. U.S. Natl. Herb. 35, 2: 69. 1961. Type. Brazil. Amazonas: São Paulo de Olivença, Rio Solimões, −3.46, −68.93, 6 Feb 1937, *A. Ducke RB 30134* (lectotype, **designated here**, RB 00539080; isotypes, K 000407350, RB 00570496, U 0002499, US 00101238). Fig. 9.4C.

Large trees, the young branches glabrous, shiny. Leaves with petioles 3–4 mm long, thick, glabrous; laminas obovate-elliptic, 8–11 × 4.5–6 cm, coriaceous, glabrous, cuneate at base, obtuse to rounded at apex, margins entire; midrib prominulous above, prominent beneath; primary veins ca 10 pairs, inconspicuous or slightly prominulous on both surfaces, venation obscure. Inflorescence a cymose panicle, the branches yellow in pubescent. Calyx 1–2 mm long, puberulous, the lobes suborbicular. Petals oblong, 8 × 2–3 mm, ochre colored, densely pubescent on exterior, glabrous within. Stamens more than 100, 5–8 mm long, connate at base, anthers ovoid 0.8 mm wide, connective thick at base, pointed upward. Disc cupular, glabrous, surrounding ovary, 1 mm high. Ovary ovoid, 3–4 mm high, densely hirsute; style 4 mm long. Drupe ellipsoid-oblong, 6.6–8 × 4.6–6 cm, rounded at both ends; exocarp 8–10 mm thick, coriaceous when dry; endocarp woody, 5 cm long, 3.5 cm broad, deeply anfractuose-rugose, with ten irregular cavities, 5 in the valves alternating with 5 larger ones, the 5 opercula longitudinal, oblong, acute at apex, usually with 1–3 seeds developing.

Distribution and habitat. Known from a single collection in the upper Solimões region (Fig. 10.25).

Phenology. Fruiting in February.

Conservation. Data deficient (DD).

Illustrations. Spichiger et al. 1990: 34, Fig. 9.

This dubious species is only known from the type of collection. The collection cited in the protolog is *Ducke RB 30135*, which is actually the type of *Vantanea micrantha*. The correct number is *RB 30134* as cited above. There are two specimens of that number at RB, and I have chosen one as the lectotype.

Numerical List of Taxa

1. **Duckesia**

    1.1 *Duckesia liesneri* (Cuatr.) K. Wurdack & C. E. Zartman
    1.2 1.2 *Duckesia verrucosa* (Cuatr.) Cuatr.

2. **Endopleura**

    2.1 *Endopleura uchi* (Huber) Cuatr.

3. **Hylocarpa**

    3.1 *Hylocarpa heterocarpa* (Ducke) Cuatr.

4. **Humiria**

    4.1a. *Humiria balsamifera* (Aubl.) St. Hil. var. *balsamifera*

    b. var. *coriacea* Cuatr.

    c. var. *floribunda* (Mart.) Cuatr.

    d. var. *guianensis* (Benth.) Cuatr.

    e. var. *imbaimadaiensis* Cuatr.

    f. var. *laurina* (Urb.) Cuatr.

    g. var. *pilosa* (Steyerm.) Cuatr.

    h. var. *guaiquinimana* Cuatr.

    i. var. *savannarum* (Gleason) Cuatr.

    j. var. *subsessilis* (Urb.) Cuatr.

    4.2. *Humiria crassifolia* Mart. ex Urb.
    4.4. *Humiria fruticosa* Cuatr.
    4.4. *Humiria parvifolia* A. Juss
    4.5. *Humiria wurdackii* Cuatr.

G. T. Prance, *Humiriaceae*, Flora Neotropica 123,
https://doi.org/10.1007/978-3-030-82359-7

5. **Humiriastrum**

5.1 *Humiriastrum colombianum* (Cuatr.) Cuatr.
5.2 *Humiriastrum cuspidatum* (Benth.) Cuatr.
5.3 *Humiriastrum dentatum* (Casar) Cuatr.
5.4 *Humiriastrum diguense* (Cuatr.) Cuatr.
5.5 *Humiriastrum excelsum* (Ducke) Cuatr.
5.6 *Humiriastrum glaziovii* (Urban) Cuatr.
a. var. *glaziovii*
b. var. *angustifolium* Cuatr.
5.7 *Humiriastrum mapirense* Cuatr.
5.8 *Humiriastrum melanocarpum* (Cuatr.) Cuatr.
5.9 *Humiriastrum mussunungense* Cuatr.
5.10 *Humiriastrum obovatum* (Benth.) Cuatr.
5.11 *Humiriastrum ottohuberi* Cuatr.
5.12 *Humiriastrum piraparanense* Cuatr.
5.13 *Humiriastrum procerum* (Little) Cuatr.
5.14 *Humiriastrum purusensis* Prance
5.15 *Humiriastrum subcrenatum* (Benth.) Cuatr.
5.16 *Humiriastrum villosum* (Fróes) Cuatr.

6. **Sacoglottis**

6.1 *Sacoglottis amazonica* Mart.
6.2 *Sacoglottis ceratocarpa* Ducke
6.3 *Sacoglottis cydonioides* Cuatr.
6.4 *Sacoglottis gabonensis* (Baill.) Urban
6.5 *Sacoglottis guianensis* Benth.
6.6 *Sacoglottis holdridgei* Cuatr.
6.7 *Sacoglottis maguirei* Cuatr.
6.8 *Sacoglottis mattogrossensis* Malme
6.9 *Sacoglottis ovicarpa* Cuatr.
6.10 *Sacoglottis perryi* K. Wurdack & C.E. Zartman
6.11 *Sacoglottis trichogyna* Cuatr.

7. **Schistostemon**

7.1 *Schistostemon auyantepuiense* Cuatr.
7.2 *Schistostemon densiflorum* (Benth.) Cuatr.
7.3 *Schistostemon dichotomum* (Urban) Cuatr.
7.4 *Schistostemon fernandezii* Cuatr.
7.5 *Schistostemon macrophyllum* (Benth.) Cuatr.
7.6 *Schistostemon oblongifolium* (Benth.) Cuatr.
7.7 *Schistostemon reticulatum* (Ducke) Cuatr.
7.8 *Schistostemon retusum* (Ducke) Cuatr.
7.9 *Schistostemon sylvaticum* Sabatier

8. Vantanea

8.1 *Vantanea bahiensis* Cuatr.
8.2 *Vantanea barbourii* Standl.
8.3 *Vantanea compacta* (Schnizl.) Cuatr.
a. Subsp. *compacta*
b. Subsp. *microcarpa* Cuatr.
8.4 *Vantanea deniseae* W. Rodrigues
8.5 *Vantanea ovicarpa* Sabatier
8.6 *Vantanea depleta* McPherson
8.7 *Vantanea guianensis* Aubl.
8.8 *Vantanea macrocarpa* Ducke
8.9 *Vantanea maculicarpa* Sabatier
8.10 *Vantanea magdalenensis* Cuatr.
8.11 *Vantanea micrantha* Ducke
8.12 *Vantanea minor* Benth.
8.13 *Vantanea morii* Cuatr.
8.14 *Vantanea obovata* (Nees & Mart.) Cuatr.
8.15 *Vantanea aracaensis* Prance
8.16 *Vantanea occidentalis* Cuatr.
8.17 *Vantanea paraensis* Ducke
8.18 *Vantanea parviflora* Lam.
8.19 *Vantanea peruviana* Macbr.
8.20 *Vantanea spichigeri* Gentry
8.21 *Vantanea spiritu-sancti* (Cuatr.) K. Wurdack & C. E. Zartman
8.22 *Vantanea tuberculata* Ducke

List of Exsiccatae

Abraham, A. A., 152 (4-1c)
Abreu, I. S., 16 (6-5); 45 (8-14)
Abreu, M. M. O., 27 (6-5)
Absy, M. L., 74 (4-1a)
Acevedo, D., 1151 (6-11)
Acevedo-Rodrigues, P. et al., 13 (2-1); 1614 (8-18); 8008, 8114 (7-5); 8257 (5-12); 10279 (6-5); 10420 (4-1d); 13720 (2-1); 14632 (5-2); 14699 (4-1d); 14703 (4-1a)
Acosta, L., 69 (6-11)
Adjimang, E. O., 4849 (6-4)
Adorno, H., 78 (4-4)
Afonso, E. A. L. 191 (4-1i)
Afzelius, A., s.n. (6-4)
Agra, M. F., 629, 1547 (6-8)
Aguiar, O. T., 1121 (8-3a); ESA105453 (5-3)
Aguilar, R. et al., 186 (8-2); 232 (5-4); 453 (8-2); 601, 740, 1576 (5-4); 1862 (8-2); 2208, 2638 (5-4); 3138 (8-2); 3590, 3794, 4134(5-4); 4139 (8-2); 4287 (5-4); 4537 (8-2); 8091 (6-11)
Aguirre S., J., 876 (4-1d)
Akpabla, G. K., 766 (6-4)
Alcazar, E. et al., 123 (6-11)
Albuquerque, B, W, P (Byron))., 11 (2-1); 1084, 1094 (4-1c); 1148 (5-2); 1166 (4-1a); 1185 (8-7); 67-33 (6-2); 67-77 (7-5); 386 (7-6); INPA20756 (7-5)
Albuquerque, B. W. P. & Elias de Paula, J., 67-33 (6-2)
Albuquerque, N. A. et al., 355 (6-8)
Alencar, M. L., 104 (7-5); 184 (4-1a); 194 (4-1d); 493 (5-2)
Alhayde, S. F., 376 (6-5)
Allemão, F. & Cysneiros, F., 255 (4-1c)
Allen, P. H., 3243 (5-12); 5812 (5-4); 6681 (8-2)

G. T. Prance, *Humiriaceae*, Flora Neotropica 123,
https://doi.org/10.1007/978-3-030-82359-7

Almeida, A. I. S., 25 (6-8)
Almeida, C. de, 1935 (4-1c)
Almeida, E. B., et al., 51, 311, 334, 366 (6-8)
Almeida, J., 2239 (8-1); 6989, (6-8)
Almeida, J. C. de, INPA204, INPA3270, INPA3996 (5-2); INPA5924 (8-18)
Almeida, S. & M. Cordeiro, 647, 733 (4-1a)
Almeida, T. E., et al., 1028 (8-14); 3894 (5-14); 4202 (4-1a)
Altson, R. A., 159 (4-1d); 545 (4-1b)
Altson R. A. & Lutz, B., 169 (4-1g)
Aluísio, J., 78 (2-1); 95 (8-11); 118 (2-1); 135 (7-5); 143 (1-2); 170 (8-18); INPA71646 (2-1)
Alvorado G., M., s.n. (8-16)
Alvarenga, D., 249 (6-5)
Álvarez, E. 157 (6-1); 930, 1182 (8-19)
Alverson, W. S. et al., 101 (5-1)
Alves, L. J. et al., 186 (8-14)
Alves, M., 1084 (8-14); 1305 (4-4); 2014 (6-8); 2305 (4-4)
Alves, R. P., 101 (4-1a)
Alves-Araújo, A., 856, 964 (6-8)
Alvorado G., M., s.n. (8-16)
Amaral, C., 43 (6-8)
Amaral, I. L. do, 860, 943 (4-1a); 1379 (8-18); 1428 (6-8); 1541 (8-15); 1543 (4-4), 1607, 1692 (8-18); 1703 (4-1f); 1705 (4-1i); 1755 (8-7); 1807, 2562 (4-4); 2668 (4-1d); 2712 (8-7); 2987 (8-7); 3144, 3145(6-8); 3253 (6-5); 3260 (7-5); 3996 (4-4);
Amaral.M. C. E. do, s.n. (5-6); CFSC8377 (5-6a); CFSC9265 (4-4)
Amaral-Santos, C., 1537, 3069, 3256 (6-5)
Amman, S., 62 (7-5)
Amorim, A. M. et al., 435 (8-3a); 575 (4-1c); 1184, 1319 (6-8); 1391 (8-21); 1414 (4-4); 1482 (6-8); 1563 (7-8); 1651, 1807 (4-4); 2048 (6-8); 2118 (8-1); 2562 (4-4); 3054 (8-14); 3232 (8-3a); 3260 (6-5); 3996 (4-4); 4048, 4084 (6-8); 4346 (6-5); 6769 (4-4); 8601 (7-8); 8630 (8-15); 9246, 9256 (4-1c)
Amorim, B. S., 2027 (8-1)
Andel, T. R. van et al., 2347 (6-3); 2393 (6-5); 4993 (6-3)
Anderson, A. B., 129, 137 (4-1c); 215 (4-1c)
Anderson, C. W., s.n. (5-3); 154 (5-10); 506, 559 (4-1d)
Anderson, W. R. et al., 7450, 8380 (4-4); 8837 (8-14); 8839 (4-4); 10795 (4-1c); 13867 (4-1b); 26223A (8-14); 35109, 35449 (4-4); 35910 (5-6b); 36130 (4-4); 36233, 36781 (8-14)
Andrade, A. C. S., 95 (4-4)
Andrade, M. J. G., 364 (4-4)
Andrade-Lima, D., 49-336, 49-371, 50-452 (6-8); 51-821 (4-4); 53-1332 (6-8); 58-3156 (4-1a); 58-3155 (6-8); 61-3781, 62-4085 (4-4); 66-4572 (4-1c); 69-5644 (6-8); 71-6558 (8-13); 72-6872 (8-14); 72-7097 (4-4); 76-9145 (6-8); 79-8074 (6-8); 8760 (4-4)

Andrade-Lima, D. A. & Madeiros-Costa 109 (6-8); 177 (4-1c)
Andre, E. F., 210 (4-1j)
Angel E., D. C., 19, 31 (4-1d)
Anjos, A. C. dos & Fonseca Neto, F. P., 12 (4-4)
Antezana, A. et al., 361, 386, 447, 745 (5-7)
Antônio, L. C., PSACF691 (6-5)
Antoniazzi, S. A., 10-239 (6-8)
Anunciação, E. A., 926 (4-1c); 1064 (7-5)
Appun, C. F., 37 (4-1a)
Araque, J. & Barkley, F., 18Va021 (4-1j)
Araújo, A.A.M., 170, 189 (6-8)
Araujo, A. P. de 348 (4-1a)
Araújo, C. M. L. R., 29 (6-8)
Araújo, D. 117 (4-4); 320 (8-1); 341, 7563 (4-4)
Araújo, G. B., 414 (6-8); 537, 543, 546 (8-18)
Araújo, J da S. s.n. (4-4)
Araujo, J. M. P., 6 (6-1)
Araujo-M., A., 2394, 2397 (8-3b)
Araújo, S. 30 (5-6a); 496 (4-4)
Arbeláez, M. V. et al., 398 (4-2); 977 (5-12)
Arbeláez, M. V. & Sueroque 543 (4-2)
Árbocz, G. F. 3817 (6-8)
Archer, W. A., 7915, 7964 (6-1)
Arévalo M, F., 530 (7-5)
Aragão, I. 83, 190 (4-1c)
Argent, G. C. G. in P. Richards 6793 (6-5)
Arias G., J. C., 1641 (6-5)
Ariati, V. et al., 855 (8-3a)
Aristegueta, L., 2174 (4-1b)
Ariza, W. & Villa, C., 49 (6-5)
Arraes, M. G. M.et al., 31879 (4-4)
Arroyo, L. et al., 360 (6-8)
Assis, A. M. et al., 2866 (5-3)
Assis, C. K., 18 (5-3)
Assis, J. S. de, 201 (4-4); 308 (8-14); 346 (4-4)
Assis, L. C. S., 114 (5-6); 523 (5-6a)
Assunção, P. A. C. L., 75 (8-18); 184 (6-5); 185 (8-11); 191 (6-5); 367 (8-8); 486, 496 (6-8); 529, 530 (4-1c); 605 (2-1); 886 (6-5); 1077a (4-1c); 1096, 1175 (4-1c); 1182 (6-5); 1199 (4-1c)
Atkins, S., PCD4591 (4-4)
Attims, Y., 265 (6-4)
Aublet, J. C. B. F., s.n. (8-7)
Aubréville, A., 92 (6-4)
Aubry-Lecompte, C. E., s.n. (6-4)
Aubrya, F., (6-4)

Aulestia, C. et al., 548 (5-13); 603 (8-16); 827 (5-4); 1035 (5-13); 1209 (5-4)

Austin, D. F. et al., 7059 (4.1a); 7070 (6-5)Aymard, G. et al., 5932 (6-5); 5965 (4-1c); 6036 (6-5); 6317 (4-1i); 6384 (6-5); 6470 (7-6); 6497 (6-5); 6505 (4-1f); 6508 (6-5); 7656 (6-3); 7949, 8320 (6-3); 8930 (6-3); 9100 (7-8); 9116 (4-3); 9161 (4-1f); 9986 (6-5); 10955 (5-1); 12674 (6-2); 14214 (8-8); 14357 (7-5)

Aymard, G. & Delgado L., 7949, 8320 (^-3)

Aymard, G. & Gómez, C., 11026 (5-1)

Azevedo, A. B., 32 (8-18)

BAFOG 35M (6-3); 48M, 102M (8-18); 124M, 131M (6-3); 220M (8-18); 228M (4-1a); 247M (8-18); 347 (6-3); 1083 (4-1a); 7587 (4-1a); 7622 (6-5); 7656 (6-3)

Bailey, I. W., 115 (4-1a)

Baitello, J. et al., 1799 (8-3a)

Baker, B. O., 58, MG9401 (2-1)

Baldwin, J., 3187 (7-6)

Balée, W. L., 369 (6-8); 3127 (6-5)

Bamps, P. R. J., 4575 (6-8); 5473 (6-5)

Barbosa, C. 7558 (5-2); 7700 (4-1a)

Barbosa, C. E. & Madriñán, S., 8364 (4-2)

Barbosa, E. M., 24 (8-18); 126, 155 (2-1); 207 (4-1c); 297 (8-18); 302 (2-1)

Barbosa M. R., 1346, 1365 (6-8)

Barbosa, R. I., 156 (4-1d); 174 (6-5)

Barbosa da Silva, M. 73 (4-1a)

Barbosa-Silva, D., 48 (6-5)

Barbosa-Silva, R. G., 261, 944 (8-14)

Barbour, W. R., 1018 (8-2)

Barford, A. S., 41076 (5-13)

Barona, A., 854 (4-4); 914 (6-1); 962 (4-4)

Barra, N. de la et al., 847 (6-8)

Barrère, P. la, s.n. (8-18)

Barreto, A. C., s.n. (4-1a); ASE74 (4-4)

Barreto, H. L. Mello., 1259 (5-3); 8505 (4-4); 8804 (5-6b); 9761 (4-4); 9849 (8-14); 9872, 9938 (5-3)

Barros, A. A. M., 2161 (4-4)

Barros, F. de B & Martuscelli, P., 1272 (8-3a)

Barroso, C. M. 265 (4-1c)

Barter, C., 68 (6-4)

Bastos, A. M., 104 (4-1a); 244 (8-7)

Bastos, M. N. C., 39 (6-8); 100, 174, 252 (4-1a); 341(6-5); 508 (4-1a); 527 (6-8); 575 (4-1a); 1039, 1189, 1255 (4-1a); 1273 (6-5); 1274 (4-1a); 1307, 1397 (6-5); 1456 (6-8); 1473 (6-5); 1521, 1557 (6-8); 1729 (6-5); 1699, 1914, 2026 (4-1a)

Baumgratz, J. F., 189 (4-1a); 1456, 1457 (4-4)

Bautista, H. P. et al., UMS20 (4-4); UMS60 (8-13); 67 (4-1a); 577 (4-1c); 1453 (4-4)

Bautista, H. P. & L. S. S. Faria, 1158 (4-4)

Bautista, H. P. & T. Jost, 1713 (6-8)

Bayma, I. A., 71 (6-8)
Beccari, N., s.n. (4-1d)
Beck, H. T. 31, 202 (6-1); 205 (8-18); 386, 523 (6-1); 2181 (5-13)
Beck, S., 29027, 29502 (5-7)
Begazo, N., 193 (6-8)
Belém, R. P., 221 (8-3a); 2418 (8-1); 2421 (4-4); 2442 (6-8); 2484, 2487, 2490 (4-4); 2496 (8-1); 2576 (8-14); 3215 (8-1); 3322 (8-21); 3830 (4-4)
Belém, R. P. & Pinheiro, 2408 (4-4); 2410 (4-4); 2418, (8-1); 2442 (6-8); 2484, 2487 (4-4); 2490 (4-1c); 2496 (8-1); 2576 (8-14); 3215 (8-1)
Belém, R. P. & Magalhães 221 (8-1); 747 (4-4); 748 (8-1)
Belém, R. P. & Mendes, J. M., 221 (8-3a)
Beltran, G. & Díaz, J. 1027 (5-2)
Bena, P., 1319 (5-5)
Benoist, R., 1239 (4-1a); 1530 (8-7)
Benti, P. 1083 (4-1a)
Bequart, J. C. C., 1649 (6-4)
Berg, C. C., P18795 (6-2); P18806, 19474 (8-18)
Bernardi, L. C., s.n. (8-22); 1751 (6-5); 2601 (4-1a); 2603 (4-1b); 2613 (8-12); 2813 (6-5); 2814 (5-10); 3033 (6-3); 3435 (8-3a); 6547, 6634 (6-5); 6762 (6-3)
Berry, P. E. et al., 780 (7-6); 5032 (4-1d); 5194 (4-1d); 5317, 5362, 5489 (4-5); 5507 (7-6); 5626 (4-1d); 5711 (4-1d); 5810 (4-1f); 5887 (4-5); 5360 (4-1d); 6028 (4-5); 6739, 7091 (7-8)
Berry, P. E. & Aymard, G., 7531 (5-1)
Berry, P. E. & Rosales C., J., 6499 (4-5)
Berti, L. M., 85 (7-2); 320 (6-3); 520 (7-2); 2614 (4-1j); 134-981 (4-1c)
Bertoluzzi, R. C. C., 441 (5-3)
Bertoncello, R. & Pansonato, M., 424 (5-3)
Betancour, J., 13381 (4-1d); 13412, 13746 (6-2)
Bigio, N. C., 647, 833, 865 (4-1a); 891 (4-1c); 897 (4-1a); 1167 (5-2); 1238 (2-1)
Billiet, F. & Jardin, B., s.n. (4-1); 1126 (6-3); 6392 (4-1a)
Binuyo, A., FHI41418 (6-4)
Black, G. A., 47-1001 (2-1); 47-1276 (4-1a); 47-1756 (4-1c); 48-2512, 48-2514 (7-8); 48-2589 (4-1d); 48-3249 (4-1a); 48-3453 (6-8); 48-3555 (6-5); 49-8313, 49-8369 (4-1a); 50-8685, 50-8831 (6-5); 50-9058 (6-8); 51-12776 (4-1a); 51-12954 (4-1j); 51-13231 (4-1a); 51-13454 (6-5); 51-13843 (6-8); 55-18577 (4-1a); 57-19306 (2-1); 57-19590 (4-1c)
Black, G. A. & Foster, M. B., 48-3393 (6-1)
Black, G. A. & Ledoux, P., 50-9810 (6-1); 50-9811 (5-5); 50-10371 (4-1c); 50-10553 (6-5); 50-10783 (5-2)
Black, G. A. & Magalhães, D., 51-11790 (4-1g); 51-12072 (6-5); 51 12954 (4-1a)
Blanchet, J. S., s.n. (4-1g); 85 (804a); 1005(4-4), 1686, 2810 (4-4), 3144a (4-1g); 3305, 3362 (8-3a); 3422, 3570 (4-4); 3805, 3837 (8-3a)
Blanco, C. A., 1229 (5-12)
Boldingh, I., 3886 (4-1d)
Bomfin, M., 35 (4-4)

Bonadeu, F., 213, 218 (4-1a)
Bonaldi, R., 592 (5-2)
Boom, B. M. et al., 5722 (5-1); 7124 (4-1d); 8525 (6-5); 8542 (6-8); 8571 (6-2); 8579 (2-1); 8580 (8-18); 8613 (8-17); 8742 (7-6); 8780 (8-4); 9364 ((4-1c)
Boom, B. M. & Gopaul, D., 7644 (4-2); 7669 (4-1c)
Boom, B. M. & Mori, S. A., 1683, 1714 (5-5)
Bordallo, MG22630 (4-1c); MG22641, 22646 (4-1a)
Bordenave, B., 780 (6-3)
Bortoluzzi, R. L. C., 441 (5-3)
Bos, J. J., 3533, 4039, 4725, 6137 (6-4)
Boschwezen Surinam (BW), 36a (4-1d); 117 (6-5); 177, 212 (4-1a); 531 (4-1d); 1120 (4-1d); 1166 (6-5); 1490 (6-5); 1547 (4-1a); 1935 (4-1a); 1936 (6-3); 2068 (7-2); 2232, 2471 (4-1a); 2599 (4-1d); 2765 (4-1a); 2816 (4-1d); 2885, 2918 (4-1a); 2974 (6-5); 3010 (4-1a); 3040 (4-1d); 3079, 3125(6-5); 3646 (4-1d); 3934 (4-1a); 3947 (4-1d); 3961 (6-5); 4033 (7-2); 4270, 4296 (6-3); 4469 (6-5);4669 (4-1a); 4673 (6-5); 4684 (4-1a); 4720 (6-3); 4770, 4810 (4-1a); 4933, 4960 (7-2); 5412 (4-1a); 5430 (6-5); 5486 (4-1d); 5527 (8-4); 5827, 5850, 5858 (4-1a); 6234 (4-1a); 6495 (6-3); 6010, 6068, 6670, 6907 (4-1a)
Bovini, M. G., 148, 1736 (6-8); 1738 (4-4); 4155 (4-4)
Boyan, R., 46 (FD7870) (4-2); 78 (FD7902) (6-5)
Boyle, B., et al., 2044, 2070, 2071 (5-13)
Brade, A. C., 13627, 20090, 121454 (4-4); RB28585 (8-3a)
Braga, B., 8, SPSF5519 (8-3a)
Braga, J. M. A., 433 (4-4)
Bragança, V. A. N., 179 (6-5)
Bresolin, A., 1407 (8-3a)
Breteler, F., 4798 (6-8); 4956 (6-3)
Britez, R. M. 1800 (5-3)
Brito. M. F. M., 74 (6-8)
Brito de Souza, A., 3191 (4-1c)
Britton, N. L., s.n. (6-1)
Broadway, W. E., 8475 (6-1)
Browne, R., 78 (6-5)
Buchtien, O., 1518 (5-7)
Burgos, J. A., 37, 85 (5-5)
Cabral, F. N., 267A (4-1d); 505 (6-5); 547, 956 (4-1c); 1233 (8-18); 1337 (4-1a)
Cabrera, I., 578 (8-16); 883 (8-10); 2724 (6-2); 2778 (6-8)
Cadorin, T. J. et al., 1787 (8-3a)
Caetano, V. J., 40 (6-5)
Calazans, C., 305 (6-5)
Calderón, C. E. et L., 2200 (4-1a); 2529 (4-1c); 2700 (4-1i)Callejas, R. et al., 1678 (4-4)
Callejas, R & Marulanda, O., 6974 (4-1c)
Campbell, D. G., P20892 (6-8); 21964(6-2); 22536 (4-1a); 22547 (4-1c)
Campos, M. M., 44 (4-1c)

Campos, P., 1387 (4-1c); 1390 (6-5)
Caño, A., 51 (7-6)
Capucho, P., 430 (2-1); 483 (8-18)
Carballo, G., 139 (6-11)
Carcerelli, C. s.n., 72 (5-3)
Cárdenas, D. et al., 3085 (8-10); 4123 (5-2); 4351 (4-2); 5564 (7-6); 6599 (4-1j); 6885, 6940 (4-1a); 9154 (8-19); 9181 (4-5); 10020 (5-12); 10604 (4-1j); 10659 (5-12); 11406 (5-12); 13487, 13649 (4-5); 14761 (4-1d); 14773, 14793, 14837, 14988 (4-5); 15017 (4-1d); 15037 (4-5); 15070 (4-1d); 15116, 15181, 15187, 15337 (4-5); 15391 (4-1c); 16278, 16361 (4-5); 16374 (4-1d); 16375 (6-5); 16413 (4-5); 16488 (4-1c); 16506 (4-1c); 16528, 16800, 16894 (4-5); 16942, 16957 (4-1d); 16984 (4-1c); 17178, 17227 (4-5); 17326 (4-1d); 17347 (4-1c); 17368 (4-5); 17475, 17791 (4-c); 17493 (4-1d); 18017 (5-12); 19016 (4-1a); 20386 (7-8); 20392 (4-1d); 20491 (4-5); 21072 (4-1j); 21199 (4-1c); 21392 (4-1a); 21581 (6-2); 21957 (4-5); 22185 (6-2); 22248 (5-11); 22407 (6-2); 22508 (4-1j); 23417, 23419 (7-8); 23639 (6-2); 24205 (4-1c); 24292 (4-1c); 24524 (4-1a); 42494 (4-1c); 42903 (6-2); 43061 (4-1c); 43111, 43778 (4-1a); 43998 (4-5); 44025 (4-1c); 44063 (5-2); 44066 (4-1j); 44135, 44163, 44343 (4-5); 45541 (4-1a); 47011 (4-1c); 48997 (8-19); 49064, 49071 (4-1j); 49375 (4-1d); 49658 (8-18); 49845 (4-2); 50222 (4-1a); 58137 (4-1j)
Cardiel, J. M. & Pedrol, J., 4665 (5-2)
Cardona, F., 774 (4-1b); 965, 1112 (4-1h); 1205 (4-1b); 1768, 1823 (4-1b); 1912 (8-12); 2139 (4-1c); 2269 (4-1b); 2362 (8-12); 2533, 2877, 2942 (4-1b)
Cardoso, D., 2025 (6-8); 3237 (7-5); 3311, 3321 (3-1); 3392 (4-1d)
Cardoso, G. L., 12 (5-2))
Carneiro Filho, A., 6, 48 (4-1d)
Carneiro, E. M., 27, 142, 176, 736 (4-4)
Carreira, L. M. M., 62 (4-1c); 117 (7-5); 604 (6-8)
Carvalho, A. S. R. 46, 66 (4-4)
Carvalho, A. M., 88 (4-4); 157 (4-1c); 1062 (8-14); 1275 (4-4); 2507, 2906, 2907 (8-14); 4346 (6-5); 6159 (8-3a); 6531 (4-4); 6777 (7-8)
Carvalho, A. M. & Gatti, J., 484 (8-1)
Carvalho, A.M. & Lewis G. P., 926 (4-4)
Carvalho, A.M. & Sant'Ana 6239 (7-8)
Carvalho, A. M. & Saunders, J., 2906 (8-14); 2907 (8-1)
Carvalho, C. 162 (4-1j)
Carvalho, D. A., 59, 112 (4-4)
Carvalho, F. de 2660, 2755, 2944, 4761 (6-4)
Carvalho, F. A. de, 1_Virua2006 (4-1c); 952, 964 (4-1d); 1051 (6-5); 1413, 1428, 1434, 1453, 1455, 1456 (4-1d); 1548, (4-1c); 1716a, 1716b (4-1d); 1791 (4-1d); 1870, 2058, 2068, 2081, 2096, 2175, 2177, 2186 (4-1c); 2188 (4-1d)Carvalho, L d'A. F., 184 (4-1a)
Carvalho, V, de, 125 (6-5)
Carvalho-Sobrinho, J. G. de, 785 (6-8); 834 (4-1d); 838 (5-2); 843 (4-1a); 1048 (8-18); 1250 (4-1d); 1648, 1649 (8-18); 1686 (6-5); 1700 (4-1c); 2635, 2667, 2725 (6-8)

Casaretto, G., s.n. (5-3)

Castaño A., N. et al., 1727 (8-8); 3446 (5-2); 10536 (8-18); 10593 (4-2)

Castillo, A., 2000 (5-12); 4102 (7-6); 5533 (5-12); 5787 (7-6); 6816 (6-5); 7143 (7-6); 7206, 7209 (6-5)

Castilho, C. V. de, 54, 709 (8-18); 920, 979 (6-5); 1014 (1-2); 1016 (8-8); 1078 (6-8); 1147, 1180 (8-18); 1193, 1226 (2-1); 1314 (8-18); 1364 (8-8);

Castro, F. 4271 (4-1a); 10694, 10837 (5-5); 17918 (4-5); 18072 (7-6)

Castro, S. Y., 1693 (6-1)

Castro, W., 3058, 3085, 3159, 3169 (2-1); 3202 (5-3)

Castroviejo, S., 12056, 12060 (4-1a)

Cavalcante, P., 1987 (4-1j); 2291 (2-1); 2549, 2550 (4-1c); 2761 (1-2); 3310 (6-5); 3349 (6-5); 3382 (4-1c); 3383 (6-5)

Cavalcanti, A.D.C., 193, 614 (6-8)

Cavalcanti-Ferreira, D. M. 614 (6-8)

Celedonia Ruiz, J. 661 (5-5)

Cerati, T. M. et al., MG249 (4-4)

Cervi, A. C. 58074 (4-4)

Cestano, L. A., 174 (6-8)

Chacón, I. A. 10 (5-4); 900 (6-11); 1124 (8-2)

Chagas, J. C. 204, 1700 (5-2); INPA86 (4-1a); INPA244, INPA1243 (7-5); INPA1638 (5-2); INPA1773 (4-1c); INPA3093, INPA3472, INPA3669 (7-5); INPA3720 (5-2); INPA3996 (5-2); INPA5954 (8-18); INPA 6721 (7-5)

Chagas, J. C. & Dionisio INPA3472 (7-5)

Chanderbali, A. & Gopaul D., 98 (5-10)

Chautems, A., 185 (8-14); 191 (4-4)

Chavarria, U., 742 (6-11)Chaves, S. R., 38 (7-5)

Chaviel, A., 177 (4-1c); 197 (6-5)

Chevalier, A., 16195 (6-4)

Chota, M., 5/93, 6/21 (8-20); 6/240 (8-18)

Christenson, G. M., 1395 (7-8); 1421 (5-12)

Cid Ferreira, C. A., 194 (6-5); 235, (4-1d); 263 (4-1d); 344, 776 (7-6); 1000 (6-5); 1141 (6-8); 1199, 1457(4-1c); 1493 (4-1a); 1860 (4-1a); 1976 (6-2); 2307 (4-1a); 2478 (4-1c); 2815 (4-1a); 3187, 3740 (5-2); 3762 (6-5); 3870 (4-1c); 3980 (4-1d); 3985 (5-2); 4095 (4-1c); 4205, 4261, 4264, 5205 (5-2); 5386 (6-5); 5386 (6-5); 5446 (6-8); 5457 (4-1c); 5593 (4-1a); 5810 (4-1i); 5852 (8-18); 6031 (8-8); 6124 (6-8); 6165 (6-5); 6208 (6-8); 6326, 6475 (6-5); 6686 (8-7); 6845 (7-5); 7021 (4-1d); 7030 (6-8); 7084(6-8); 7415 (6-5); 7429 (6-8); 7536 (2-1); 7563 (4-1d); 7578 (6-5); 7615 (5-2); 7625 (8-7); 7718 (5-2); 7797 (4-1c); 7864 (8-7); 7867, 7870 (8-5); 8103 (4-1c); 8108 (5-2); 8195 (5-2); 8459 (8-4); 8484 (8-7); 8574 (8-18); 8867 (4-1a); 9080 (4-1d); 9086 (5-2); 9143 (4-1d); 9148 (4-5); 9167, 9193(4-1a); 9202 (4-1i); 9272 (7-5); 9394 (4-1c); 9510 (2-1); 9521 (6-2); 9550 (4-1c); 9565 (8-18); 9740 (5-2); 9768 (4-1c); 9792 (4-1a);9805 (5-2); 9818 (4-1c); 10130 (4-5); 10143A (6-3); 10160 (4-1j); 10672 (5-2); 10936 (4-1j); 11004 (4-1d); 11027 (5-2); 11222 (4-1c); 11523 (4-1d); 11532 (6-5); 11584 (4-1a); 11645 (4-1a); 11689 (4-1c); 11870 (7-8); 11871 (6-5); 11888, 11897 (4-1d); 11917 (4-1d); 11989 (4-1a); 12057

(7-5); 12059 (4-1c); 12075 (4-1b); 12093 (4-1a); 12341, 12455 (4-1d); 12623 (8-18); 12787 (4-1a); 12901 (4-1b); 12933 (6-8); 12973 (6-5); 13009 (5-2); 13020 (6-5); 13178 (4-1d); 13933 (6-8); 19256 (6-5)

Cid Ferreira & Lima, J., 3762 (6-5)

Clarke, H. D., 656, 774, 776 (4-1d); 863 (4-2); 868, 885 (8-12); 976(4-2); 1274 (6-5); 1829 (6-3); 2912 (6-5); 2913 (6-1); 3175 (2-1); 3936 (6-5); 4497 (6-3); 5243 (4-1c); 6944, 7009 (4-1d); 10859 (5-10)

Clark, H. L. 6906 (4-1d); 6912 (4-1c); 7048 (4-1d); 7078 (7-8); 7235, 7252 (4-1d); 7343, 7346 (8-18)

Clark, H. L. & Maquirino, P., 6905 (4-1d); 7812 (7-8); 7902 (7-6)

Clark, K. & Christenson G. M., s.n. (5-12)

Clavijo R., L., 383, 538 (5-12); 1383 (6-1)

Clavijo R., L., & Montes 1383 (6-2)

Clavijo R., L & Tanimuka, P, 703 (6-2)

Coelho, D. F., (Dionisio)783 (4-1c); 870 (2-1); INPA3646 (2-1); INPA3928 (7-5); INPA4101 (2-1); INPA4138 (6-2); INPA8481 (8-18); INPA25932 (8-18); INPA52396, 870 (2-1)

Coelho, L. F., (Luiz) s.n., 33, 160 (4-1c); 546, 650 (7-5); 880 (4-1d); 4064 (4-1a); INPA1056 (4-1c); INPA1422 (8-18); INPA1684 (4-1c); INPA1814 (5-2); INPA3146 (6-2); INPA3255 (7-6); INPA3670 (4-1c); INPA3928 (7-5); INPA4064 (4-1a); INPA4272 (6-2); INPA5135, INPA5136, INPA5137, INPA5139 (6-8); INPA5196, 31 (8-8); INPA5227 (8-18); INPA5234 (6-5); INPA5240 (1-2); INPA6361 (6-5); INPA6490 (8-7); INPA7252 (7-5); INPA18752 (6-8); INPA202860 (4-1c); INPA36008 (6-5); INPA42102 (4-1c); INPA60331 (8-7)

Coelho, L. F. & Mello, F., INPA3255 (7-5)

Coêlho, L. S., 24 (6-8)

Coélho, R. et al., 41 (5-3)

Colchester, M., 2352 (4-1j)

Colella, M. et al., 1673 (6-5)

Colque, O. & Mendoza, L. 443 (6-8)

Conceição, A. A. 1355 (8-13)

Cooper, G. P., 68, 274 (6-4)

Coradin, L. et al., 1009 (6-5); 6088 (4-4); 6409 (8-14)

Cordeiro, I et al., 16 (4-2d); 105, 127 (4-1d); 133 (6-8); 136 (7-8); 166 (6-8); CFCR783 (4-4); 2269 (4-1a); 8204 (4-4)

Cordeiro, M. R., 1151 (4-1a); 1294 (4-1c); 4880 (6-8)

Córdoba, M. P. et al. 67 (5-2); 167 (4-2); 181 (4-1a); 246 (8-19); 529 (4-1c); 785 (7-8); 913 (6-5); 1862 (4-1a)

Cornejo, M. et al., 384 (5-7)

Correa, A. L. 411 (8-18); 431 8-5); 458 (6-5)

Correa A., M. D. & Dressler, R. L., 712 (6-11)

Correa-Gomez, D. F., 137 (6-2)

Correia, C. M. B., 32 (8-3a)

Correia, V. M., 3 (5-9)

Cortés B., R. et al., 201 (4-2); 243 (6-8); 247 (4-2)

Costa, A. L., s.n., 569, 941, 1125 (4-4)
Costa, C. S., 498 (4-1a)
Costa, D. S. 137 (4-1a)
Costa, G., 36, 93 (8-14); 167 (4-4)
Costa, F. M., 1922 (4-1i)
Costa, F. N., & Fiaschi, P., 210 (8-14); 401 (4-4)
Costa, J. 109 (4-1c); 772 (8-13)
Costa J. C., 3436 (4-1c)
Costa, M. A. da S., 118 (6-8); 1044 (6-5); 1117 (2-1); 1118 (6-8)
Costa Lima, P. A., 529 (4-1d)
Costich, D. & Cardoso, R., 1071 (8-4)
Couto, A. P. L., 143 (8-14)
Cowan, R. S., 2145 (4-2); 38700 (4-1a); 39263, 39266 (4-1d)
Cowan, R. S. & Maguire, B., 38034 (4-1c)
Cowan, R. S. & Soderstrom, T., 1820 (4-1b); 2145 (4-2)
Cowan, R. S. & Wurdack, J., 31090, 31301 (4-1b); 31472, 31502 (4-1j); 32022 (6-5)
Croat, T. B., 54008, 54053 (4-1c); 54158 (4-1b); 59251 (4-1c); 62194 (4-1c); 96568 (5-7); 101798, 101896 (4-1a); 102098, 102236 (4-1d)
Cruz, E. D., 461 (6-8); 677 (6-5)
Cruz, J. S. de, 2110, 2227, 2645 (4-1d)
Cuatrecasas J., 7203 (7-8); 14418, 14956 (5-4); 16615, 17186 (5-13); 17226 (6-9); 19727 (4-1j); 19909 (5-8); 19927 (6-9); 19937 (8-16); 19998 (6-9); 19989 (5-8); 26650 (8-7); 26651 (8-18)
Cuevas, E & Pariamo, H., 7637 (6-8)
Cunha, L., 195, 197, 201 (6-8)
Cunha, N. M. L., 1024 (2-1); 1030 (8-8)
Curran, H. M., 159 (4-1g)
Custodio Filho, A., 1536, 1741, 2685 (5-6)
Custodio Filho, A. & Gentry, A. H., 4682 (5-6a)
Daguerre, N. et al, 219 (6-9)
Dalcomo, G., RB 86212 (8-21)
Daly, D. C., 806(4-1a); 964 (4-1a); 3815 (5-2); 3916 (8-17); 3967, 3972 (6-5); 5459 (6-5); 5469 (4-1a); 5602A (7-5); 5703 (6-2); 9405 (6-2); 12193 (2-1)
Damasco, G. 565 (4-1a); 1304 (6-5)
Damazio. L. s.n. (RB82987) (5-3)
Damião, C, 2899 (7-5); 2903 (6-5); 3073 (4-1a)
Damião, C. & Mota, A. de, 608 (8-11); 617 (8-4); 630 (8-18); 719 (6-5)
Daneu, L., 477 (4-4)
Daramola, B. O., s.n. (6-4)
Daranda, B. O., FHI46926 (6-4)
Dário, F. R. et al., 1130A (6-8)
Davidse, G. et al., 4850 (4-1b); 16892 (4-1f); 17057, 17063, 17411 (4-1c); 17530, 17533, 17698, 17861 (4-1a); 17889 (6-8); 17891 (4-1a); 22688, 22879, 23037 (4-1b); 27755, 27769 (6-5)

Davidse, G. & Herrera C, G., 31480 (6-11)
Davidse, G. & Huber, O. 22804 (7-1)
Davidse, G. & Miller, J. S., 26550 (4-1d); 26643 (7-6); 26712 (7-8); 27040 (5-1); 27046 (5-1); 27352 (6-8)
Dávila, D. 15 (8-19); 505 (5-2)
Dávila, N. C., 5807 (4-1d)
Davis, D. H. 141 (4-1d); 334, 358 (4-1e); 635, 647 (4-1d)
Davis, T. A. W., 256-FD2247 (4-1d); 600 FD 2720 (5-10); 643 FD 2055 (4-1d)
Davis, D. M. 141 (4-1d)
Dayton, W. A. & Barbour, W. R., 3004 (6-11); 3129 (8-2)
De Bruijn, J., 1645, 1684 (6-3)
Defler, S., 177 (6-5); 179 (8-18)
Deighton, F. C., 2677 (6-4)
Del'Arco, & Arrais, G. 1012 (4-1c)
De La Cruz, J. S., 2202, 2210, 2227, 2644, 2645 (4-1d)
Delascio C., F. et al., 9361 (4-1c); 9494 (4-1d); 11911 (4-1c)
Delascio C., F. & Liesner, R. L., 13560 (4-2)
Delgado, L., 279 (6-5); 805 (4-1i); 911 (4-1c); 941 (7-6); 1080 (4-1c)
Delázkar et al., 5 (5-1)
Dcmarchi, L. O., 110, 426 (4-1c); 621 (4-1c); 768 (5-2); 820 (4-1d); 889 (7-5)
Denys, see Loubry
Devia A, W. et al., 3944 (6-??); 4184, 4273 (5-1); 4330 (5-13)
De Wilde, J. J. F. E., 808, 10963, 11404 (6-4)
De Wilde, W. J. J. O., 661 (6-4)
De Wilde, W. J. J. O., & B. E. E. De Wilde, D2134 (6-4)
De Vries, W. II., s.n. (4-1d)
Dias, H. M., 132, 209 (4-4); 492 (8-1)
Dias, S. S. 1 (4-4)
Díaz S., C., 2035 (4-4); 7736 (8-18)
Díaz, W. A., 908 (6-5); 3268 (4-1b); 5243 (4-1e); 7275 (7-6)
Dick, C., 91 (6-3); 115 (8-17); 147 (8-8); 151 (8-7)
Dinklage, M. J., 2973 (6-4)
Dinuyo, A., FHI41418 (6-4)
Dionizia, F. et al. 28 (5-2)
Dombrowski, L. T., 12635 (5-3)
Donant Herb. 1686 (4-1c)
Dressler, R. L., 4458, 4467 (6-6)
Duarte, A. P., 97, 2295 (6-8); 2508 (8-14); 2629 (4-4); 3280 (8-14); 3717 (4-4); 5967 (6-5); 6553 (6-5); 6632 (6-8); 6671 (4-4); 6827 (8-1); 8028 (6-8); 8551, 9020, 9336 (8-14); 9750 (4-4); RB70631 (8-14)
Dubs, B., 1681 (6-5); 2508 (4-1c)
Ducke W. A., s.n. (5-2); s.n. (8-11);11 (6-8); 12 (6-2); 16, 16a (6-2); 77 (8-8); 87 (4-1c); 98 (8-8); 157 (8-18); 200 (8-7); 241 (2-1); 243 (5-2); 255 (7-5); 265 (3-1); 267 (8-18); 305 (2-1); 324 (8-7); MG363 (6-8); 416 (8-8); 440 (4-1c);, 519 (6-8); 541 (4-1c); 751 (8-11); 752 (8-17); 781 (8-18); 1055 (6-1); 1174 (6-2); 1175 (7-5);

Evans, R. J. et al., 2610 (4-1d); 2866 (2-1); 3220, 3232 (4-1a)
Evans, R. J. & Lewis, G., 1832, 1835 (4-1d)
Faber-Langendoen, D. et al., 432, 459 (6-9); 530 (5-8); 551 (6-9); 547, 876 (5-8); 1005, 1078 (6-9)
Flakenberg, D de B., 11023, 11032 (8-3a)
Fanshawe, D. B., 715 (4-1a); 724 (7-2); 942 (6-5); 1715 (5-10)
Farah, F. T. et al., 2159 (5-3)
Farias, D., 6797 (4-1d)
Faria, J. E. Q., et al., 5348 (8-14); 5428 (4-4); 6049 (8-14)
Farias, E., 421 (4-1c)
Farias, G. L., 65 (8-1); 629 (5-9)
Fariñas, M et al., 472 (4-1f); 478 (6-5)
Farinaccio, M. A., 780 (6-8); 840 (6-5)
Farney, C., 373 (4-4); 1765 (4-1d); 1776 (6-8); 1882 (4-1b); 1313, 1388 (4-4); 1776 (6-5); 1882 (4-1b); 2091 (4-1c); 2108 (5-2); 2109 (4-1a); 2367, 2616, 2708, 2826 (4-4); 4598 (5-9); 4605 (4-4)
Faustino, T. C. 26, 27 (8-13); 32 (8-3a)
Félix, H. C., 2 (4-1d)
Félix-da-Silva, M. M., 22 (4-1a); 44 (6-8)
Fernandes, M. G. C. et al., 1607 (5-6)
Fernandes-Bulhão, C. & Stefanillo, D., CFB527 (6-8)
Fernández P., Álvaro, 185 (5-4); 201 (8-2); 2084 (4-1d); 2142 (7-8); 2148 (4-1d); 2228 (6-8); 2276 (7-4); 7063 (6-5)
Fernández Ángel, 2277 (4-1c); 5587, 5698, 5795 (6-5); 5887, 6963 (4-1c)
Fernández, E. P., 163 (4-4)
Fernández, V., 5 (6-5)
Fernández, Y. & M., 451, 4561 (4-1c)
Ferraz, J., HST18451 (4-4)
Ferreira, A. M., 40 (8-8); 154, 58-302 (8-7)
Ferreira, C. et al., 369 (4-4); 577 (6-5)
Ferreira, E., 57-10 (6-2); 57-25 (8-18)
Ferreira, G. C., 130 (6-5)
Ferreira, J. D. C. Arouck, 375 (4-4); 396, 402 (8-13)
Ferreira, L. V., 07PNJ (4-5); 47PNJ (4-1d); 25 (6-5); 45 (7-5); 72 (4-1c); 136, 257 (7-5)
Ferreira, M. C. et al., 577 (6-8); 622, 754 (4-4); MG206523 (4-1j)
Ferreira, P. 31 (6-8)
Ferreira Junior, C. A. et al., 659 (8-14); 662, 754 (4-4)
Ferreira-Pinto, M. A., 519 (5-9)
Ferreira, S., 302 (8-7)
Ferreira, V. B. R., 12 (8-1); 39, 55, 79 (4-4); 59 (8-1); 82, 86, 89, 92, 108 (4-4); 124 (8-1); 162 (4-4)
Ferreira, V. S., 95 (8-13)
Feuillet, C., 4049 (4-1c)
Fiaschi, P.et al., 332, (8-14); 878 (4-4); 1444 (8-1); 2472 (4-4)

Fidalgo de Carvalho, M., 2944 (6-4)
Figueirêdo, L. 260 (8-13)
Figueiroa, R., 35 (6-8)
Filho, J. P. L. HUEFS84595 (5-3)
Filho, S. J., 175 (4-4)
Finotti, 305, 306 (8-3a)
Fittkau & Coelho, D., INPA12793 (4-1d)
Fleury, M., 33166 (6-4)
Flores, B. M., 12 (7-5)
Flores, J. L. T. P., 20141009_05H01 (8-18)
Flores, M., 530 (7-5); 531 (4-1c)
Flores, T. B. & Romão, G. O., 1241 (4-4)
Focke, H. C., 1018, 1286 (4-1d)
Foldats, E, et al., 3671 (4-1f); 3673, 3822 (4-4); 9005 (6-5); 9061 (4-1d); 9088 (4-1c); 9378 (4-5)
Folli, D. A., 816 (8-3a); 878 (4-4); 886 (8-3a); 983 (4-4); 1121 (6-8); 1299 (5-9); 1377 (6-8); 1393 (5-9); 1558 (4-4); 1582, 1683 (8-1); 2095 (8-21); 3422 (8-3a); 4527 (8-1); 4639 (8-21); 5268 (6-8); 5376, 5530 (4-4); 5389 (6-8); 7300 (4-4)
Fonnegra G., R. et al., 4711 (5-1)
Fonseca, A. R., s.n. (8-14); ASE511, ASE629 (4-4)
Fonseca, M. C., 662 (4-4)
Fonseca, M. R. & Guedes, M. L., 508, 1066 (4-4)
Fonseca, W. N., 443 (8-14)
Fontella, J., 3112, 3190 (4-4)
Fontes, C. G. et al., 234 (5-3)
Forbes, 325 (6-10)
Forest Department British Guiana 404A (7-2); 600 (8-7); 663 (4-1e); 931 (5-10); 2055 (4-1d); 7957 (4-1d)
Forzza, R. C., 303 (6-2); 3884 (8-3a); 6537 (4-1d); 6863 (4-1d); 6983, 7029, 7036, 7917 (7-5); 7947 (7-8); 7955a (4-1d); 7994 (4-1b)
Foster, P. F., 520 (6-8)
Foster, R. B., 4126 (6-6); 14537 (6-11)
Fox, J. E. D., 34, 99 (6-4)
Fraga, C. N., 2815 (4-1a); 3533 (8-14)
França, F. et al., 545 (5-3); 1004 (8-14); 1180 (4-4); 3151 (8-3a); 3233 (6-8); 4143 (4-4); 4144 (8-13); 4434 (4-4)
França, F. & Melo, E., 997, 1180 (4-4)
França, G. S. & Raggi, F. 545 (5-3)
Francisco, INPA1809 (8-18); INPA2044 (7-5); INPA2084 (5-2)
Franco R., P. et al., 3235 (4-1c); 3255 (5-12); 3288, 3698 (5-2); 4261 (4-1c)
Frazão, A. RB8188 (6-8)
Freire, G. Q., 47 (8-1)
Freire, L. & Figueiredo, M. B., 29 (6-8)
Freitas, C. A. A., 178 (5-2)

Freitas, M. A. de, 243 (8-18); 771 (6-8); 778 (8-8); 860 (8-4); 980 (5-5); 1771 (6-8); 51689 (8-4)

Fróes, R. L. 11742 (6-8); 11813 (4-1c); 19933 (8-14); 20480 (4-1c); 20803 (5-2); 21090 (7-6); 21192 (6-2); 21338 (4-1d); 21342 (4-1d); 21346 (4-1c); 21370 (7-7); 21411 (7-8); 21437 (7-6); 22458 (6-8); 22472 (7-5); 22587 (6-5); 22644 (5-16); 22703 (7-5); 22738 (4-1a); 22747 (7-8); 22760, 22838, 22842 (4-1d); 22857, 22940 (6-5); 23494 (6-5); 24137 (7-6); 24256 (6-5); 24297 (6-8); 24916 (7-5); 24924 (5-12); 24934 (6-2); 24936 (8-7); 25185 (8-18); 25369 (6-5); 25459 (8-18); 25463, 25480 (5-2); 25565 (2-1); 25783 (6-3); 26071 (7-5); 26428 (8-17); 26812 (4-1c); 25438 (5-2); 26636 (6-3); 27985, 28407 (5-12); 28454, 28895, 29486, 29854 (4-1a); 30093 (4-1c); 30284 (6-5); 30416 (6-8); 30670 (6-5); 32391 (6-8); 32478 (6-5)

Fróes, R. L. & Addison, G., 29096, 29102 (4-1d); 29119 (7-5); 29211 (4-1d); 29144 (5-16)

Fróes, R. L. & Black, G. A., 27572 (4-1c)

Fuentes, A. F. et al., 1661 (6-5); 6506 (8-3b); 7637 (5-7); 17102, 17136 (6-8)

Funch, R., 60, 104 (8-13); 174, 586, 619, 824 (8-14); PCD827 (4-4); 1082 (8-14); 2033 (4-4); 2060 (8-14)

Fundación Biológica Puerto, CHI-66 (5-10)

Furlan, A. et al., CFCR450, CFCR841, 2546 (4-4)

Gadelha-Neto, P da C., 23, 1052, 1914, 2005, 2405, 3411 (6-8)

Galdames, C. et al., 1296 (6-9); 1430 (5-4); 1462 (6-9); 4404 (8-6); 4441, 6221 (5-4)

Galeano, M. P. et al., 1377 (6-2)

Gancy, W. 3201 (8-14)

Ganev, W. 90 (8-14); 348 (8-3a); 1019 (8-14); 1329 (8-3a); 1760 (8-13); 2020 (8-14); 2652 (8-3a); 2786 (4-4); 2949 (4-1a); 3008, 3201 (8-14); 3380 (8-3a); 3541 (4-4)

García, J. D. 1534 (5-2)

García, R. J. F. et al., SPSF07070 (5-3)

García, S. 148, 167 (5-5)

García-Barriga, H., 13681 (6-5); 14287 (5-12)

Gardner, G. A., 1146 (6-8); 1263 (4-4); 3047 (6-5); 4452 (8-14), 4452bis (4-4)

Garetta, A. O. et al. 304, 754 (5-3)

Garetta, A. O. & Faria, M. B. 55 (5-3)

Garetta, A. O. & Martins, R. F. A., 201 (5-3)

Garvizu, M. & Fuentes, A. F. 366 (6-8)

Garwood, N. & Foster R. B., 440 (8-6)

Gasche, J. & Desplats, J., 212 (6-8)

Gaspar, A. L. de & Assis, F. C., 2787 (4-4)

Gasson, P., PCD6191 (8-14)

Gates, B. & Estabrook, G., 199 (4-4)

Gaul, T. D., 214, 215 (2-1); 216 (6-8); 217, 218 (8-18)

Gély, A., 383 (6-5)

Gentry, A. H. et al., 1931 (8-6); 1938 (5-4); 7406 (8-6); 8843 (6-10); 12939 (4-1d); 12956 (8-17); 24054 (8-16); 24274 (6-9); 28611 (5-4); 35388 (8-16); 35464

(5-13); 35500, 40421 (5-8); 40437 (6-9); 42837, 42891 (6-1); 47902 (5-4); 50012 (4-1c); 53294 (8-16); 53677(5-4); 56138 (8-20); 56473 (8-19); 56630 (8-20); 56970 (5-4); 69996 (5-13); 70895 (6-8); 77753 (2-1); 78796 (8-2); 80778 (5-7); 10832 (7-6); 10867 (4-1f)

Gentry, A. H. & Brand, J. 36787 (5-8)

Gentry, A. H. & Pilz, G., 32860 (6-4)

Gentry, A. H. & Sánchez, M., 65253, 65287, 65288 (5-12)

Gentry, A. H. & Stein, B. A., 46400, 46450 (4-1d; 46898 (5-1); 46932 (8-7); 46980, 47295 (6-2)

Gentry, A. H. & Tillett, S. S., 10832 (7-6)

Germano, C. M., 54 (8-7)

Giacomin, L. L., 1913 (4-1a); 2121 (4-1i)

Giaretta, A. O. et al., 531 (8-1)

Gillespie, L. J. 2505 (6-8); 2637 (4-1d); 2696 (4-1e); 2810 (6-10); 2856 (4-2)

Giordano, L. C. S. et al., 787, 1183 (4-4); 1186 (4-1c); 1230 (4-4); 1237(4-1c); 1603, 1604, 1656 (4-4); 1684, 1834 (4-4); 1884 (4-1c); 1893 (4-4); 1961 (4-1c); 1962 (4-4); 2037, 2115, 2041 (4-4); 2213, 2220 (5-6b); 2316, 2382 (4-4); 2383 (4-1c); 2394 (5-6b); 2416, 2489, 2493, 2501, 2636 (4-4); 2638 (5-6b); 2664, 2669, 2670 (4-4); 2678 (8-21); 2679 (4-4); 2680 (4-4); 2681 (8-1); 2682 (8-13); 2683 (5-9); 2713, 2714, 2715, 2716, 2717, 2718, 2720 (4-4); 2721 (8-14); 2722, 2723 (8-13); 2724, 2725 (4-4); 2808 (5-3)

Giraldo-R, A. & Corona-T, C., 2015 (5-13)

Giulietti, A. M., et al., CFCR1343 (8-14); 1958 (8-13); PCD2783 (4-4); CFCR4288, CFSC7352 (8-3a); SPF20179 (8-14)

Glaziou, A. F. M., s.n. 63 (4-1g); 731 (4-1g); 7766 (5-5); 6196, 7765 (4-4); 7766 (5-6); 8286 (4-4); 10078(8-7); 10342 (4-4); 10437 (4-1c); 11828, 11829 (8-3a); 12515, 12919 (4-4); 13574 (6-5); 14640, 16723 (8-3a); 16724 (5-6b); 18064 (5-6); 18178 (5-3); 18179 (5-6); 18180 (4-4); 18181, 18182 (8-3a); 18962 (4-4); 18963 (8-14);18964 (5-6); 19573 (5-2)

Gleason, H. A., 729 (5-10)

Godinho, R. & Macedo, M., 122 (6-5)

Goes, B.T. P. M. et al., 193 (4-1c)

Gomes, A. L., s.n. (4-4)

Gomes, F. S. et al., 228, 677, 805, 1164 (6-8)

Gomes, J. M. L., 1315 (4-4); 2209, 2210, 3049 (8-21)

Gomes, J. S., 238 (6-8)

Gomes, M., 537, 563 (2-1); 622 (4-4); 643, 685, 812, 970, 1234, 1383, 1435, 1489 (2-1)

Gomes, P. S. et al., 228, 883 (6-8)

Gomes, R., 4 (5-2)

Gomes-Costa C. A., 108 (4-1d)

Gómez P. L. D., 3298, 6934 (6-6); 22031 (5-4)

Gómez-Laurito, J., 6934, 6966 (6-6)

Gonçalves, D., 28 (4-4)

Gonçalves, F. B., 92 (6-5)

González, J., 1149 (6-6)
González, M. 516 (4-1a); 633 (4-1c); 639, 670 (5-2)
González R., J. A., 1149 (6-5)
Gordano, L. C. S., 2683 (CVRD)
Gordillo R., E., et al., 313 (4-5)
Gorts van Rijn, A. R. A. 465 (4-1a)
Gossweiler, J. 751, 8182, 8707, 8751 (6-4)
Gottsberger, G. K. & Posey, D. A., 24-22183 (6-5); 32-19183 (6-5)
Goulding, M., 2180 (7-5)
Grández, C. A. et al., 2570 (5-1)
Granville, J. J., et al., 1753 (8-5); 9121 (6-5); 10792 (8-5)
Grayum, M. H. et al., 7662, 8299 (6-11)
Graziela (G. M. Barroso) RB178995 (4-1a)
Grenand, P., 2513 (4-1c); 3062 (8-9); 3086 (5-15)
Grijalva, A. et al., 598 (5-4)
Grings, M. et al., 1760 (5-3)
Grogan, J., 161 (4-1a); 162 (6-5); 481 (4-1a)
Groppo, M., 762, 997, 2233 (4-4)
Gaul, T. D., 216 (6-8)
Guánchez M., F., 11 (6-5); 353 (4-5); 883 (4-1b); 3239 (6-2); 3286, 3529, 3555 (4-1f); 4623, 4651 (6-5)
Guánchez M., F., & Huber, O., 4558 (6-5)
Gudiño, E. & Moran, R. C., 1268 (5-13)
Guedes, T., 58, 80 (6-2)
Guedes, M. L. et al., 109(8-14); 252 (8-3a); PCD300, 358, 381, 684, 709, 757, 758, 967, 1072 (8-14); 1202, 1387 (4-4); 1503, 1504, 1921, PCD2094 (8-14); 2106, 2119, 2126, 2141, 2368, 2456 (6-8); 2541 (4-4); 2709 (8-3a); 3375, 4006, 4778, PCD4805, 5136 (4-4); 5323 (6-8); 5454 (4-4); 5456 (8-1); 5499 (4-1c); 5517 (8-3a); 7030 (6-8); 9507 (4-1c); 11820 (8-3a); 12129 (6-8); 12407 (8-14); 14197 (8-3a); 14352, 17133 (8-14); 17642, 23206 (6-8)
Guilding, L., s.n. (6-1)
Guillaumet, J. L., 5731 (7-2)
Guillemin, A., 205 (4-4)
Guillén V., R. & Marmañas, M. N., 4622, 4638 (6-8)
Guillén V., R. & Soliz, P., 3845 (6-5)
Guimarães, E. F., 921, 1004 (4-4)
Guppy, N., 308 (6-5)
Gurgel, E. S. C., 303 (8-18); 735, 751 (8-18); 884 (6-8)
Gusmão, E. F. de et al., 493 (6-8)
Gutiérrez R., A., 64, 82 (5-5); 134 (5-1)
Hage, J. L. & Santos, T. S., 906 (8-1)
Hahn, W. J. et al., 3830 (4-1d); 4544 (4-2)
Haidar, R. F. & E. R. Santos, 1138 (6-5)
Hamaguchi, J. O., 38 (6-5)
Hamilton, C. W. & Davidse, G., 2817 (8-6)

Hammel, B., 8444, 9075 (6-11); 12954 (8-6); 13192 (6-11); 15335 (8-16); 17026, 17030, 17904 (5-4); 17917 (8-2); 18014 (5-4); 18171 (8-2); 18173 (5-4); 18390, 20631 (6-11)

Hammel, B. & Koemar, S., 21768 (4-1j)

Harley, R. M. et al., CFCR6129 (8-14); CFCR7029 (8-13); 10539 (4-1c); CFCR14058, CFCR14958 (8-14); 15558 (8-14); 15649, 15914, 16008, 16664 (4-4); 17364 (8-1); 17917 (8-2); 17960, 18245, 18533 (4-4); 18577 (8-13); 18651 (8-14), 18683(8-13); 19249 (4-4); 19327 (8-3a); 20554 (4-4); 20556, 20811(8-14); 20846 (8-3a); 22273, 22286, (8-14); 22718 (4-4); 24305 (8-3a); 25363 (4-4); 25834 (8-14); 26408, 27832 (4-4); 50165 (8-14); 50243 (4-4); 50322 (8-3a); 50400 (4-4); 50715 (4-4); 51537, 53956 (8-3a); 54461, 54508 (8-13); 54662 (8-13); 54757, 54926, 55684 (8-3a)

Harmon, P., 182, 324 (5-4)

Haroldo, INPA58199 (8-18)

Harrison, S. G., 1052, 1734 (4-1d)

Hart, J. H., s.n. (6-1)

Hartshorn, G. S., 1275 (6-11); 2139, 2251 (8-2)

Hatschbach, G. 12500, 16515; 20893, 21265, 21270 (8-3a); 27454, 28075 (8-14); 32199 (8-3a); 41620 (4-4); 42069, 46529 (8-14); 46551, 47031, 47491, 50243, 50400 (4-4); 54916, (8-3a); 56145 (5-3); 56751 (4-4); 59751 (5-3); 65644 (6-8); 67007 (6-8); 67164 (4-4); 68608 (8-3a); 69499, 69746, 72126 (4-4); 75577 (6-8)

Hawes, J., 1085 (1-2); 1202 (8-8); 1301 (8-7); 1473, 1516 (8-8); 1649 (1-2)

Hawkins, T. 1971 (4-1c); 2166 (4-1a)

Heloisa, H. V. de A. et al., 18 (6-8)

Henderson, A., 933 (1-1)

Henkel, T. W, 266 (4-1c); 600 (4-1a); 827 (6-5); 3279 (4-1c); 3937 (4-1a); 5698 (4-1b)

Herb. J. Miers 6167 (8-18); 8915 (4-1g)

Herb. Richard s.n. (4-1a); s.n. (4-1g); s.n. (5-3)

Heringer, E. P. & Eiten, G. 15003 (5-3)

Heringer-Sales, A., 4242 (4-4)

Hernández, L., 25 (6-5); 345 (4-1b); 356 (8-12); 391 (4-1g); 394 (8-12)

Herrea C., G., 3114 (6-11); 4585 (5-4)

Herrera, H. et al., 1726 (6-9)

Hitchcock, A. E., 16938 (4-1d)

Hoehne, F.C., 3021 (5-6b); 4970 (4-4); 7970 (4-1g); 29281 (8-3a)

Hoff, M., 5363 (4-1j)

Hoffman, B., 702 (4-1d); 1002 (6-8); 1012 (6-5); 1600 (6-10); 1624 (4-1e); 1683 (4-1d); 1713 (4-2); 1745, 1755 (6-10); 1873 (4-2); 1875 (4-1c); 1996 (4-1i); 3008 (6-8); 3415 (4-1e); 3669 (4-1d); 4564 (7-3); 6079 (6-5); 6145 (6-5)

Hoffmann, S., 90-3-43 (4-1d); 90-5-127 (4-1c)

Hohenkerk, s.n. (6-5); FH404A (7-2); 663 (4-1f)

Holanda, A. S. S., 452 (4-1d); 467 (4-1a); 472 (7-5); 473 (4-1d); 475 (4-1c); 477 (4-1j); 478 (4-1a); 479 (6-5); 482 (4-1j); 483 (6-5); 485 (4-1a); 486 (4-1d); 487

(4-1a); 496 (6-5); 497 (4-1d); 499 (4-1c); 501, 516 (6-5); 536, 537, 538, 539, 540, 541, 542, 543, 544 (4-1d) 545 (4-1a); 547 (4-1c); 549 (4-1a); 550, 551 (4-1c); 552 (4-1j); 553 (4-1a); 555, 575 (4-1c); 562, 563, 564, 565 (4-1d); 662 (4-1j); 663 (4-1a); 664 (4-1j); 785, 786, 788, 790 (4-1c); 791 (4-1d); 793 (4-1a); 794, 795, 796, 797, 798, 815, 816, 817, 818, 819, 820 (4-1c); 821 (5-2); 822, 823, 824, 825, 826 (4-1c); 828 (5-2); 865, 866, 867, 868, 870, 871, 872, 873, 874, 875, 876, 877, 878, 879, 880, 881 (4-1c)

Holdridge, L. R., 5164 (6-6); 5216 (6-11)

Holle, J.L., s.n. (6-4)

Holst, B. et al., 2181 (8-12)

Holst, B. & Liesner, R. L., 2350 (8-12); 2630 (6-5); 2791 (5-10)

Hopkins, M. J. G., 966 (4-1c); 1421, 1450 (8-18); 1777(4-1j); 1942, 2088 (7-5); 2096 (4-1c); 2256, 2261 (6-5)

Hostman. W. R., 793 (4-1d)

Householder, J. E., 2182 (4-1d); 2417 (5-16); 2466 (5-2); 3019 (7-5)

Houtmonster, 541A, 542A, 543A (4-1a)

Huashikat, V., 2712 (8-18)

Huber, J. E., 96 (4-1c); 239, 940, 1260 (2-1); 1850 (6-1); 2785 (4-1a); 6011 (4-2) 6992 (6-8); MG9401 (2-1); MG9583 (8-18); 10446 (6-5)

Huber, O., 1072 (4-1i); 1857(4-1i); 2026 (4-3); 2344 (4-1f); 2677 (4-1i); 2823 (4-3); 2832 (4-1i); 2840 (4-1a);3082 (5-12); 3089 (4-1c); 3155 (6-5); 3231 (4-1a); 3277 (4-3); 3300 (4-1d); 3308 (4-1i); 3418 (4-1d); 3429 (4-1i); 3478 (4-1f); 3731 (4-1c); 3834, 3931 (4-1f); 4083 (4-1f); 4223 (6-5); 4738 (4-1a); 4842, 4888, 4922 (4-5); 4926 (4-1b); 4930 (6-5); 5009 (4-1f); 5080 (4-1d); 5174, 5196 (6-5); 5241, 5332 (6-5); 5361 (4-1f); 5572 (4-5); 6011 (4-2); 6035 (4-1b); 6085 (4-5); 6179 (4-1b); 6222 (4-1a); 6294, 6657 (6-5); 6745 (8-12); 7275 (4-1b); 7441 (8-12); 7473 (4-1e); 7559 (8-12); 7619 (4-1); 7912 (8-12); 7990 (4-3); 8016 (4-1b); 8212 (4-1c); 9319 (6-5); 9322 (4-1c); 9379 (4-1b); 9619 (4-2); 9667 (8-12); 9860 (4-1e); 9882 (4-1b); 10329 (4-1b); 10497, 10498, 10620 (8-12); 10669 (4-1d); 10709 (4-1j); 10723 (4-1d); 10748 (4-1i); 10770 (4-1d); 10943, 11236 (4-1b); 11315 (8-12); 11331 (4-2); 12103 (4-1c); 12386 (6-5); 12539 (4-1c)

Huber, O. & Alarcón, C., 6656, 6743 (4-1b); 7619 (4-2); 9683 (8-12); 10485 (6-5)

Huber, O. & Medina E., 5813 (4-1f); 5958 (4-5)

Huber, O & Tillett, S., 2832 (4-1i); 2885 (4-1a); 5332 (6-5); 5524 (4-1f); 5938 (4-5)

Humbert, H., 27440 (4-1d); 27422 (4-1c)

Humbert, H. & Schultes, R. E., 27363 (5-16); 27364 (4-1j)

Hunt, D. R. & Ramos, J. F., 5753 (4-1c)

Hurtado, P., 132 (5-4)

Icárcel, V. & Chota, M. 5/93 (8-18)

Idrobo, J. M. et al., 6869 (4-1j); 11468 (5-12)

INPA 86 (4-1c); 204 (5-2); 620 (4-1b); 1056 (4-1b); 1243 (7-5); 1638 (6-2); 1684 (4-1d); 1700 (5-2); 1773 (4-1c); 1814 (6-2); 2044 (7-5); 2084 (6-2); 177655 (6-5)

Im Thurn, E. F., s.n. (4-1d); s.n. (7-2)

Irvine, F. R., 2321 (6-4)

Irwin, H. S. et al., BG7 (4-1d); 246 (4-1a); 5011 (4-1a); 6010 (6-8); 9292, 12624 (4-4); 16311 (4-1c); 18227 (6-8); 20039, 20115 (8-14); 20652 (4-4); 22574 (4-1c); 22718, 22719, 22767 (4-4); 22926 (5-6b); 23371 (4-4); 23429 (8-14); 23567 (4-4); 27516 (8-14); 28081, 28546 (4-4); 28626 (8-14); 31059 (8-13); 32354 (4-4); 32395 (8-13); 54482, 55274, 55420 (4-1a); 57533 (4-1a); 57565 (4-1d)

Isacksson, J. G. L., 87 (8-18)

Ivanuskas, N. M., 452 (8-3a); 909f (5-3); 2213, 4074, 4076, 4125, 4140, 4367 (6-8)

Jansen, J. W. A., 1088 (6-4)

Jansen-Jacobs, M.J. et al., 1100 (4-1d); 1397 (4-1d); 1898. 1990 (6-5); 1954 (2-1); 3229, 5399 (4-1c)

Janssen, A., 466 (4-1d)

Jardim, A. B., 224 (4-4); 234, 743 (6-8)

Jardim, J. G., 478 (6-8); 529 (8-21); 792 (7-8); 2199 (4-1c); 2497(8-14); 5149 (6-8); 5287, 5422 (8-1); 6329 (6-8)

Játiva, C. & Epling, C. C., 1014, 1055, 1135, 1147 (5-13)

Jenman, G. S., s.n. (5-10); 287 (7-2); 299 (4-1d); 478 (7-2); 542 (4-1d); 1023 (4-1b); 1281 (4-1d); 2386 (5-10); 2489 (7-2); 3912 (4-1a); 4136 (4-1j); 4719 (7-2); 4851 (5-10); 4883 (4-1d); 5561, 5562 (4-1d); 5672 (4-1a)

Jesus, F. P. R. et al., 18 (6-5); 337 (5-6a)

Jesus, J. A. de, 89 (6-8); 595 (8-1)

Jesus, M. C. F., 428 (5-3)

Jesus, N. G. de, 11, 14, 623, 931 (6-8); 1264 (4-4); 2117 (6-8)

Jiménez M., A., 133 (6-6)

Jiménez, Q. et al., 961 (6-11); 1131, 1315 (8-16)

Jiménez-Saa, H., LBB14361 (6-5); LBB 14396 (6-3)

Jobert 362 (8-7)

Johnstone, A. T., 164/31 (6-4)

Joly, A. B. et al., s.n. (4-4); 1202 (4-4)

Jordan, H. D., 2057 (6-4)

Jost. T. 149, 205 (4-4)

Jost, T. & Bautista, H. P., 263 (6-8)

Jost, T. & Ferreira, M. C., 136 (4-4)

Julião, G. R., 62, 63 (5-2); 64, 65 (8-11); 66 (8-18)

Junker, N. W., 5525 (4-1a)

Junqueira, A. B. & Ramos, J. F., 1087 (1-2)

Kajekai, C. et al., 279 (5-7)

Kajekai, C. & Wisum, A., 1257 (5-7)

Kanu, K. M., 86 (6-4)

Kappler, A., s.n. 2144 (7-3)

Kawasaki, M. L et al., 71 (7-5); 142 (4-1c); 148 (4-1j); 275 (4-1c); 377 (8-7); CFCR8341, CFSC9106 (4-4)

Keay, R. W. J., FHI28077 (6-4)

Keel, S. et al., 233 (4-1a); 244, 311 (7-5)

Kelloff, C. L., 611 (4-1d)

Kennedy, H., 2437 (6-9)
Kennedy, J. D., 396A, 1688 (6-4)
Killeen, T. J. et al., 6158, 4520, 6916, 6929 (6-8)Killip, E. P. & Smith, A. C., 28681 (4-1a)
King, H. L., 278 (6-4)
King-Church, L. A., s.n. (6-4)
Kinup, V. F. et al., 3928 (6-8); 4489 (2-1)
Kirkbride, J. & Lleras, E., 2836 (6-5)
Klaine, R. P. 271 (6-4)
Klein, R. M., 117 (8-1); 37b (8-3a); 1225, 1244, 1650, 7345, 9644(8-3a)
Klein, R. M. & Bresolin, A. 7298 (8-3a)
Klein, V. L. G., 490 (8-1)
Klug, G., 1091, 1130 (8-19); 1315 (4-1a); 1564 (7-7); 2846 (4-1a); 3706 (4-4)
Knab-Vispo, C., 269 (6-5)
Knapp, S. et al., 7457 (4-1c)
Knob, A. 853 (6-5); 1105 (4-1a)
Koechlin, J., 2632 (6-4)
Kollman, L., 707 (8-13); 1282 (8-14); 1806 (4-1a)
Komura, D. L., 1806 (4-1a)
Konno, T., 774 (4-4)
Korning, J. et al., 57591 (5-4)
Kozera, C. 826 (5-3)
Krahl. A. H., 248 (5-2)
Kral, R., 72051 (4-1e); 72081 (4-1b)
Krieger, L., CESJ10381 (4-1c); 10400 (4-4); 10401, 10455 (4-1c)
Krukoff B. A., 121 (6-4); 1483 (4-1f); 4956 (8-18); 6371 (8-17); 6506 (6-8); 6653 (6-5); 7082 (6-5); 7120 (8-18); 7182 (8-17); 7926 (4-1c); 7928 (4-1c); 8227 (8-18); 8757 (6-5); 11270 (5-7)
Kubitzki, K., 75-76, 79-201 (5-2); 79-96 (4-1c); 84-144 (5-2); 84-116 (4-1c); 84-299, 84-344 (4-1c); 87-23 (7-5); 87-37 (5-2); 88-35 (6-5); 88-68 (5-2); 88-114 (6-5)
Kuhlmann, J. G., 2 (6-8); 150 (4-1d); 170 (7-5); 171 (6-8); 179 (4-4); 463 (4-1a); 813 (6-5); 1873 (4-1c); 1995 (4-1a); RB3509 (4-1c); RB2128 (6-8); RB2894 (4-1d); RB3510 (6-5); 6674 (4-4); 6682 (8-1); RB20453 (5-6b); RB21029 (7-5); RB21038 (6-5); RB46769 (5-6a); RB47414, RB72803 (8-3a)
Kuhlmann, M. & Jimbo, 38 (8-7); 142 (6-8); 197 (2-1); 223, 245 (8-18)
Kukle, H. & Boom, B. 26 (5-2); 34 (8-18)
Kukle, L., 26 (5-2)
Kurtz, B. C., s.n. (4-4)
Kuypper, J., 33 (4-1d)
Labiak, P. H., 5715 (4-1b)
LBB 117 (6-5)
Ladrach, W. E., (6-9)
Laguna, A., 159 (8-6)
Lamb, F. B., 133 (8-10); 141, 145, 170 (5-1); 350 (6-5c)

Landim, M., 400, 633, 746, 933, 1206 (6-8)
Landrum, L. R., 4262 (5-6)
Lands Bosbeheer Suriname, 9465 (4-1a)
Langsdorff, G. H., s.n. (8-14)
Lanjouw, J., 195, 334 (4-1d); 1253 (4-1a)
Lanjouw J. & Lindeman, J. C., H8 (4-1a); 267, 268 (4-1d); 573, 652 (4-1a); 911 (4-1d); 968 (4-1c); 1797, 1798 (4-1d); 2194 (6-3); 2869 (6-5); 3259, 3289, 3317 (4-1d)
Lanna, J. P., 1402 (4-4)
Lao, E. A. et al., 5, 195 (6-11)
Lao, E. A. & Gentry, A. H., 530 (8-6)
Lao, E. A. & Holdridge, L., 193 (6-11)
La Rotta, C., 552 (8-18)
Larpin, D., 962 (6-5)
Lasser, T. & Vareschi, V., 3888 (4-1b)
Laurênio, A., 508, 979 (6-8)
Lautert, M., 407 (2-1); 2031 (6-8)
Leal, C. G. & Silva A. da, 86 (6-8)_
Leblond, J. B., 402, 441 (4-1c)
Le Cointe, P., RB11357 (4-1c); RB20424 (6-5)
Leeuwenberg, A. J. M., 5488 (6-4)
Leitão Filho, H. F., 650 (5-3); 7892, 17182 (4-4)
Lemée, A. s.n. (6-3)
Lemos, C., s.n. (8-3a)
Lemos Filho, J. P., 4057 (5-3)
Lemos, M. C., 79 (5-2); 80 (6-5); 81 (6-8); 82 (8-18)
Lems, K., s.n. (4-4); 5043 (5-4)
Lent, R. W., 1923, 2220 (5-4)
Lépiz, E., 328 (6-6)
Le Prieur, M., 253 (6-5); s.n1838, s.n.1840 (4-1a)
Lepsch-Cunha, M. M. et al., 1030 (8-8)
Lescure, J. P., 668 (7-9); 879 (6-5)
Letouzey, R., 15099 (6-4)
Lewis, G. P. et al., CFCR 7075 (4-4)
Libiak, P. H., 5715 (4-4)
Liesner, R. L. 1224 (5-4); 3708 (4-1f); 3855, 3905 (4-1d); 4102 (5-2); 5500 (6-8); 6081 (4-1d); 6253 (7-6); 6905 (4-1d); 7169 (7-6); 7542 (4-1d); 7547 (7-8); 8823 (4-1c); 15876 (5-1); 15920 (4-2); 17619 (4-1b); 17929 (4-1j); 18144 (4-1b); 19087 (4-1c); 19152, 19472 (8-12); 19779, 20487 (4-1c); 23974 (8-12); 24017 (6-5); 24032 (8-12); 24038, 24083 (4-1b); 24115 (8-12); 24354 (6-5)
Liesner, R. L. & Carnevali, G., 22410 (4-2); 22880 (4-1d); 22589 (1-1); 22633 (4-2)
Liesner, R. L. & Clark, H. L., 3356 (4-1d); 9072 (7-6)
Liesner, R L. & Holst, B. K., 20628, 20661, 21181(6-5); 21758 (4-1b); 21812 (6-5)

Lima, B. & Lima, J., 350 (2-1)
Lima, D de, 49-336 (6-8); 53-1273 (4-1a); 53-1332 (6-8); 1623 (4-1g)
Lima D. de A. L., 68 (6-5); 72 (7-5); 104 (6-5); 169 (6-2)
Lima, D. P. 12697 (8-1); 12764 (4-4)
Lima, E. S., 143 (6-8)
Lima, H. C. de 1064 (8-14); 1685 (8-1); 1897 (8-1); 2918 (8-1); 2957 (8-14); 2975 (4-4); 2976 (8-1); 3175 (5-2); 3180 (7-7); 3184 (5-2); 3307 (4-2); 3375 (8-18); 3678 (8-3a); 3914 (8-3a) 7282, 7288 (8-15); 7239 (4-1i); 7259 (4-1b); 7339 (4-1a); 8027, 8030, 8032 (4-4); 8074 (6-8); 8095, 8096 (4-1c); 8176 (4-1d); 8191 (4-1a); 8192 (6-5); 8217 (7-5)
Lima, J., 3 (7-5); 456 (4-1d); 524 (8-18); 568 (8-18)
Lima, J. & Zimmerman, B., 524 (8-18)
Lima, J. C. A. de, 3 (4-1a)
Lima, J. P., 1786, 1821 (2-1)
Lima, L. C. L., et al., 54 (6-8)
Lima, R., 2181, 2318 (6-8)
Lindeman, J. C., 258 (4-1d); 4201 (4.1c); 4202 (4-1a); 4381 (4-1d); 6354 (4-4); 6541,6861 (4-1a); 6862 (4-1b); 6880 (4-1a); 6881, 6882, 6883 (4-1c)
Lindeman, J. C. & Roon, A. C. de, 721 (5-10)
Linder, D. H., 219, 1469 (6-4)
Lins, A. et al., 292 (6-8)
Lira, S. S. et al., 333, 366, 415, 424, 463, 472, 474, 550, 572, 585, 587 (6-8)
Lisboa, A., MG2327 (4-1f); MG2330 (6-5); MG4699 (4-1c)
Lisboa, F. A., 4699 (4-1c)
Lisboa, P. et al., 97, 104, 133, 879 (4-1c); 1283 (4-1a); 1796 (6-5); 1822 (8-7); 1898 (7-8); 2117 (8-7); 2210 (2-1)
Lisboa, R., 6667 (4-1c)
Little, E., L., 6233, 6320, 6412, 6413 (5-13)
Little, E. L. & Dixon, R. G., 21093, 21148 (5-13)
Liuth, H. S., 131 (6-8)
Lleras, E., P17474 (2-1); P19657 (5-2)
Lobato, C. 513 (4-1a)
Lobato, D. A., 12 (4-1a)
Lobato, L. C. B., 25 (4-1c); 90 (6-8); 217, 513 (4-1a); 1855 (8-18); 2069 (4-1c); 2448 (4-1a); 2540 (4-1c); 2951 (6-8); 3022 (6-5) 3027 (2-1); 3445 (6-5); 3670, 3714 (4-1a); 3991 (4-1c)
Lobato, L. C. B. & Oliveira, J., 251 (5-2)
Lobato, L. C. B., & Smith, N., 4548 (4-1a)
Lobo, M. G. A., 113 (6-8); 133 (6-5)
Loizeau, P. A., 304 (5-5); 540 (8-14); 566 (4-4)
Lohman, L. G. & Peckham, H., 147 (4-1a)
Lombardi, J. A., 2834, 3409 (5-3); 3522 (8-21); 4072, 4075 (5-3); 4300 (8-14); 4321, 9054, 9647 (4-4); 10177 (4-1a)
Londoño, A. C. et al., 42, 542, 979, 995, 1045, 1068, 1102, (8-19); 1133 (8-18); 1248, 1264, 1367, 1387 2518 (8-19); 2537 (8-20); 2710, 2760 (8-19)

Lopes, L. C., 149 (8-1)
Lopes. L. K. C. & Sousa. R. S., 8 (4-1a)
Lopes, M. 130 (5-2); 882 (4-4)
Lopes, P.M., 2 (4-4)
Lopes, W. de P., 171 (5-3)
López C., R., 5605 (7-5); 6348 (6-1); 7915 (6-5)
Lorenzi, H., 183, 1361 (5-3); 1732 (8-18); 1740 (2-1); 5274 (8-3a); 6894 (6-8)
Loubry, D (Denys on some labels), 706 (8-9); 1074 (7-9); 1789 (5-15); 1951 (6-3); 2211 (4-1c)
Loureiro, A. A. (Arthur), INPAW 6148 (8-11); INPA15524 (6-8); INPA16138 (8-18); INPA37819, INPA37850, INPA37969, INPA39501 (7-5); INPA47977 (4-1d); INPA47997 (5-2); INPA48008 (4-1d); INPA48018 (6-5); INPA48176 (8-7); INPA48377 (4-1d); INPA48486 (8-18); INPA 50632 (8-11)
Loureiro, R. L. de 153 (4-1a)
Lourenço, A. R., 102, 109 (6-8)
Lourival, INPA60922 (8-18)
Lovo, J. et al., 324 (8-14)
Lowe, J., 4248 (4-1c)
Loza, I et al., 1435 (5-7)
Lucas. E. J., 897 (4-4); 929 (8-1); 984 (4-4)
Lucas, F. C. A. 301, 419, 602 (6-5); 613 (4-1a)
Lucena, M. F. A., 740 (6-5)
Luetzelburg, P. von, 40 (8-14); 22561, 22575, 22627 (4-1d); 24014 (4-1j)
Luiza, A. s.n. (8-21)
Luize, B. G., 117, 130, 231, 297, 514, 521 (5-14)
Lughadha, E., 5959 (8-3a)
Luke, Q., 13083 (6-4)
Lunt, W., s.n., 5984 (6-1)
Luteyn, J. L. et al., 6292 (4-1b)
Lutz, B., 681 (4-1g)
Luz, A. A. da, 342 (8-1)
Luz, G. O., 13 (4-1d)
Lyra Lemos, R. P. de, 1059 (4-4)
Maas, P. J. M., 2294 (4-1d); 5649 (4-1); 5753 (6-10); 6577, 6633 (7-5); 6804 (7-6); 6894 (6-5); 6905 (5-11); 7325 (4-1a); 9872 (8-1); P12746 (7-8)
Maas, P. J. M & Westra, L., 3564 (4-1d); 3602 (5-10); 4293 (4-1e)
Macêdo, A., 1293 (6-8); 3922 (6-5); 4034 (4-1a)
Macedo, M. 35 (4-1c); 3905, 4128 (4-1a)
Macedo, M. & Assumpção, S., 1854 (6-8); 1916, 2605 (4-1a)
Macedo. M. & Godinho, R., 4128 (4-1a)
Machado, J. W. B., 54 (5-3)
Machado, M. A. B. I., 200, 262, 423 (8-18)
Macia, M. J. et al., 6348, 6443, 6999, 7032 (6-8)
Maciel, U. N. et al., 173 (5-2); 195, 246 (6-5); 262 (6-8); 577 (6-5); 1517 (8-7); 1659 (1-2); 2106 (8-7)

Madison, M. T., 6139 (5-16); 6227 (4-1d); 6690 (4-1a)

Madriñán, S. & Barbosa, C. 977 (6-10); 1046 (4-1d)

Maduro, C. B., 4 (4-1c)

Magalhães, C. M., 198 (6-8)

Magnuson, W. E. 4665 (6-5); INPA212038 (6-8)

Maguire, B. et al., 24223, 24443, 24707, 24789 (4-1b); 24836, 24844 (6-5); 29337, 29416 (6-5); 29416a (4-1j); 29541, 29697 (4-1b); 29769 (4-1j); 30018 (4-1b); 30453 (4-1d); 30483, 30561 (4-3); 30622 (4-1b); 30693 (6-7); 30791 (4-1d); 30844 (6-5); 30885, 30918 (4-1b); 30967 (6-5); 30987 (4-1c); 32022 (6-5); 32686 (4-1b); 32763 (4-1h); 33099 (4-1h); 33242 (4-1b), 33388 (4-1b); 33576, 33583 (8-12); 34913 (5-11); 35040 (4-1c); 35140, 35453 (4-1b); 35882 (4-1j); 36210 (4-1j); 36295 (4-1d); 36354 (4-1f); 36456 (4-1d); 36580 (4-3); 37632 (4-1i); 40105 (4-1b); 40159 (4-1b); 41640 (8-7); 41821 (4-1f); 41917 (4-1d); 43893 (4-2); 47026 (4-1a); 51695 (8-18); 53526, 53688 (4-1c); 56006 (8-7); 56047 (6-5); 56068 (8-7); 56542 (5-2)

MAIA 184 (4-1c)

Maguire, B. & Fanshawe, D. B., 23233 (4-2); 23295 (4-1b); 23450, (4-1d); 32158 (4-1e); 32250, 32619 (6-10)

Maguire, B. & Politi, L., 27627 (4-1h); 27695 (4-1b); 27974 (4-1c); 28828 (4-1d)

Maguire, B. & Stahel, G., 23654, 23696 (4-1d); 24957 (4-1a)

Maguire, B. & Wurdack, J. J., 34677, 35579 (4-1f)

Mahecha, G., 4 (5-13)

Maia, L. A., 104 (7-6); 184 (4-1a); 194 (4-1d); 493 (6-5)

Malme, G. D., 2237 (6-8)

Mann, G., s.n., 925, 1417 (6-4)

Mansano, V. de F., 07-435, 661, 723 (4-4)

Mantone, L., 146 (4-4)

Mantovani, W. & Rocha, D. M. J., 12735 (4-1c)

Marin, E., 330 (5-10); 405 (4-1c); 410, 1061 (6-5)

Marín, J., 155, 162 (5-4); 330 (5-10); 1189, 1190 (4-1c)

Marín C., N. L. 632 (4-1d); 676 (4-5); 1678 (4-1c); 1681 (4-1a); 1749 (6-2)

Marinho, T., 5 (1-2); 197 (7-5)

Marinoni, 114 (6-8)

Markgraf, 3034, 3424 (4-4)

Marques, M. C., 255 (8-3a)

Marquete, N., 494, 613 (4-4)

Marquete, R., 425, 642, 4132, 4213 (4-4)

Mars, L. 3304.2567 (8-18)

Martin J., s.n. (4-1a); s.n. (5-15); s.n. (6-3)Martins, L. H. P., 12 (4-1a); 13 (4-1c)

Martins, M. V., 129 (4-1d)

Martinelli, G. 287, 1911, 1354 (4-4); 1976 (4-1a); 4407 (4-4); 5365 (8-14); 5374 (4-4); 5410 (8-13); 5889 (8-14); 6860 (4-1c); 6881, 7316, 7321 (4-1c); 9190, 9651 (4-4); 9668 (4-1c); 10675 (5-6a); 11032 (8-1); 11232 (8-14); 11358 (4-4); 12099, 12276 (4-1c); 12288 (6-5); 14497 (1-4d); 14499 (5-2); 16983 (4-5); 17169 (7-5); 17276 (4-1b); 17327, 17331 (8-15); 18787 (8-14)

Martins da Silva, R. C. V., 31 (8-7)
Martius C. F. P von., s.n. s.n. (4-1a); s.n. (4-1c); s.n. (4-2); s.n. (5-2); s.n. (6-1); s.n. (8-14)
Martyn, E. B., 136 (4-1d)
Matos, E. N. de, 203 (6-8); 368, 372 (4-4)
Matos, F. B., 1742, 1964, 1969 (6-8)
Matos, F. D. de A., 3 (8-8)
Matos, I. S., 27, 45 (6-5)
Matos, G. M. A., 276 (6-8)
Matos, M. Q. et al., 62 (6-5)
Matta, A. B. da, 218 (8-18); 225 (6-5)
Mattos, F. D. de A., 28 (5-2)
Mattos Silva, L. A. et al., 208 (6-8); 403 (8-1); 449 (4-4); 622 (8-1); 704 (4-4); 859 (6-8); 932 (6-8); 960 (8-1); 1193, 1282 (4-4); 1436 (8-21); 1613 (8-13); 1965 (8-3a); 2441 (4-1c); 2485 (4-4); 2643 (8-1); 2645 (4-1c); 3021 (4-4); 3895 (6-8); 3931, 4227, 4350, 4630 (4-4); 4749, 5085 (8-1)
McDaniel, S. T., 15445 (4-1c)
McDowell, T., 1731 (4-1d); 2325 (4-1c); 2326 (5-10); 2509 (4-1j); 2920, 2993 (6-10); 3124 (4-1c); 4066 (4-1a); 4600 (4-1e); 4612 (4-1c); 4734 (6-10)
McDowell, T. & Gopaul, D., 2538 (4-1d); 2730, 3005 (4-2)
McPherson, G., 7884, 8116, 8299 (6-11); 9511 (6-11); 10892, 11008, 11296 (8-6); 13714 (6-4); 20471 (8-6); 20501 (5-4); 20561 (6-9); 20885, 20914 (5-4)
Medeiros, A., RB483175 (4-1a)
Medeiros, H., 2990 (4-1j)
Medina, B. & Quizhpe, W., 145 (5-7)
Meer, P. P. C. van, 957,987 (6-4)
Mehlig, U., 411 (6-5)
Meikle, R. D., 525 (6-4)
Meinich, L., 25 (5.2)
Meio, M. R. F., 643 (5-3)
Meireles, J. E., 568 (4-4); 570 (8-1); 609 (8-21); 611, 612 (4-4); 613 (8-1); 621 (4-1c); 628, 629 (4-4); 640 (6-8)
Melinon, M., s.n. (4-1a); (4-2); s.n. (6-3); s.n. (6-5); 48 (4-1a); 100 (8-18); s.n. (8-18); 377 (4-1a); 584 (6-5); 1863,1864 (4-1a)
Melito, M., 110 (8-1)
Melo, A. et al., 898 (7-5)
Melo, E. 1136, 1295, 1646 (4-4); 8070 (6-8)
Melo, M. F. F., 338 (6-8); 479 (6-5); 498 (4-1c); 746, 785 (6-8)
Mello, A. T., 58 (8-5)
Mello, F. C. de, 26 (5-2); 41 (8-18); 50 (8-11); 72 (1-2); 84 (8-18); 92 (2-1); 101 (6-8); INPA3341 (8-8); INPA3908, 41 (8-18); INPA4182 (6-2); INPA55232, (1-2); INPA55257, 55378 (2-1); INPA55448, 55449 (1-2); INPA55452 (2-1); INPA55449 (1-2); INPA58196 (2-1); INPA58197, INPA58918 (1-2); INPA60169 (8-18)
Mello Filho, L. E., 1186 (4-1g); 3044 (4-4)

Mello Silva, R. 2751 (8-14); 3005 (4-4); CFCR9627 (4-4); CFCR9890 (8-13); CFCR10165 (8-3a)

Melnyk, M. et al., 1, 15, 34, 90 (6-5)

Melo, E. de 1136 (4-4); 1326 (8-14); 2820 (4-4)

Melo, E. & França, F., 1295 (4-4)

Melo, G. A. R. de, VIC10305 (8-14)

Melo, M. F. F., 746, 785 (6-8)

Menandro, M. S. 47 (1-2); 53 (2-1); 277 (5-9)

Mendes, M. S.436 (8-14)

Mendes-Magalhães, G., 2117, 15489 (8-14)

Méndez, P. & Aulestia, M., 336 (5-4)

Mendonça, M. J. A. de, 56 (6-2)

Mendoza, C. & Arroyo, L. 1573 (6-8)

Mendoza, H., 9614 (4-4); 9817 (4-1c)

Menezes, C. M., 381 (6-8)

Menezes, L. F. T. et al., 1552 (5-9); 1620 (5-3); 1654 (6-8); 2006 (5-3)

Mennega, A. et al., 5649 (4-1b)

Merker, C. A., et al., 3041 (6-11)

Mesquita A. L. et al., 845 (6-5); 847 (8-18); 849 (7-5); 853 (4-1c)

Mesquite, C. G. & Cohnhaft, M., 1865 (4-1d)

Mesquita, M. R. et al., 332, 842 (8-11)

Mexia, Y., 5815 (4-4); 6049 (8-7)

Meyer, G., 107 (2-1)

Miereles, J. E., 609 (8-19); 613 (8-1); 616 (5-9); 629 (4-4); 631, 640 (6-8); 646, 677 (4-4)

Miller, R., 179 (6-5)

Milliken, W. et al., 569 (4-1a); 758 (4-1j); 1984 (4-1a)

Minorta C., V., 1042 (5-2)

Miquel s.n. (4-1d)

Miranda, A. M. & Ferraz, J., 6173 (6-8); 6246 (4-4)

Miranda Bastos, H. de 380 (6-5)

Miranda, C. A. B. de, 308 (6-8)

Miranda, E. B., 425 (4-1c); 999 (8-14)

Miranda, F. E. L. de, 445 (2-1); 704 (4-1a); 830 (4-1c); 964 (6-2)

Miranda, I. 2026 (4-1j)

Miranda I. P. de A. 19 (4-1d); 115 (6-5)

Miranda, I. S. et al., 139 (4-1a); 1019, 1089 (4-1c)

Miranda, J. P., 390 (4-1c)

Miranda, M. C. C., 229 (4-1c)

Miranda, T. B. et al., 206, 682 (8-3b)

Moema, Y., s.n. (6-8)

Monsalve, M., 689 (5-13); 1121, 1123 (6-9); 1336 (4-1c)

Monteagudo, A., 5877 (5-5)

Monteiro, M. M. et al., 162 (8-1)

Monteiro, M. M. & Oliveira, A. G., 72 (5-9)

Monteiro, M. T., 23646 (4-4)

Monteiro, O. P. (Osmarino), s.n. (INPA50057 (2-1); s.n. INPA50048 (6-5); s.n. (INPA50891) (8-4); 971 (4-1a); INPA50895 (5-2); 72-275 (6-5); 76-1299 (6-5); INPA75831 (6-5); 1210 (7-2); 1271 (8-18)

Monteiro, O. P. & Lima, J., 136 (8-8)

Monteiro, O. P. & Ramos, J., 26 (4-1d); 733 (8-7); 791 (8-4)

Monteiro da Costa 281 (6-5)

Montoya J., M., 1694, 2924, 2852 (4-5)

Moore, R. A., 20, 24, 44 (2-1)

Moraes, A. et al., 11 (8-14)

Moraes, M. A., 181 (4-1d); 211 (4-1b)

Morales, C. et al., 1607, 1969 (5-7)

Morales, J. F., 1549 (6-11)

Morawetz, W., 27-30878 (4-5

Moreno, P. P., 25540, 25636, 26822, 27308, 27309 (6-11)

Mori, S. A. et al., 4153 (6-11); 4707 (5-4); 8019 (4-1d); 8549 (6-5); 9622 (4-4); 10215 (7-8); 10478 (4-4); 10874 (5-9); 10893 (4-4); 10906 (5-9); 11421 (4-1c); 11983 (8-1); 12156 (6-8); 12583, 12783, 12954 13138 (4-4); 13181 (8-13); 13197, 13198 (8-14); 13343 (8-14); 13825 (7-8); 14070a (4-1a); 14337(8-14); 14391 (8-14); 14981 (8-18); 16348, 16417 (6-5) 16113, 16161, 16308, 16348, 16371, 16417, 16429, 16489 (6-5); 16520 (6-8); 17286 (6-5); 17423 (2-1); 17429 (8-17); 17538 (8-9); 17667 (5-2); 18722 (8-7); 19278 (5-2); 20196 (6-5); 20481(8-18); 20717 (8-5); 20843 (6-5); 21841 (8-18); 22457 (5-2); 23358 (5-5); 23441, 23572 (5-15); 23436, 23584, 23788 (6-5); 23793 (5-5); 25329 (6-3); 27275 (6-8)

Mori, S. A. & Bolten, 8196 (6-3); 8303 (4-1d); 8317 (4-1d); 8521, 8549 (6-5)

Mori, S. A. & Boom, B., 13924 (4-4); 14337, 14391 (8-14); 14778 (8-7); 14981 (8-18); 15303 (8-7)

Mori, S. A. & Kallunki, J., 4114 (6-11)

Mori, S. A. & Pennington, T. D., 18096 (5-5)

Mori, S. A. & Smith, N. P., 25047, 25159 (8-5)

Morokawa, R., 338 (5-9)

Mosén, H., 3475 (5-3); 3477 (6-8)

Moss, M., 13, 57 (4-1a)

Mostacedo, B. et al., 1660 (6-8)

Mota, C. D. A. da et al., 6 (1-2); 210 (4-1a); 608 (8-11); 686, 1435, 1489 (2-1); 2899 (7-5); 2903 (6-5); 3121, 3193 (4-1c); INPA60470, 60488(4-1i); INPA60633 (4-1c); INPA61036 (4-1d); INPA61399 (4-1a); INPA61555 (4-1i)

Mota, R. C. da 1154 (5-3)

Moura, M. O., et al., INPA171653 (8-18)

Moura, O. T., 687, 1088, 1353, 1432, 1433, 1456, JPB22701, JPB22702, JPB22704, JPB24088 (6-8)

Munhoz, C. B. R., et al., 7385 (4-1a)

Murillo, J. et al., 1038 (6-2); 1075 (6-5)

Museu Goeldi, 1260 (2-1); 9419 (4-1a); 9583 (8-18); 9664 (8-7); 9670 (8-18); 9672 (5-5); 9680, 9723 (8-18); 10130 (6-8)

Mutchnick, P., 37 (6-5); 1167 (4-1d); 1208 (6-3)
Nadruz, M., 517 (8-3a)
Nascimento, A. F. S., 98 (4-4)
Nascimento, F. H. F., 158 (4-4); 383, 384, 575, 610, 658 (8-3a)
Nascimento, J. R., 3 (8-8); 5 (8-18); 17, 108 (2-1); 109 (8-11); 199a, 217a, 321a (2-1); 570 (8-11); 595, 595a (2-1); 613 (8-18); 630 (6-5)
Nascimento, O. C. do, 422 (4-1c); 599 (7-6); 619 (4-1d); 694 (6-8); 812 (4-1a); 857 (6-8); 874 (6-5); 1169 (6-5)
Nascimento, T., 22 (6-5)
Nave, A. G., 1433 (6-8)
Nee, M. 31124 (4-1a); 34494 (4-1a); 42620 (8-7); 42758 (8-18); 42847 (8-7); 46502, 48403 (6-8)
Negrelle, R. R. B., 740 (8-3a)
Neill, D. et al., 11755 (8-16); 11784, 11829 (5-13); 14652, 14653, 14954, 14962, 14969, 15163 (5-7); 15185, 15186, 15187, 15189, 15190 (4-4) 15227, 15441, 15442, 15504, 15526, 15880, 16366, 17405 (5-7)
Neill, D. & Quizhpe, W., 14653, 14954, 14962, 14969, 16271 (5-7)
Nelson, B. W. et al. 310 (6-8); 1263 (6-5); 1272 (7-5); 1506 (7-5); 1514 (4-1c); P21063, P21082 (4-1d)
Nevers, G. de et al., 4162 (5-4); 4375 (6-11); 4585 (5-4); 4801, 4818 (6-9); 5291 (6-11); 5533 (5-4); 5589 (6-11); 5844 (6-9); 6141 (6-11); 6597 (8-6); 6874 (6-9); 7075, 7225, 7480, 7525, 7562, 7575 (6-11); 7941 (6-9)
Neves, D. M., 1562 (8-14)
NicLughadha, E. N. et al., PCD5959 (8-3a)
Nitta, A., 17588 (4-1a)
Noblick, L. R., 1266, 2173, 2175 (4-4)
Norberto, F. et al., 142 (6-5); 167 (4-1c)
Nunes, N. L. et al., 29 (4-4)
Nunes, T. S., 792 (8-1)
Occhioni, P., 1164 (4-1c)
Oldeman, R., T304 (6-3); 1725 (6-8)
Oldenburger, F. H. F. et al. 261 (6-5)
Oliveira, A. A. de et al., 29 (5-3); 91 (8-18); 95 (8-14); 182, 183, 235, 402 (6-8); 811, 1101c (4-4); 2778 (8-3a); 3679 (5-3)
Oliveira, A. C. A., 236 (8-4); 246 (2-1); 278, 279 (8-18); 358 (6-8); 359 (8-18); 425 (4-4); 428 (8-3a); 2234 (4-1a)
Oliveira, A. M. & Stehmann, J. R. 100, 160 (5-3)
Oliveira, A.P. P., 87 (6-8)
Oliveira A. R. de (Adair), INPA59638 (8-16: INPA59777, INPA59965, INPA60520 (6-8); INPA72909 (2-1)
Oliveira, E. de, 3562, 3609, 3745, 3813 (8-18); 3845 (2-1); 3934, 3962 (8-18); 4053 (8-18); 4063 (2-1); 4115 (8-18); 4220, 4230, 4314, 4455, 4476 (6-8); 4485 (8-18); 4569 (2-1); 4605 (6-8); 4797 (2-1); 4808 (8-18); 4823 (2-1); 6252, 6733 (6-8); 6784 (5-2)

Oliveira, J. et al., 101 (6-8); 233 (8-7); 523 (8-18); 655 (2-1); 699 (8-7); 701, 1000 (4-1j); 1005 (6-5)

Oliveira, J. B., 74 (6-8)

Oliveira, J. C. L., 369 (2-1)

Oliveira, J. D. P. et al., CFCR6562 (4-4)

Oliveira, J. M., 3032, RB197659 (4-1a)

Oliveira, P. V., 11-254 (6-8)

Oliveira, R. P. de et al., 226 (8-14); 266 (4-4); 470 (8-14)

Øllgaard, B. et al., 57591 (5-4)

Onishi, E. & Fonseca, S. G., 1263 (4-1)

Ordones, J. et al., 1843 (5-6)

Oren, D. C., 4 (4-1a); 5 (4-1c)

Orlandi, R. P., 736 (8-3a); 869 (6-5)

Ortiz, O. O., 557 (8-6)

Pabón E., M. et al., 314 (5-12)

Pacheco, M. et al., 85 (6-8); 143 (1-2)

Paciencia, M. B., 2335 (8-1)

Paes, L. E., 112 (5-3)

Paixão, J. L., 1738 (6-8)

Palacios, E. & Rodrigues, A., 103 (6-5)

Palacios, P. et al. 2609 (5-12); 2705 (5-2) Palacios, W. 2616, 2655 (5-13); 9371, 9394 (6-1); 11433 (5-4); 16486 (5-13)

Palheta, E., 3304.814.2 (6-5)

Pangaio, L. (4-4)

Pardo, C. S., 597 (8-3a)

Passos, B. C. dos, 1095(6-5)

Passos, L., 5380 (8-3a)

Pastore, M., 448 (4-1a)

Patiño, V. M., 12 (8-16)

Patmore, P. N. & Dufour, D. L. 162 (4-1j)

Paula-Souza, J. 6119 (8-14)

Pearce, R., s.n. (8-3b)

Peixoto, A. L. 323 (8-1); 340 (4-4)

Pena, B. S., s.n. (2-1); 702 (4-1c)

Pena, M. A., 788 (8-14)

Peña, R. M. & Martinez, J. P., 6, 30 (5-13)

Pennington, R. T. et al., 5 (6-8); 393 (4-1d)

Pennington, T. D. et al., 9950 (6-1); 12742 (8-18); 13855 (8-7); 17414 (8-18); 18235 (4-1a); P22772 (8-18)

Pennington, T. D. & Chango, N. D., 12311 (5-5)

Peñuela, M. C., 1600 (4-1a)

Perea, J., 3144, 3163, 3477, 3502 (5-7)

Perea, J. & Flores, V., 3176, 3186 (4-4); 3144, 3163, 3477 (5-7)

Perdiz, R. de O. et al., 812 (4-4); 1496, 1515 (7-5); 1985 (4-1j); 2072, 2236 (7-5); 2643 (6-5); 2830 (4-4)

Peron, M., 657 (5-6b)
Pereira, A., 12 (4-1c); 83 (4-1a)
Pereira, B. A. S. & Alvarenga, D. 3335 (5-3); 3659, 3699B (4-4)
Pereira, Edmundo, 1701, 1709 (4-4); 1788 (4-1a); 1829 (4-1c); 2044 (4-4); 2165 (8-14); 3560 (4-1c); 3964 (4-1c); 9617 (4-1c)
Pereira, E. da C., 232 (8-11); 401 (6-5)
Pereira, O. J. et al., 139, 141 (4-4); 232, 270 (8-1); 271, 303, 320, 372, 949, 1138, 2927 (4-4); 3000, 3049 (8-1); 3092 (4-4); 3134 (6-8); 3161 (4-4); 3171, 3257 (8-1); 3295, 3523 (4-4); 3604 (6-8); 3667(6-8); 3792 (4-4); 3836 (6-8); 3838 (5-3); 3874 (5-9); 3970, 4071 (4-4); 4112 (5-9); 4165, (6-8); 4244 (4-4); 4360 (6-8); 4463 (4-4); 4568 (4-4); 4771 (5-9); 4780 (4-4); 4858 (5-9); 4861 (4-4); 5145 (5-9); 5394(5-3); 5940, 6095 (6-8); 9510 (6-8)
Pereira, L.A. & Chagas, E. C. O., 101, 110, 212 (6-8)
Pereira, M. C. A., 119 (4-4)
Pereira, M. J., 3304.2832 (6-8)
Pereira, N., MG30277 (4-1c)
Pereira, P. A., 62 (4-1c); 564 (4-1j); 565 (6-8); 566 (4-1j); 568 (4-1d); 580, 588 (7-8); 607, 608, 609(4-1d)
Pereira, T. C., 55, 68 (6-8)
Perreira, H. S. 132 (2-1); 0110209-0 (6-5)
Perreira, B. A. S. & Alvarenga, D., 3335 (5-3)
Pereira-Silva, G., 14742, 14808 (6-5); 15635 (8-18); 15879 (6-5); 15977 (1-2); 16029, 16030 (6-5); 16041 (1-2); 16193 (4-1a); 16307 (2-1)
Pérez, L. P., 1271 (6-2)
Pérez, L. P. & Arévalo, R., 1259 (6-5)
Pérez, R. et al., 1099 (8-6); 1248 (6-11)
Peron, M., 441, 657 (5-6b)
Persaud, A. C., 102 (7-2); 191 (4-1a); 288 (7-2)
Pessoa, E., 725 (4-1d)
Pessoal do L. P. F. Brasília, 976 (2-1); 967 (8-18); 1013, 1040 (2-1); 1043, 1063 (8-18); 1080, 1094 (2-1); 1101 (8-18); 1143 (2-1); 1159(8-18); 1167, 1182, 1213 (2-1); 1235 (8-18); 1252, 1263 (2-1); 1287, 1309, 1332 (8-18)
Pessoal do CPF INPA6108 (1-2)
Phelps, K. & Hitchcock, C., 508 (4-1b)
Philcox, D. et al., 3060 (6-8); 3131 (6-5); 3573 (8-18)
Philcox, D. & Ferreira, A., 3545, 4149 (6-5)
Pickel, D. B.,591 (6-8)
Pinheiro, R. S., 2102 (4-4)
Piñeros, P. 72 (5-2)
Pinho-Ferreira, M. A. M., 519 (5-3)
Pinto, G. C. P., 143 (8-13; 304 (4-4); 377 (8-14)
Pinto, L. J. S., 76 (4-4)
Pipoly, J. J., 6781 (6-5); 6834 (4-1b); 7155 (4-2); 7456 (4-1d); 7655 (4-2); 7692 (4-1d); 7773 (4-1e); 7808 (4-2); 7856 (4-1b); 7990 (6-10); 8460 (4-1d); 9132 (4-1d); 9348 (4-1b); 9534 (4-1a); 9697 (4-1d); 10190 (4-1b); 10254, 10479 (4-1b); 10484 (8-12); 11426 (5-10); 11497 (6-3); 13625 (6-5)

Pirani, J. R. CFCR2091 (8-14); 1906 (4-4); 2003 (8-14); 2066 (8-13); 2870 (4-4); 5038 (5-3); CFCR7277, H51430 (8-14); CFCR 10882 (5-3)

Pires, J. M., s.n. (8-18); 41 (6-5); 365 (7-6); 588 (7-6); 683 (6-5); 708 (3-1); 754 (4-1d); 989 (4-1a); 1029 (4-1j); 1030 (5-12); 1410 (6-8); 1510 (6-8); 1785 (6-5); 3877 (4-1f); 4017 (6-8); 4518 (8-7); 6140 (6-8); 6936, 6944 (5-5); 14028 (6-5); 14111 (6-5); 14135 (6-5); 14196, 14209 (4-1d); 51067 (4-1c); 51553 (6-3); 51695 (8-18); 51943 (2-1)

Pires, J. M. et al., s.n. (INPA13242) (7-5); 4954, 5105 (8-18); 5380 (8-7); 6209 (4-1f)

Pires, J. M. & Black, G. A., 31 (6-1); 2961 (4-1g)

Pires, J. M. & Furtado, P. P., 17129 (6-8)

Pires, J. M. & Cavalcante, P. B., 52159 (4-1a); 52266 (4-1a)

Pires, J. M. & Coêlho, L. INPA13242 (7-5)

Pires, J. M. & Leite, P., 14809 (4-1d)

Pires J. M. & Silva, N. T., 1409 (8-18); 1753 (4-1a); 1754 (8-18); 1763 (2-1); 1767 (4-1a); 1783 (6-5); 1888 (6-2); 1974 (6-5) 2063 (4-1); 2070 (6-5); 2078 (4-1a), 2087, 2088 (6-5); 2089 (4-1c); 4192 (4-1c); 4624, 4629 (4-1a); 4702 (4-1c)

Pires (O'Brien), M. J., 577 (4-1j); 683 (6-5); 897 (6-1); 1763 (2-1); 1785 (6-5); 1888 (6-2); 1974, 2087, 2088 (6-5); 2216, 2278, 2366, 2386 (8-18)

Pires, M. J. & Silva, N. T. 577 (4-1j); 683 (6-5); 1409, 1751 (8-18); 1753 (4-1a); 1754 (8-18); 1762, 1763 (2-1); 1766 (8-18); 1767 (4-1a); 1785 (6-5); 2063 (4-1a); 2078 (4-1a); 2164, 2194 (2-1); 2195, 2210, 2278(8-18)

Pires, O. & Honda, M., 195 (5-2)

Pittier, H. 16260 (6-6)

Pivari, M. O. D. et al., 2349, 2542 (5-3)

Plowman, T. C. et al., 9604 (4-1c); 9734 (4-1a)

Plowman, T. C. & Davis, E. W., 12644 (7-5)

Plowman, T. C. & Almeida, G. E. M., 10047 (4-1a)

Plowman, T. C. & Carvalho, A. M., 12830 (4-1c)

Plowman, T. C. & Guanchez, F., 13522 (6-5)

Poeppig, E., s.n., 81, 2932, 3011 (4-1c)

Pohl, J., s.n. (4-1g)

Poiteau, A., s.n. 1824 (4-1a)

Polak, M. et al., 310 (6-5); 526 (6-3)

Poland, C., 6671 (4-4)

Pontes, A. F., 228, 234, 308, 309, 311, 312, 316, 320, 322, 325, 335, 337, 340, 371, 693(6-8).

Poole, J. M., 1743, 1784, 1793 (4-1d)

Popovkin, A. V., 461, 511 (6-8)

Posada, A & Matapí, D., 477 (8-19)

Poulain, S., 72 (6-5); 83 (8-20)

Prado, L. F., 682 (4-2)

Prance, G. T. et al., 1292 (4-1a); 1648 (6-5); 2003 (4-1d); 2179 (2-1); 2217 (8-18); 2276 (6-2); 2646 (7-5); 2836 (8-18); 3051 (8-18); 3062 (6-8); 3283 (8-7); 3614 (6-8); 3687 (5-2); 3710 (8-18); 3719 (5-2); 3740, 3819 (4-1d); 4114 (6-5);

4404 (4-1b); 4663 (7-5); 4736 (6-10); 4750 (4-1d); 4778 (5-2); 4830 (4-1a); 4881, 5147 (4-1d); 5560 (4-1a); 8282 (8-18); 8869 (2-1); 9540 (6-5); 9795 (4-1g); 9966, 10258 (6-8); 10386 (6-5); 10393 (6-8); 10424 (6-5); 10427, 11472 (7-5); 11541(7-6); 12600 (7-7); 13195 (6-5); 13340 (4-1c); 14032 (5-14); 14853 (4-1a); 14879 (5-2); 15029 (8-18); 15046, 15123 (7-5); 15175 (7-6); 15458 (6-8); 15509 (7-5); 15914 (7-8); 17533 (8-5); 17722 (6-2); 17770 (8-18); 17904 (4-1c); 17982 (5-5); 22598 (4-1d); 22602 (5-2); 22621 (6-2); 22734 (8-18); 22808 (6-2); 22974 (2-1); 23036 (8-17); 23391, 23484 (4-1c); 24878 4-1c); 24892 (8-17); 24976 (4-1c); 24980 (6-5); 28055 (5-8); 28833 (4-1d); 28855 (4-1i); 28974 (4-1b); 28996 (8-15); 29496 (6-5); 29610 (8-18); 29611 (4-4); 29748 (4-1d); 30392 (6-5)

Prance, G. T. & Huber, O., 28298, 28332, 28414 (4-1b)

Prance, G.T. & Pennington, T. D., 1965 (6-8); 2003 (4-1d)

Prance, G. T. & Silva, N. T., 58544 (6-8); 58585 (4-1a); 58722 (8-7); 58838 (6-5); 58954 (6-5); 59189 (6-8)

Prata, A. P. do N. et al. 229 (6-5), 2291 (6-8)

Prata, E. M. B., 777 (8-11); 792, 931 (4-5)

Prescott, M., 181 (6-8)

Prévost, M. F., 3233 (8-7)

Prieto, A. et al., 631 (6-5); 5302 (6-2); 5337, 5357 (4-5); 5650, 5695 (6-5)

Procópio, L. C. 377 (6-5)

Proença, C. et al., 2514 (4-1c)

Pruski, J. F., 3260 (8-18); 3517 (6-8)

Pulle, A., 52, 150 (4-1d)

Queiroz, L. P., 195 (8-14); 541 (4-4); 583 (8-14); 626 (4-4); 664 (8-14); 1819 (4-4); 1958 (8-14); 2456 (4-4); 3044 (6-8); 3944 (8-14); 3987 (4-4); 4214, 5059 (8-14); 5601, 5622 (8-13); 9532 (4-4); 9921 (8-3a)

Quelal, C. et al., 675 (5-4)

Quesada, F. J., 466, 1152 (5-4)

Quevedo, R. C. et al. 993 (6-8); 2609 (4-1a)

Quinet, A., 1629 (4-1j); 2667 (4-1j); 2721 (4-1b); 2718 (8-15); 2723 (4-1b); 2732 (8-15); 10160 (4-1j)

Quiroz-Villarreal, D. K. 9 (6-4)

Quizhpe, W. et al. 580 (5-4); 612 (5-7); 1992, 2060, 2785, 2853, 3130, 3197, 3238 (5-7)

Quizhpe, W. & Luisier, F. 612 (5-4); 1992, 2060 (5-7)

Rabelo, B. V., 547, 588, 744 (4-1a); 2923 (4-1a); 2991, 3036 (6-5); 3224, 3233 (2-1); 3252, 3254, 3296, 3297 (6-5); 3538 (8-18); 3668 (6-1)

Ramage, G. A., s.n. (6-8)

Ramalho, F. B., 46 (4-4)

Ramírez, J. G. et al., 7215, 7318 (6-8)

Ramírez, N., 3751 (4-1c); 4032 (8-12); 4033, 4295, 4641 (4-1b)

Ramos, A. E., et al., 99(6-8)

Ramos, C., 370 (5-4)

Ramos, J. F., 386 (6-8); 878 (4-1c); 1148 (4-1a); 1163 (2-1); 1666 (401a); 1785 (8-17); 2839 (8-18); 2953 (4-1c); INPA62149 (4-1d)

Ramos, G. & Patiño, V. M., s.n. (8-16)
Ramos, M. B., 446 (6-8)
Rankin, J. et al., 48, 100 (8-18)
Ratter, J. A., et al. 1139 (4-1c); 1398 (4-1c); 1434 (4-1a); 1814 (6-1); 2709 (4-2); 5730 (4-1a); 6700 (6-8); 6801 (4-1a); 6810 (6-8); 7815 (6-5)
Rech, s.n. (7-3)
Redden, K. M., 1489 (8-12); 1574 (4-2); 2008 (6-10); 2145 (4-2); 4236 (7-2); 6582 (6-10); 6678, 7121, 7188 (4-1e); 7193 (4-1); 7264 (6-10)
Reis, I. N., 9 (4-1a)
Reis. L. Q., INPA57836 (8-18)
Reis, R., 9 (8-21)
Reitsma, J., 1273, 2336 (6-4)
Reitz, P. R.., 3353, 4595 (8-3a)
Reitz, P. R. & Klein, R., 1589, 1730, 1744, 1836, 3024, 4595, 9128 (8-3a)
Restrepo, D. et al., 140 (5-12)
Restrepo, J. F., 940, 955 (6-1); 1023 (8-18)
Revilla, J., 1225 (8-19); 4049, (4-1c); 4169 (7-5); 8528, 8535 (4-1a); 12210 (2-1); 12706 (6-8); 12707 (4-1a); 12720 (6-8); 12742 (6-5)
Revilla, J. & Forero, J., 4049 (4-1c); 4121 (4-1a); 4199 (7-5)
Revilla, J. & Mota, C. D. A. da, 4135 (4-1c); 4169 (6-5)
Reyes, D. & Morales, C., 1214 (5-7)
Rezende, S. G., 1730 (4-4); 1731 (8-14)
Ribamar, J. & Ramos, J., 108 (2-1); 109 (8-11); 217 (2-1); 234 (6-5); 238 (5-2); 321 (2-1); 370 (4-1a)
Ribas, O. S. et al. 7715, 7731 (8-14)
Ribeiro, A. J., 28 (4-4); 52 (8-3a)
Ribeiro, B. C. S. & Pinheiro, G. S., 1111 (8-7)
Ribeiro, J. E. L. S., 929 (8-8); 1119 (8-18); 1134 (6-2); 1163 (6-5); 1428 (6-2); 1640 (4-1c); 1881 (6-8); 2691 (4-1c); 2857 (4-1d); 2875 (4-1c); 2880 (5-2)
Ribeiro, M. & Oliveira, A.G., 29 (5-3)
Ribeiro, R. D., 1279 (6-1)
Ribeiro, T., 208 (4-4); 466 (4-1f)
Ribeiro Filho, A. A., 17 (4-4); 44 (8-14); 219 (8-13); 297 (8-14)
Richard, C. Herb. s.n. (4-1c); s.n. (8-18)
Riedel, L., s.n. (4-1b); s.n. (4-1g); s.n. (5-6); s.n. (5-9); s.n. (6-5); s.n., 1271 (8-14); 1437 (6-5); 3570 (4-1g)
Rimachi, M. 4110 (4-4); 4577, 8070 (8-19)
Rios, M., 666 (6-5)
Rizzini, C. T., 1511 (4-4)
Roa, A., 257 (8-18); 307 (6-5); 317 (6-2); 394, 436, 445 (6-5); 605 (6-2); 636 (5-2); 686 (5-1)
Roberto, INPA58193 (8-18)
Robles, R et al., 2079 (6-11)
Rocha A. E. S. & Lobato, L. C., 305 (4-1a); 837 (4-1j)
Rocha, F. F., 13, 25 (8-14)

Rocha, K. D., 106 (6-8)
Rocha, R. F. A. 133 (4-4)
Roderjan, C. V. et al., 1411 (8-3a)
Rodrigues, E., 110 (7-5)
Rodrigues, M. N. 272 (6-5); 1992 (4-4)
Rodrigues, M dos S., 504 (6-8)
Rodrigues, W. A. et al., INPA86 (4-1c); 132 (4-1b); INPA213 (4-1a); INPA244 (7-5); 296 (5-2); 470 (4-1c); 526 (6-2); 527 (8-18); 734 (8-7); 772 (6-5); INPA620 (4-1c); 1056 (4-1c); INPA 1243, 1407 (7-5); INPA1422 (8-18); INPA1638 (6-2); INPA1684 (4-1c); INPA1700 (5-2); INPA1773 (4-1c); INPA1809 (8-18); 1814 (6-2); INPA2044 (7-5); INPA2084 (6-2); 2221 (2-1); 2391a (8-18); 2449a (4-1c); 2484a (7-5); 2934, 2935 (6-5); 3024 (6-2); 3152 (6-2); 3284, 3359, 4792, 5001, 5005 (7-5); 5178 (6-8); 5259 (8-7); 5241(8-11); 5409 (8-18); 5427 (2-1); 5439 (8-18); 5446 (6-5); 5452 (8-11); 5456 (6-5); 5459 (8-18); 5492 (8-11); 5619 (6-8); 5876 (6-8); 6043 (5-2); 6072 (4-1c); 6553 (7-5); 7081 (8-18); 7643, 7673 (2-1); 8165 (8-11); 8820 (7-5); 9143 (6-5); 9428 (4-1a); 9472 (8-18); 9620 (6-8); 9769 (4-1d); 9999 (7-5);10001 (4-1c); 10130 (4-4); 10228 (4-1a); 10478 (6-5); 10481 (4-1d); 10944 (4-1j);10945, 11034 (6-5); 11044a (4-1c); 11086a (2-1); 11127 (4-1j); 11128, 16811 (6-5)
Rodrigues, W. A. & Chagas, J., 1210 (4-1a); 1313 (7-5); 1493 (5-2); 1669 (8-18); 1675 (4-1c); 1717 (2-1); 1718 (8-18); 1723 (2-1); 1756 (8-18); 2224A (6-2); 2317 (4-1c); 2352A (6-2); 2370 (5-2); 2740 (4-1c)
Rodrigues, W.A. & Coêlho, L., 1446 (6-8); 1709 (4-1c); 1712 (4-1a); 1908 (8-7); 2037 (4-1c); 2045 (7-5); 2166 (8-7); 3919 (7-5)
Rodrigues, W. A. & Freitas, M., INPA14369 (8-18)
Rodrigues, W. A. & Lima, J., 2065, 2154 (4-1c); 2252 (7-5); 2449 (4-1c); 2537, 2559 (7-5); 2689 (4-1c)
Rodrigues, W. A. & Loureiro, A., 942 (8-18); 4943 (6-8); 5734, 5750 (8-7); 5943 (6-8); 7080 (8-18); 7150 (8-4); 7169, 7172. 9505 (8-18)
Rodrigues, W. A. & Osmarino, 5677 (8-7); 5729 (6-8); 6033. 6744 (8-18); 6964 (8-7); 6983 (6-8); 7031 (8-18); 8239 (6-2); 8313 (6-1)
Rodríguez, A., 543 (6-5); 645 (5-4); 841 (4-1a); 4476 (8-2)
Rodriguez, A. & Surubí, J., 570, 627 (6-8)
Rodríguez, J., 6506 (8-3b)
Rodríguez, M. 54 (6-1); 1518 (4-4); 3657 (8-18)
Rodríguez, R., 12 (6-2); 20 (4-1j)
Rojas, A., 293 (8-18); 3631, 3654 (6-6)
Rojas, G., 62 (6-8)
Romão, G. O., 1246, 2520 (4-4)
Rombouto, H. E., 228 (4-1a)
Romero G., G. A., 1850 (4-1d); 2130 (6-5)
Romero G., G. A. & Melgueiro, E., 2130 (6-5); 2136a (4-1c)
Romero, R. et al., 5723 (8-14)
Romero-Castañeda, R., 2721 (5-13); 3542 (7-8); 4785, 4942 (5-1); 5129 (4-1c); 5174, 5442, 5497 (5-13); 8379, 8404 (5-1)

Roosmalen, M. G. M. van, 1151 (4-1d); 1270 (6-2); 1446 (4-1j)
Roque, N., 399 (8-14)
Rosa, C. D., 6 (8-3a)
Rosa, N.A., 313 (4-1c); 1328 (5-5); 1132 (6-7); 1230 (2-1); 1568, 2067 (4-1d); 2241 (4-1a); 2728 (6-5); 2850 (5-5); 3183 (4-1a); 5485 (5-2)
Rosa, N. A. & Cordeiro, M. R., 1492 (4-1j); 1568, 1710 (4-1d); 1714 (4-4)
Rosa, N. A. & Lira, S. B., 2241 (4-1b); 2292 (8-18)
Rosa, N. A. & Santos, M. R. dos, 1892 (4-1c); 1972 (4-4); 1705, 2067 (4-1c)
Rosa, N. A. & Vilar, 2726 (6-8); 2728, 2870 (6-5)
Rosário, C. M. et al., 25 (6-8); 44 (4-1c); 898 (8-18)
Rosas Jr., A., 293 (8-18)
Ross, A. F. 232 (6-4)
Rosselli, A. P. & Rodríguez, A., 124 (8-18)
Rossi, L., s.n. (4-4); 1043 (5-3); 1985 (4-4); CFSC7284 (4-4)
Roswear T. S., (6-4)
Roubik, D. W., 54 (4-1c)
Rozemiro, 24 (2-1); 48 (8-7)
Rubio, D. et al., 1092, 1680 (5-4)
Rudas L., A., 2607, 2677 (6-2); 3681 (8-7); 7151 (4-5); 7164 (6-5); 7171 (4-5); 7238 (6-5)
Rudge, E., s.n. (4-1a)
Rueda, R. M. et al., 1596, 4708, 7467, 16966, 17210, 17860, 17828 (6-11)
Ruíz, J., 511 (8-19); 1351 (8-20)
Rylands, A. B., 55 (4-1a)
Rylands, A. & Hage, J. L., 68 (8-14); 84 (7-8); 125 (8-14)
Saar, E. et al., PCD5380 (8-13); 2652, 5604 (8-13)
Sabatier, D., 38-3 (7-9); 807 (5-15); 815, 836b, 842 (7-9); 989, 1013 (7-3); 1020 (5-15); 1071, 1130, 1172 (8-9); 1692 (8-7); 1765, 1779, 1988 (5-15); 2304 (5-5); 2373 (6-5); 2384, 2923 (8-18); 4360 (5-5); 4374, 4669 (8-18); 4898 (8-9); 5043 (6-3); 5574 (8-9)
Sabatier, D. & Molino, J. F., 5342 (8-9)
Sabatier, D. & Prévost. M. F., 2269 (5-15); 2808 (6-3); 4066 (5-5); 4067, 4351, 4672 (7-9); 4911 (8-9)
Sabatier, D. & Smock, J.-L., 5782 (6-3)
Saborio, J. C. et al., 130 (8-2)
Sacramento, A., 24, 66, 105, 162, 353, 550, 581 (6-8); 643 (4-4)
Saddi, E. M., 717 (6-8); 912 (8-14)
Saé-Silva, I. M. M., 71 (6-8)
Sagot, P. A., s.n. (4-1a)
Sakagawa, S., 418 (6-8); 434 (4-1a); 716, 728(4-1j)
Sakuragui, C. M., CFCR13981 (4-4)
Saldias, M. et al., 3609 (6-8)
Sales, A. B., 32 (6-8)
Sales, J. & Rosário, C., 1994 (4-1a)
Salinas, N. R., 721 (4-1a)

Salomão, R. P. de, 898 (4-1a); 976 (4-1c)
Salzman, P., s.n. (4-1b); s.n. (4-4)
Sampiao, A. B. et al., 556 (4-1c)
Sampaio, P. S. P. et al., 105, 183 (5-3)
Samuels, J. A., s.n. (4-1d)
Sánchez S., M. 841 (4-1a); 971A (6-10); 3813 (8-20); 4407 (5-2); 5907 (6-2)
Sandeman, C., 2699 (4-1d)
Sandino, J. C., 3411, 5005, 5006, 5007, 5008, 5010 (6-11)
Sandwith, N. Y., 374 (7-2); 399 (4-1a); 545 (4-1b)
Sanjoa, E et al., 2864 (6-5); 5587 (6-5)
Sano, P. T. et al., 14410 (8-14)
Sant'Ana, S. C., 74, 947, 1034 (4-4)
Santana, D. L. et al., 309 (8-14); 716 (4-4)
Santana, J. P. 450 (6-8)
Santos, A. L. S., 25,156 (6-8)
Santos, E. 2087 (4-4)
Santos, E. B. dos, 311 (4-4)
Santos, E. R. 1139, 1789 (6-5)
Santos, G. dos et al., 493 (2-1); 563, 605 (6-8); 611 (6-5); 716 (4-1i); 741, 801 (4-1a)
Santos, J. L. dos, 685 (4-1d); 716 (4-5); 718 (6-5)
Santos, J. U. et al., 246 (6-8); 802 (6-8)
Santos, L. A. S., 95 (6-8); 1440 (6-8)
Santos, L. O., 523 (4-1c)
Santos, M. F & Serafim, H. 328 (5-6)
Santos, M. G., 251, 676 (4-4)
Santos, M. M., 100 (8-14)
Santos, M. R dos, 100 (6-8); 272 (6-5); 676 (2-1)
Santos, O. A. dos, 473 (4-1c)
Santos, R. B. 62 (8-13)
Santos, R. M. et al., 1452 (6-8)
Santos, S. O., 211 (5-2); 282 (6-8)
Santos, T. R., PCD5648 (8-14)
Santos, T. S. dos et al., 2290 (8-1); 2298, 2342, 2343, 2626, 2741 (4-4); 2737(8-1); 2871 (4-1c); 2983 (8-1); 3114 (8-14); 3287 (8-1); 3432 (8-3a); 4540 (6-8); 5648 (8-14)
Santos, T. S. dos & Mattos-Silva, 2287 (8-1)
São-Mateus, W., 15 (4-4)
Saraiva, S. C., 16 (6-8); 17 (2-1); 708 (7-5)
Sardelli, L. F. et al., 6-228, 6-258 (6-8)
Sarmento, A. C. & Bautista, H. P., 841 (8-14)
Sarquis, R. S. F. R. et al., 41 (4-1c); 45 (6-5)
Sartori, A. C. B., 236 (4-4)
Sasaki, D., 222, 1320, 1452 (6-8); 1982 (4-1a); 2100, 2158 (2-1); 2242 (4-1a)
Sastre, C. et al., 3362 (4-1j); 3433 (4-1a); 4125 (7-3); 8425 (4-1b); 8454 (4-1c)

Sastre, C. & Echeverry, R., 655, 657 (6-1)
Sastre, C, & Raichel D., H., 5211 (5-12)
Scarponi, T.M., 67 (4-4)
Schmidt, S., 247 (4-4)
Schomburgk, R., s.n. (4-1e); s.n. (8-7); 10 (8-12); 47 (8-7); 135 (5-10); 164 (4-1b); 166 (5-10); 175 (4-1f); 270 (4-1d); 346 (4-1f); 543 (7-2); 560 (4-1f); 571, 574 (6-5); 576 (4-1i); 584 (5-10); 628 (4-1f); 825 (5-10); 842 (6-5); 845 (4-1i) 968 (4-1f); 982 (8-7); 1359 (5-10); 1552 (8-12); 1581 (8-7)
Rob 166, 574, 576, 628, 825, 982
Rich 135, 270, 560, 584, 842, 1359, 1552; 1581
Schmidt, M. & Pabst, G., 9681 (4-1a)
Schultes, R. E., 9431a, 9435 (4-1a); 23131 (5-2)
Schultes, R. E. & Cabrera, I., 15034 (4-2); 15511 (4-1a); 15922 (5-12); 16893 (4-1a); 17045 6-2); 17231 (4-1j); 17253 (6-2); 18319, 18371 (4-1j); 18480 (4-1a); 19290a, 19519 (6-2); 19951, 19963 (4-1f)
Schultes, R. E. & López, F., 8881 (6-5); 9267 (8-18); 9298d, 9363 (4-1d); 9510 (4-1j); 9701 (5-2); 10339 (4-1c)
Schultes, R. E. & Pires J. M., 9103A (4-1j)
Schultes R. E. & Rodrigues, W. A., 26203A (4-1c)
Schutz Rodrigues, R. et al. 2295 (7-5)Schwacke Herb 3093 (4-4); 3974 (4-1c); 3975 (6-5); 3976, 3977 (7-5); 3978 (5-2); 5571, 8344 (4-4)
Sebastião, P. & Assumpção, 14 (4-1c)
Secco, R. & Cardoso, O., 583 (6-5)
Segadas-Vianna, F., et al. I-310, I-369, I-385, I-439, I-821, I-907, I-945, I-1383, 3506 (4-4)
Sellow, F., s.n. (4-1c); s.n. (4-4); 171 (4-1c); 180, 2212, 2228 (4-1g)
Semir, J. et al. 4361 (5-6b)
Sette Silva, E., s.n. (1-2); s.n. (5-2); 6 (4-1b); 196 (7-5); 207 (5-16); 501 (4-1b); 838 (8-7); INPA/WWF1301.421.2 (8-4, 8-16); 1301.1780.2 (8-18); 1301.2810.2 (1-2); 1301.2908.2 (5-2); INPA/WWF2303.333.2 (6-16); INPA/WWF3304.4480.2 (8-18); INPA/WWF3402.4075.2 (8-8)
Setz, E. L., 36 (8-18)
Setz, E. Z. P., F36 (8-7); 1162 (2-1)
Shoko, S., 618 (8-6)
Sick, H. 557 (6-5)
Sidney 1263 (4-1a)
Silva, A., 59 (6-8)
Silva, A. E. da, 145 (6-5)
Silva, A. F. da, 346, 1986 (4-4)
Silva, A. P. da, s.n., 1301.3307.2, 1301.4455.2 (2-1)
Silva, A. S. L. da 303 (4-1c); 308, 590 (8-18); 768 (6-5); 2421 (6-1); 1398 (4-1c); 3895 (4-1c); 3911 (6-5); 3922 (4-1a); 4042 (6-5); 3034 (6-8); 3882 (6-8); 3895 (4-1c); 3922 (4-1c)
Silva, A. R. 35 (6-5)
Silva, B. M., 26 (4-1c); 153, 163 (6-8)

Silva, C. 10 (8-1); 15 (4-4)
Silva, C. A. 409 (7-8)
Silva, C. F., 3402.888.2 (8-8)
Silva, D. B., 48 (6-5)
Silva, F. et al., 40, 169, 230 (6-5)
Silva, H. C. H. 337 (6-8)
Silva, I. A., 81 (5-9)
Silva, I. R., 146 (5-2)
Silva, J. A. C. da, 205 (7-5); 224 (4-1d); 648 (4-1d); 782 (4-1i); 876 (4-1d); 899 (4-1c); 936 (6-2); (4-1d); 1024 1165 (6-5); 1243 (4-1c); 1298 (8-8); 1364 (4-1c); 1413 (7-5); 1462 (6-5); 1472, 1498 (8-18); 1520 (4-1c); 1693 (8-7); 1783 (4-1c); 1801 (8-18); 1844 (6-5); 1860 (5-2)
Silva, J.B. D., 54 (6-8); 55 (8-8); 56 (6-8)
Silva, J. F., 416 (1-2)
Silva, J. M. & Abe, L. M., 3164 (8-3a)
Silva, J. M. & Motta, J. T., 8691, 8698 (5-3)
Silva, J. M. & Campos, R. P., 8451 (5-3)
Silva, L.A. M. et al., 208, 422 (6-8); 449 (4-4); 622 (8-1); 859, 932 (6-8); 960 (8-1); 1193 (4-4); 1415 (7-8); 1436 (8-21); 2643 (8-1); 3895 (6-8); 4749 (8-1)
Silva, Luan, 562 (4-4)
Silva, L. N., 39 (6-8); 72 (4-1c); 73 (6-8)
Silva, L. R., s.n. (6-8); 422 (6-8)
Silva M. Aparcida da, 3226, 7679 (4-4)
Silva, M. F. da et al. 273, 685 (6-2); 807 (8-18); 838 (8-7); 1057 (7-6); 1135 (7-6); 1161, 1194 (7-5); 1522 (4-1c); 1586 (6-8)
Silva, Manoel, INPA27679 (1-2)
Silva M. F. F. da 6, 208, 1060 (4-1c)
Silva, M. G. da 482, 718 (8-7); 948 (7-5); 1548 (5-2); 1796 (6-5); 2544 (6-5); 2697 (8-7); 3356 (4-1c); 4046 (6-8); 4194 (6-5); 5588 (4-1a); 5728 (6-1); 5923 (8-7); 7160 (8-15); 7163 (4-4)
Silva, M. G. & Lima, R. P., 604 (6-8)
Silva, M. G. & Rosário, C., 3855 (2-1); 4046, 4846, 4987 (6-8); 5223 (4-1c); 5884 (6-8)
Silva, M. J., 61, 81, 112 (6-5)
Silva M. M. da, 6 (4-1a); 368 (5-3)
Silva, M. N., 187 (4-1c)
Silva, N. T., 426 (5-5); 981 (2-1); 1018 (6-5); 1040 (8-18); 1135 (7-5); 1291, 1401, 1405 (8-18); 1754 (8-18); 4517 (7-2); 4564 (4-1d); 4692 (4-1a); 4722 (1-2); 4726 (5-5); 5248 (4-1c); 5260 (8-7); 5403 (8-18); 5493 (2-1); 5510 (6-5); 58410 (8-1); 60947 (6-2)
Silva, N. T. & Brazão, U., 60947 (6-2)
Silva, R. M., 185 (4-1a)
Silva, S. B. da 165 (8-14)
Silva S. F. da 213A (1-2)
Silva, S. J. R. da, DTS110 (6-5)

Silva T. C. S., 932 (5-5)
Silva, V. G. da, 241, 302 (2-1)
Silva, W. S. da 55 (5-2); 103 (7-5)
Simon, M. F., 1505, 1533 (8-4); 1617, 1844 (4-1a); 2359(6-8)
Siniscalchi, C. M., 54 (4-1d); 547 (8-14)
Siqueira, D. R., 103 (6-8)
Siqueira, G. S., 51, 218 (5-9); 847 (8-21); 1068 (8-3a)
Siqueira, R., MG8281 (6-5); MG8735, MG8775 (8-7)
Siqueira-Filho, J. A., 3559 (4-4)
Small, D., 99 (6-4)
Smith, A. C., 2176 (4-1d); 2423 (4-1a)
Smith, D. A., 106 (6-11)
Smith, D. N. et al., 13680 (6-8)
Smith, L. B., 6406, 6694 (4-4)
Soares, C. R. A. et al., 639 (6-8)
Soares, E., 95 (4-1a); 169 (6-5); 170 (8-17); 229 (6-5); 231 (8-18)
Soares, M. P., VIC19647 (8-3a)
Soares-Silva, L. H. et al., 1178, 1529 (8-14)
Sobel, G. L., 4768 (6-5); 4785 (5-5)
Sobral, M., 3593 (4-1a); 3989 (4-4); 3994 (8-1); 7333 (5-3); 9720 (6-8); 11238 (4-1a)
Soejarto, D. D., 2863 (5-1)
Soejarto, D. & Lockwood 2392, 2396 (5-12)
Solis, F. et al., 28 (5-4)
Solomon, J. C., 9560 (5-7)
Sothers, C. A., 130, 191 (8-18); 192 (6-5); 304 (2-1); 305 (6-5); 841, 831, 843 (8-8); 847 (1-2)
Sosef, M. S. M., 1648 (6-4)
Sousa, A. B., 129 (4-1c)
Sousa-Silva, S. et al., 6 (4-4); 622 (8-1); 628 (4-4)
Souza, C. S. D., 38 (4-4); 65 (4-1c); 73, 114 (4-4)
Souza, E. B., 1957 (4-1d)
Souza, F. M. et al., 558 (5-3)
Souza, G. P. P. de, & Gusmão, E. F. de 403 (6-8)
Souza, J. A. de, 78 (2-1); 95 (8-11); 118 (2-1); 135 (7-5); 143 (1-2); 170 (8-18); INPA61899, 61934, 71646, 71647, 71648, 72908 (2-1)
Souza, L. A., 253 (4-4)
Souza, M. A. D. de, 1160 (7-5)
Souza, M. F. L. 229 (6-5)
Souza, S. A. da M., 921 (6-5)
Souza, V. C., 8316 (4-4); 15474 (6-8); 17909 (6-5); 30024 (6-2); 32525 (8-14)
Spada, J., 8 (8-21); 97 (8-21); 233 (8-3a)
Sperling, 6154 (2-1)
Spichiger, R. E. et al., 1738, 1741, 1742, 1743 (8-20); 1744 (8-19); 1745 (5-5); 1746 (8-20); 4560 (8-20); 4562 (6-5)

Splitgerber, F. L., s.n. (4-1d)

Spruce, R., s.n. (4-1c); s.n. (4-1d); s.n. (6-1); s.n. (6-5); s.n. (7-6); 13, 164, 181 (4-1c); 763 (6-5); 928 (4-1c); 1009 (6-5); 1499 (4-1c); 1714 (7-5); 1715 (5-2); 1969, 2419 (7-6); 2424 (5-2); 2426, 2443 (5-2); 2454, 2457 (4-1j); 3073, 3094, 3194 (7-6); 3409 (4-1d); 3419 (4-1f); 4335 (4-1g) 5963 (6-5)

Stahel, G., 18 (6-5); 90 (4-1a); 263 (6-3)

Stahel, G. & Gonggryp J. W., 3570 (4-1a)

Stancik, J. F. 155 (6-8); 397, 426 (8-18); 491 (1-2)

Stannard, B. et al., CFCR7033 (8-13); H51758, H52747, H52765 (8-14)

Stapf, M. N. S., 462 (8-1)

Stefanello, D. et al., 63, 87, 102 (6-8)

Stehmann, J. R. et al., 2663 (8-14); 4643 (8-1); 4666 (4-4); 4760 (5-9)

Stergios, B. et al. 8327 (5-11); 10472 (6-5); 11045 (5-10); 11254 (6-5); 13283 (6-5); 15304 (8-7); 15534 (5-11); 16283 (6-5)

Stergios, B. & Aymard, G., 9215 (7-6)

Stergios, B, & Delgado, L., 12961 (6-1)

Stergios, B. & Velazco, J., 14421, 14666 (6-5)

Stergios, B. & Yánez, M., 15106 (4-1c)

Stevens, W. D. 8991, 23445, 23812 (6-11)

Stevenson, D. W., 1007 (4-1d); 1059 (4-2)

Stevenson, P., 1707 (6-1); 1759 (8-20)

Steward, W.C., 78 (4-1d); 326 (7-5); 411 (7-6); P12746 (7-8); P20245 (4-1d); 21062 (8-4)

Steyermark, J. A., 57817 (4-1f); 57880 (4-1j); 58228 (4-1b); 59186 (4-1b); 59621 (4-1g); 60192 (4-1j); 60289 (4-1g); 60756 (6-5); 75695, 75995 (4-1b); 86679, 86872 (6-3); 90262 (4-1b); 93180 (4-1f); 93218 (4-1b); 94197 (7-4); 98192 (4-1f); 103165, 104530 (4-1b); 105121 (4-2); 105483, 106651 (8-12); 108916, 109248 (4-1b); 109745, 109806 (4-1b); 112828 (4-3); 112857 (4-1f); 113557 (4-1b); 115135, 115190 (6-1); 117640, 117842 (8-12); 125730 (4-1j); 127537, 127957 (4-1c); 130091, 130235, 130286, 130343 (4-1b); 130398 (6-5)

Steyermark, J. A. & Bunting, G., 102442 (5-11); 103255 (4-3); 105483, 105490 (8-12); 109806 (4-1b); 113235 (4-1c); 123896 (4-1b); 125733 (6-5); 126141 (4-1b); 128612 (4-1b)

Steyermark, J. A. & Wurdack, J. J., 1109 (4-1b); 1313 (6-5)

St. Hilaire, A. de, s.n. 1145 (4-4); B1705, 1984 (8-14)

Stone, J. et al., 3283 (6-4)

Stropp, J. & Assunção, P., s.n. (3-1); 45 (5-2); P75 (6-5); 336, 475 (7-5); 475a (5-2); 629 (3-1)

Strudwick, J. J., 3491, 4248, 4341 (4-1a)

Sucre, B. D., 1381, 3871, 5392, 5423, 7937, 8374 (4-4); 8394, 8734 (8-1)

Take, K. I., 5 (6-10)

Talbot, P. A., 1744 (6-4)

Tamashiro, J. Y., 18684 (5-3)

Tamayo, F., 3123 (8-12); 3585 (4-1d)

Tameirão-Neto, E., 3192, 3213, 5410 (8-14); 5136 (5-3)

Tate, G. H. H., 50 (4-1a); 142 (7-6); 209, 283, 286 (4-1f); 330 (4-1i); 331 (4-1j); 733, 1113 (4-1b)
Tavares, A. S., 252 (6-1); 523 (8-7)
Tavares, A. S. & Silva, M. G., 10, 29, 30, 109 (8-15)
Taylor, E. L., 1277 (4-1c)
Teixeira, E., RB43676 (5-3); RB60812 (5-6a); RB83977 (5-3)
Teixeira, L. O. A., 10 (6-5); 1178 (4-1c); 1338 (6-8); 1648 (4-5)
Teixeira, G., 113 (6-5)
Tejera, E. & Braun, A., 5 (6-5)
Terra-Araujo, M. H., 474 (4-1d)
Teunissen, P. A. et al. 11339 (6-5)
Teunissen-Werkhoven, M., LBB16254 (6-3)
Thomas, W. W. et al., 493 (5-2); 3209 (6-2); 3223 (5-1); 3235 (8-7); 3433 (6-2); 3510 (4-1j); 3877A (6-8); 4156 (4-1a); 4793, 5088, 5316 (5-2); 5318 (8-7); 6188 (4-1j); 8129 (8-1); 8170 (8-7); 8932 (4-4); 9224 (6-8); 9398, 9466, 9788 (7-8); 9878, 9978 (8-1); 9988 (4-4); 10063, 10083 (8-1); 10384 ((8-1); 10955 (4-4); 11610 (4-4); 12967 (8-14); 16069 (6-8)
Thomaz, L. D., 7 (4-4); 1339 (8-21)
Thomsen, K. 170 (8-2); 441 (5-4); 519 (8-2); 1617 (6-11)
Thornwill, A. S., 239 (6-4)
Tipaz, G. A. et al., 1443 (5-4); 2567, 2590 (5-13)
Tillett, S. S. et al., 751-32 (4-1b)
Tillett, S & C., 45358 (4-2); 45600 (4-1b); 45678 (8-12); 45679 (4-1b)
Tirado, M. et al. 471, 639, 825 (5-13)
Todzia, C. A., 2341 (8-5)
Toledo G., M et al., 1148 (2-1)
Torke, B. M., 285 (4-1d)
Tostain, O., 815 (4-1a)
Tosto, M. G. et al., 30, 88 (6-8)
Traill, J. W. H., 80 (4-1d); 81 (7-5)
Tripp, E., 2984 (6-10)
Trujillo, W. et al., 1903 (6-5)
Tsugaru, S. & Sano, Y., B-921 (7-5)
Tsuji, R. 2319 (5-3)
Tutin, T. G., 83 (4-1a); 421 (6-3)
Ule, E. H. G., 6142 (4-1c); 7625 (4-1c); 8801 (8-12)
University of Guyana Bio, 41, 106 (4-1d)
Urban, I., 1691 (8-3a)
Urbanetz, C., 190, 381 (8-3a)
Urrego, L. E., 760 (4-1j)
Uslar, Y. et al. 601 (6-5)
Utrera, A. & Cordero, Y., 85 (4-1b)
Valcárcel, J. & Chota, M., 5/93 (8-20); s.n. (8-22), 34090 (8-20)
Vale, N. B. 246(6-5)
Valedão, R. de M., 131 (4-4)

Valente, A. S. M. et al., 334 (5-3)
Valera, A., 126 (2-1)
Valiant, P. 101 (8-14)
Valle, J. R. & Valio, I. F. M. 8 (4-4)
Van Donselaar, J., 109, 1674 (4-1j)
Van Rooden, J. et al. 435 (6-9)
Varanda, E. et al., CFCR 4497(4-4)
Varejão, M de J., INPA150515 (4-1c); INPA140520 (6-5)
Vareschi V. & Foldats, E., 4563, 4573 (4-1b); 4673 (7-1)
Vargas, A. A., 129 (8-18)
Vargas, C. K. & Phillips, P., 1119 (5-4)
Vargas, I., 156 (6-1)
Vargas, O., 1406 (6-11)
Vasconcelos, J. C. de M., 2286 (6-8)
Vasconcelos, T. N. C., 437 (4-1a)
Vásquez, R. et al., s.n. (8-22); 4928, 6845 (5-5); 7696 (8-20); 8192 (6-8); 10430 (5-5); 13775 (8-18); 13858 (5-5); 15789 (8-18); 18057 (5-5); 18114 (6-1); 19949, 23497 (5-5)
Velazco, J., 830 (4-1c); 932 (4-1d); 1114 (6-5); 1826 (5-2); 1838 (4-1d)
Vélez, J., 982 (6-8)
Veloso, H., 37b (8-3a)
Verdi, M., 6942, 7047, 7116 (4-4)
Versteeg, G. M., 265 (7-3)
Vervloet, R. R., 2399 (5-9)
Viana, B. F., 41 (4-4)
Viana, G., 88 (6-5); 269 (6-8
Viana, P. L., 2885 (4-1a); 3018 (4-1a); 3038 (4-1i); 3084 (4-1a)
Vidal U., A., 2 (8-18)
Vicente, A., et al., 1, 3 (6-8)
Vicentini, A., 361 (6-4); 378 (8-18); 413 (8-7); 447 (1-2); 761 (8-7); 795 (6-2); 1298 (4-5); 1330, 1420 (4-1d); 1476, 1569, 1585, 1586, 1587, 1588 1637, 1638; 1639 (4-4); 1865 (4-1c); 2109 (4-1a)
Vieira, M. G. G., 9 (4-1c); 736, 889 (6-5); 1272 (4-1d); 1275 (5-2); 1322, 1359 (4-1d); 1369 (5-2)
Vigne, C., 2800, 2801 (6-4)
Vilhena, R. P., 49 (4-1c); 50, (1-2); 51 (5-2); 53 (2-1); 54 (1-2); 55 (7-5); 56 (8-18); 60 (4-1a); 94 (7-5); 163, 267 (6-5); 350 (6-8)
Villa, G., 1034 (6-2); 1796 (8-18)
Vincelli, P. 1160 (6-1)
Vinha, P. 1013 (6-8); 1245 (4-4)
Virillo, C. B., 217 (8-3a)
Vreden, C. LBB13669 (6-5)
Wachenheim, G., s.n., 179, 489 (8-18)
Wallnöfer, B., 12-71088 (8-18)
Warming, E., 1570 (5-6a)

Walter, B. M. T. et al., 5247 (4-4)
Webster, G. L., 25089 (4-4)
Weddell, H. A., 526, 2361 (4-1c)
Weinberg, B. s.n., 311, 580, RB261409 (4-4)
Werfe, van der, H. et al., 10022, 10152 (8-20); 24445 (5-4)
Wesenberg, J., 52, 1012 (8-3a)
Whitton, B. A., 293 (4-1c)
Wied-Neuwied, M. A. P., s.n. (8-14)
Wieringa, J. J., 1081, 1362 (6-4)
Wiggins, I. L., 18988 (6-6)
Wijinga, V., 614 (8-19)
Wilbert, W. et al., 213, 88365 (6-1)
Williams Ll., 12065 (2-1); 13868 (4-1j); 13903 (4-1d); 14310 (7-8); 14637 (7-8); 15052; (4-1f); 15061 (6-5); 15253 (7-6); n15418 (4-1f); 15753 (6-2); 15950 (7-6)
Williams, Ll & Silva, N. T. 18201 (6-8)
Williams, L. O. et al., 28677 (5-4)
Withelm, D. & Gonçalves, J., 148 (6-5)
without collector, s.n. (7-3)
Worbes, M. 1017 (7-5)
Wullschlaegel, H. R., 1393 (4-1d)
Wurdack, J. J., 293 (6-1)
Wurdack, J. J. & Adderley, L. S., 42705 (4-1d); 42760 (4-5); 43355 (5-12)
Wurdack, J. J. & Monachino, J., 40881 (5-2); 41149 (6-5); 41380 (4-1i)
Wurdack, K. J. et al. 4154 (4-1b); 4775 (7-2); 4911, 5898 (6-10)
Xavier, L. P., JPB2111 (6-8)
Yánez, M., 47 (4-1d); 256 (4-5); 467 (4-1f)
Zamora, N. 1183, 1528, 1823 (5-4); 1862 (8-2)
Zamora, N. & Castrillo, M., 1963 (8-2); 1994 (6-11)
Zanin, H. R. W. 11-68 (6-8); 6507 (4-1a)
Zappi, D. C., 759 (8-14); 1297 (4-1c); 1523 (4-4); 1583 (8-14); 2064 (5-6); 2067 (4-1c); 2899 (6-5); 3064 (6-8); 3186 (4-1j); 3188 (4-1a); CFSC9325 (4-4); CFCR9922 (8-13)
Zartman, C. E., 5722 (4-1c); 5880 (6-8); 6377 (6-5); 7190, 7326, 7336, 7337 (7-5); 7402 (5-2); 7773 (7-5); 7774, 7775 (5-2); 7778 (4-1c); 7807 (4-1j); 7814 (5-2); 8496 (7-5); 8540 (4-1c); 9593 (8-15); 9621 (4-1c); 9643 (8-4); 9660, 9697 (4-1a); 9748 (4-1d); 9749 (5-2); 9750 (7-8); 9754 (5-16)
Zarucchi, J., 1959 (4-1i); 2561 (4-1d); 3130, 3170, 3187 (5-2)
Zarucchi, J. & Barbosa, C. E. 3596, 3616 (6-5)
Zenker, G., 148, 440, 1249, 1624, 1671, 1677, 1953, 2499, 2760a, 4407 (6-4)
Zent, S., 386.39 (4-1f)
Ziburski, 88/2, 88/18 (7-5)
Ziller, S. R. & Curcio, G., 621 (5-3).

References

Acosta-Vargas, L. G. 2015. Population status of the tree *Sacoglottis holdridgei* (Humiriaceae) at Isla del Coco National Park, Costa Rica. Revista Biol. Tropical 64 (Suppl. 1): S263–S275.

Allen P. H. 1956. Rainforests of Costa Rica, p. 351.

APGIII. 2009. An update of the Angiosperm Phylogeny Group classification for the orders and families of flowering plants. J. Linn. Soc. Bot. 161: 105–121.

Araujo, P. A. M. & A. D. Mattos-Filho. 1985. Estrutura das madeiras Brasileiras de dicotyledoneas. XXVII. Humiriaceae. Rodriguesia 37: 91–114.

Aublet, J. B. C. F. 1775. Histoire des plantes de la Guiane Français 1; 564–566, 572. P. F. Didot, London, Paris.

Baillon, H. E. 1862. Aubrya gabonensis. Adansonia 2: 266–267.

——— 1873. Notes sur les géraniales et les Linacées. Adansonia 10: 368–371.

——— 1874. Série des Humiri. Histoire des Plantes 5: 51-56 t. 88–97. Paris

Bentham, G. 1853. Notes on Humiriaceae. Hooker's J. Bot. Kew Gard. Misc. 5: 97–104.

Berry, E. W. 1922. Pliocene fossil plants from Eastern Bolivia. Johns Hopkins Univ. Stud. Geol. 4: 178, Pl VIII, figs 5–11.

——— 1924a. Fossil fruits from the Eastern Andes of Colombia. Bull. Torrey Bot. Club 51: 64, figs 20–22.

——— 1924b. New tertic species of *Anacardium* and *Vantanea* from Colombia. Pan Amer. Geol. 42: 259, pls. 1–5.

——— 1929a. Tertiary fossil plants from Colombia, South America. Proc. U. S. Natl. Mus. 75, art. 24, pp1–12, pls. 1–5.

——— 1929b. Fossil fruits in the Ancon sandstone of Ecuador. J. Paleontol. 3: 300, figs 4, 5.

——— 1929c. Early Tertiary fruits and seeds from Belen, Peru. Johns Hopkins Univ. Stud. Geol. 10: 155–157, pl 1.

Bhikhi, C. R., P. J. M. Maas, J. Koek-Noorman & T. R. van Andel. 2016. Timber trees of Suriname. LM Publishers, Volendam.

Bove, C. P. 1997. Phylogenetic analysis of Humiriaceae with notes on monophyly of Ixonanthaceae. J. Comp. Biol. 2: 19–24.

——— & T. S. Melhem. 2000. Humiriaceae: Pollen and Spore Flora of the World 22: 1–35.

Calderon, E. 1998. Humiriastrum melanocarpum. The IUCN Red List of Threatened Species 1998: e.T38418A10116317. https://doi.org/10.2305/IUCN.UK.1998.RLTS.T38418A10116317.en

Camacho, R. L. & I. Montero. 2006. *Humiriastrum procerum* pp 85–90 In Cárdenas L. & Salinas, Libro rojo de plants de Colombia: Especies maderables amenazadas. Instituto Amazónico de Investigaciones Científicas SINSCI, Bogota.

G. T. Prance, *Humiriaceae*, Flora Neotropica 123,
https://doi.org/10.1007/978-3-030-82359-7

Cavalcante, P. B. 2010. Frutos comestíveis na Amazônia. Ed 7. Belém, Museu Paraenese Emílio Goeldi.

Clusius, C. 1605. Exoticorum libri decem. Lib II cap 18. Leiden, Plantiniana Raphalengii.

Cardoso, D., J. G. Carvalho-Sobrinho, C. E. Zartman, D. L. Komura & L. P. 2015. Unexplored Amazonian diversity: rare and phylogenetically enigmatic tree species are newly collected. Neodiversity 8: 55–73.

Costa Abreu, V. G. da, G. M. Corrêa, I.A. Santos Lagos, R.R. dos Silva, & A. F. de C. Alcâtara. 2013. Pentacyclic triterpenes and steroids from the stem bark of uchi (*Sacoglottis uchi*, Humiriaceae). Acta Amazonica 43: 525–528.

Cremers, G. 2001. Transfer of holotypes from CAY in French Guiana to P in France. Taxon 50: 293–296.

Cronquist, A. 1968. The evolution and classification of flowering plants. Boston Houghton Mifflin.

——— 1981. An integrated system of classification of flowering plants. Boston. Houghton Mifflin.

Cuatrecasas, J. 1961. A taxonomic revision of the Humiriaceae. Contr. U.S. Natl. Herb. 35, 2: 25–214.

——— 1964. Miscelanea sobre flora Neotropica. Ciencia (Mexico) 23: 137–139.

——— 1972. Miscelanea sobre flora Neotropica III. Ciencia (Mexico) 27: 171–172.

——— 1990. Miscellaneous notes on Neotropical flora XVII. New species in the Humiriaceae. Phytologia 68: 260–266.

——— 1991. Miscellaneous notes on Neotropical flora XX. A new species of *Humiriastrum*. Phytologia 71: 165–166.

——— 1993. Miscellaneous notes on Neotropical flora XXI. A new species of *Humiriastrum* from Brazil. Phytologia 75: 235–238.

Cuatrecasas, J & O. Huber. 1999. Humiriaceae In J. A. Steyermark et al. (eds.) Flora of the Venezuelan Guayana 5: 623–641.

Delprete, P. G., R. M. Baldini, N. Fumeaux & L. Guglielmone. 2019. Typification of plant names published by Giovanni Casaretto from specimens collected in Brazil and Uruguay. Taxon 68: 783–827.

Ducke, A. 1922–25. Plantes nouvelles ou peu connues de la région amazonienne (II partie), Humiriaceae. Arch. Jard. Bot. Rio de Janeiro 3: 175–180, 271. 1922; (III serie). 4: 99. 1925. Ducke, A. 1930. Plantes nouvelles ou peu connues de la région amazonienne (IV Serie) Humiriaceae. Arch. Jard. Bot. Rio de Janeiro 5: 142; (V serie) 6: 39–40. 1933.

———. Ducke, A. 1935–38. Plantes nouvelles ou peu connues de la région amazonienne (VII serie). Arq. Inst. Biol. Veget. 1: 205–207. 1935. (X serie) 4: 24–31. 1938.

———. 1938. Plantes nouvelles ou peu connues de la région amazonienne (X serie). Arq. Inst. Biol. Veg. 4: 24–31.

Elias de Paula, J. & J. L. de H. Alves. 1977. Madeiras nativas, anatomia, dendrologia, dendrometria, produção e uso. Brasília, Fundação Mokiti Okada: 169–170.

Endlicher, S. 1840. Genera plantarum, pp. 1039–1040. Fr. Beck, Vienna.

Engel, J. & D. Sabatier. 2018. *Vantanea maculicarpa* (Humiriaceae): A new tree species from French Guiana. Phytotaxa 338 (1): 130–137.

Gastal, M. L. & M. X. A. Bizerril. 1999. Ground foraging and seed dispersal of a gallery forest tree by the fruit-eating bat *Artibeus lituratus*. Mammalia 63: 108–112.

Gentry, A. H. 1975. 87A, Humiriaceae, in Flora of Panama. Ann. Missouri Bot. Gard. 62: 35–44.

——— 1990. Contribution to the study of the flora and vegetation of Peruvian Amazonia XX. Candollea 45: 379–380.

——— 2001. Humiriaceae in Flora de Nicaragua 2: 1149–1150. Monographs in Systematic Botany from the Missouri Botanical Garden 85.

Giordano, L. C. da S. & C. P. Bove. 2008. Taxonomic considerations and amended description of *Humiriastrum spiritu-sancti*. Humiriaceae. Rodriguesia 59: 151–154.

Gunn, C. R. & J. V. Dennis. 1972. Stranded tropical seeds and fruits collected from Carolina beaches. Castanea 37: 195–200.

Hallier, H. 1921. Beiträge zur Kentniss de Linaceae. 9 die Humiriaceeen. Beih. Bot. Centralbl. 39, Abt. 2: 56–62, 174.

Hartshorn, G. 2010. P 625 In Tomlinson, P. B. & M. H. Zimmerman (Eds.). Tropical trees as living systems. Cambridge University Press.

Heimsch, C. 1942. Comparative anatomy of the secondary xylem in the Gruinales and Terebinthales of Wettstein with reference to taxonomic grouping. Lilloa 8: 83–198.

Henry, O., F. Feer & D. Sabatier. 2000. Diet of the lowland tapir (*Tapirus terrestris* L.) in French Guiana. Biotropica 32: 364–368.

Herrera, F., S. R. Manchester, C. Jaramillo, B. MacFadden, & A da Silva-Caminha. 2010. Phytogeographic history and phylogeny of the Humiriaceae. Int. J. Plant Science 171: 392–408.

———, S. R. Manchester, J. Vélez-Juarbe & C. Jaramillo. 2014. Phytogeographic history of the Humiriaceae (Part 2). Int. J. Plant Science 175: 828–840.

Hill, A. W. 1933. The method of germination of seeds enclosed in a stony endocarp. Ann. Bot. (London) 47: 873.

Holanda, A. S. S. de, A. Vicentini, M. J. Hopkins & C. E. Zartman. 2015a. Phenotypic differences are not explained by pre-zygotic reproductive barriers in sympatric varieties of the *Humiria balsamifera* complex (Humiriaceae). Pl. Syst. Evol. 301: 1767–1779.

———, C. E. Zartman, M. J. Hopkins, J. J. Valsko & A. H. Kral. 2015b. First occurrence of *Schistostemon* (Urb.) Cuatrec. (Humiriaceae) in states of Roraima and Pará, Brazil. Check list11 (1). Article 1520.

Hooghiemstra, H., V. M. Wijninga & A. M. Cleef. 2006. The paleobotanical record of Colombia: implications for biogeography and biodiversity. Ann. Missouri Bot. Gard. 93: 297–324.

Hoorn, C. 1994. Fluvial palaeoenvironments in the intracratonic Amazonas Basin (early Miocene-early Middle Miocene, Colombia). Palaeogeogr. Palaeoclimatol. Palaeoecol. 109: 1–54.

Huber. J. 1909. Materials para a flora Amazonica. Bol. Mus. Paraense Hist. Nat. 5: 413.

Hutchinson, J. 1973. The families of flowering plants. Ed. 3. Oxford. Clarendon Press.

——— & J. M. Dalziel. 1928. Flora of West Tropical Africa 1: 274, fig. 114.

Johnson, C. D., J. Romero & E. Raimúndez-Urrutia. 2001. Ecology of *Amblycereus crassipunctatus* Ribero-Costa (Coleoptera: Bruchidae) in seeds of Humiriaceae, a new host family for bruchids, with an ecological comparison to other species of *Amblycerus*. Coleoptera Bull. 55: 37–48.

Johnson, M. & A. Colquhoun. 1996. Preliminary ethnobotanical survey of Kurupukari: An Amerindian settlement of Central Guyana. Econ. Bot. 50: 182–194.

Jupiassú, A. M. S., 1970. Madeira fossil: Humiriaceae de Irituía, Estado de Pará. Bol. Mus. Paraense Hist. Nat., Nov. Ser. Geol.14: 1–12

Jussieu, A. H. L. de. 1829. Humiriaceae, in A. St. Hilaire (ed.). Flora Brasiliae Meridionalis 2: 87–91.

Kirchheimer, F. 1951. Ueber das Vorkommen einer Gattung der Humiriaceae im Europäischen Tertiär. Planta 39: 75–90.

Kloster, A., S. Gnaedinger & K. Adami-Rodrigues 2012. A new fossil wood of the Humiriaceae family from Miocene Solimões Formation, Acre, Amazon, Brasil. XV Simposia Argentino de Paleobotánica y Palinologia. July 10–13. Corrientes, Argentina.

Kubitski, K. 2013. Humiriaceae, pp 223–228 in Families and genera of flowering plants. Springer.

Kuhlmann, M. 2012. Frutos e Sementes do Cerrado Atrativos para a Fauna: Guia do campo. 360 pp. Editora Rede de Semente do Cerrado.

Langenheim, J. H. 2003. Plant resins: chemistry, evolution, ecology, and ethnobotany. Timber Press, Portland and Cambridge.

Lima, I. P, M. R. Nogueira, L. R. Monteiro & A. L. Peracchi. 2016. Frugivoria e dispersão de sementes por morcegos na Reserve Natural Vale, sudeste do Brasil. Chapter 26 in Floresta Atlântica de tabuleiro: diversidade e endemismos na Reserva Natural Vale: 433–452.

Lorente, M. A. 1986. Palynology and paleofacies of the Upper Tertuary in Venezuela. Diss. Bot. 99: 1–232

Macedo. M. & G. T. Prance. 1978. Notes on the vegetation of Amazonia II. Dispersal of plants in Amazonia white sands campinas: The campinas as functional islands. Brittonia 30: 203–215.

Maduka, H. & Z. Okoye. 2005. Bergenin, a Nigerian alcoholic beverage additive from *Sacoglottis gabonensis* as an antioxidant protector of mammalian cells against 2,4—Dinitrophenyl Hyrazine-induced lipid protection. Internet J of Toxicology 3 (1): 1–6.

Martius, C. E. P. von. 1827. Nova genera et species plantarum brasiliensium 2: 142–148, t. 199.

Marx, F., A. A. Andrade, M. das G. B. Zoghbi, & J. S. Maia. 2002. Studies of edible Amazonian plants. Part 5: chemical characteristics of Amazonian *Endopleura uchi*. Eur. Food. Res. & Technol. 214: 331–334.

McPherson, G. 1988. A new species of Vantanea (Humiriaceae) from Panama. Ann. Missouri Bot. Gard. 75: 1148–1149.

Metcalfe, C. R. & L. Chalk. 1950. Anatomy of the dicotyledons: leaves, stem, and wood structure in relation to taxonomy with notes on economic Uses. Oxford, Clarendon Press.

Mori, S.A., G. Cremers, C. A. Gracie, J.-J. de Granville, S. V. Heald, M. Hoff, & J. D. Mitchell. 2002. Guide to the vascular plants of French Guiana. Part. 2 Dicotyledons. New York Botanical Garden Press.

Morris, D. 1889. A Jamaica drift fruit. Nature 39: 322–323.

Morris, D. 1895. A Jamaica drift fruit. Nature 53: 64–66.

Nees von Esenbeck, C. G. D. & Martius, C. E. P. von. 1824. Nov. Acta. Acad. Caes. Leop.-Carol. German. Nat. Cur. 12: 38, t 7.

Oltmann, O. 1971. Pollenmorphologisch sytematische Untersuchungen innerhalb der Geraniales. Diss. Bot. 11: 1–163.

Pesce, C. 2009. Oleaginosas da amazônia. Ed 2: 293–295. Belém, Museu Paraense, Emílio Goeldi.

Pinedo-Vasquez, M., D. Zarin, P. Jipp, & J. Chota-Inuma. 1990. Use-values of tree species in a communal forest reserve in Northeast Peru. Conservation Biology 4: 405–416.

Planchon, J. E. 1848. Sur la famille des Linnées. Hooker's J. Bot. Kew Gard. Misc. 6: 588.

Pinto, G. P. 1956. O oleo de uxi: seu estudo quimico. Bol. Tecn. Inst. Agron. N. 31: 187–194.

Pires-O'Brien. Prance, G. T. & O'Brien, C. M. 1994. Fénologia e história natural de famílias de árvores Amazônicas: Caryocaraceae, Connaraceae, Humiriaceae. Bol. Fac. Ciénc. Agrárias Pará 5: 1–20.

Pons, D. & D. D. Franceschi. 2007. Neogene woods from western Peruvian Amazon and palaeoenvironmental interpretation. Bull. Geos. Czech Geol. Survey 82: 343–354.

Reid, E. M. 1933. Notes on some fossil fruits of Tertiary age from Colombia. S. Am. Rev. Geogr. Phys. Geil. 6 (3): 210–212

Rodrigues, W. A. 1982. Duas novas espécies da flora amazônica. Acta Amazônica. 12: 295–300.

Sabatier, D. 1987. Studies on the flora of the Guianas 12: quelques nouveautés chez les Humiriaceae. Proc. Kon. Ned. Acad. Wetensch. C: 90: 203–210.

——— 2002. *Vantanea ovicarpa* (Humiriaceae), a new species from French Guiana. Brittonia 54: 233–235.

——— 2018. *Vantanea maculicarpa*, a new tree species from French Guiana. Phytotaxa 338 (1): 130–134.

Schnizlein. 1843–1870. Iconogr. Fam. Regn. Veg. 3, t. 222.

Schultes R. E. 1979. De plantis Toxicariis e Mundo Novo Tropicale XXI. Interesting native uses of the Humiriaceae in the northwest Amazon. J. Ethnopharmacology 1: 89–94.

——— & R. F. Raffauf. 1990. The healing forest: medicinal & toxic plants of the northwest Amazonia. Dioscorides Press, Portland.

Selling, O. E. 1945. Fossil remains of the genus *Humiria*. Svensk. Bot. Tidskr. 39: 258–269.

Shanley, P. & N. Rosa. 2004. Eroding knowledge: An ethnobotanical inventory in Eastern Amazonia's logging frontier. Econ. Bot. 58: 135–160.

——— & G. Medina. 2005. Frutíferas e Plantas Úteis na Vida Amazônica. CIFOR, Imazon, Belém, Brazil.

Silva, T. B. C., V. L. Alves, L. M. Conserva, E. M. M. da Rocha, E. H. A. Andrade & R. P. L. Lemos. 2004. Chemical constituents and preliminary antimalarial activity of *Humiria balsamifera*. Pharm. Biol. 42 (2): 94–97.

Silva, L. R. & R. Teixeira. 2015. Phenolic profile and biological potential of *Endopleura uchi* extracts. Asian Pacific J. Trop. Med. 8 (11): 889–897.

Silva, S. L. da, V. G. de Oliveira, T. Yano & R de C. S. Nunomura. 2009. Antimicrobial activity of bergenin from *Endopleura uchi* (Huber) Cuatrec. Acta Amazonica 39: 187–192.

Silva-Caminha, S., C. Jaramillo & M. L. Absy. 2010. Neogene palynology of the Solimões basin, Brazilian Amazonia. Paleeontographica Abt. B. 283: 1–67.

Sothers, C. A., J. M. Brito & G. T. Prance. 1999. Humiriaceae in Ribeiro, J. E. L da S et al. Flora da Reserva Ducke. INPA, Manaus.

Spichiger, R., J. Méroz & P. A. Loizeau. 1983. Las Humiriáceas y Lináceas del Arborétum Jenaro Herrera (provincia de Requena, departamento de Loretro, Perú. Contribución al estudio de la flora de la vegetación de la Amazonia peruana IV. Candollea 38: 717–732.

———, J. Méroz, P. A. Loizeau & L. S. de Ortega. 1990. Humiriaceae in Los Árboles del Arborétum Jenaro Herrera vol. 2. Boissiera 44: 23–35.

Stern, W. L. & R. H. Eyde. 1963. Fossil forests of Ocú, Panama. Science 140: 1214.

Suryakanta. 1974. Pollen morphological studies in the Humiriaceae. J. Jap. Bot. 49: 112–122.

Takhtajan, A. 1969. Flowering plants origin and dispersal. Edinburgh, Oliver & Boyd.

——— 1997. Diversity and classification of flowering plants. New York, Columbia University Press.

Tchouya, G. R. F., G. P. N. Oblang, J-B Bongui & J. Lebibi. 2016. Phytochemical study of *Sacoglottis gabonensis* (Baill.) Urb. Isolation of Bioactive Compounds from the bark. J. Amer. Chem. Sci. 11: 1–5.

Urban, I. 1877. Humiriaceae in Martius, Fl. Brasiliensis 12 (2): 435–454.

——— 1893. Humiriaceae in P. Taubert, Plantae glaziovianae novae vel minus cognitae. Bot. Jahrb. Syst. 17: 503.

Van der Ham, R., T. van Andel, L. Butcher, I. Hanno, T. Lambreschts, K. Lincicome, P. Mikkelsen, J. Mina, B. Patterson, E. Perry & M. Romance. 2015. Furrowed blister pods stranded on northern Atlantic Ocean coasts represent an undescribed *Sacoglottis* (Humiriaceae) endocarp most similar to the fossil *Sacoglottis costata*. J. Bot. Res. Inst. Texas 9: 137–147.

Weyland, H. 1938. Die fossilen Sacoglottis-Fruechte und eine neue Art der Gattung, *Sacoglottis germanica* n. sp. Dacheniana, Ser A, 98: 163–164.

White, F. 1962. Geographic variation and speciation in Africa with particular reference to Diospyros. Publ. Syst. Assoc. 4: 71–103.

White, L. J. T. 1994. *Sacoglottis gabonensis* fruiting and the seasonal movements of elephants in the Lopé Reserve, Gabon. J. Trop. Ecol. 10: 121–125

Wijninga, V. M. & P. Kuhry. 1990. A Pliocene flora from the Subachoque Valley (Cordillera Oriental, Colombia). Rev. Paleaobot. Palynol. 62: 249–290.

Winkler, H. 1931. Unterfam. IV, Humirioideae, in Engl. & Prantl, Die Naturlichen Pflanzenfamilien, zweite Auflage. Band 19a, 106: 126–129, t. 58, 59.

Wurdack, K. J. & C. C. Davis. 2009. Malpighiales phlygenetics: gaining ground on one of the most recalcitrant clades in the angiosperm tree of life. Amer. J. Bot. 96: 1551–1570.

——— & C. E. Zartman. 2019. Insights on the systematics and morphology of Humiriaceae (Malpighiaceae): androecial and extrafloral nectary variation, two new combinations, and a new *Sacoglottis* from Guyana. PhytoKeys 124: 87–121.

Xi, Z, B. R. Ruhfel, H. Schaefer, A. M. Amorim, M. Sugumaran, K. J. Wurdack, P. K. Endress, M. L. Matthews, P. F. Stevens, S. Matthews & C. C. Davis. 2012. Phylogenomics and a posteriori data partitioning resolve the Cretaceous angiosperm radiation Malpghiales. Proc. Natl. Acad.U. S. A. 109: 17519–17524.

Zamora, N. 2007. Humiriaceae in Manual de plantas de Costa Rica 6: 10–14. Missouri Botanical Garden Press.

Index of Local Names

G. T. Prance, *Humiriaceae*, Flora Neotropica 123,
https://doi.org/10.1007/978-3-030-82359-7

Index of Scientific names

New names are in **boldface** and synonyms are in *italics*. Page numbers in boldface indicate primary page reference. Page numbers with an asterisk (*) indicate pages with illustrations or maps. New lectotypifications are indicated by **LT** in **boldface** following the number of the page where published. Fossil names are followed by **F.**

G. T. Prance, *Humiriaceae*, Flora Neotropica 123,
https://doi.org/10.1007/978-3-030-82359-7

Printed in the United States
by Baker & Taylor Publisher Services